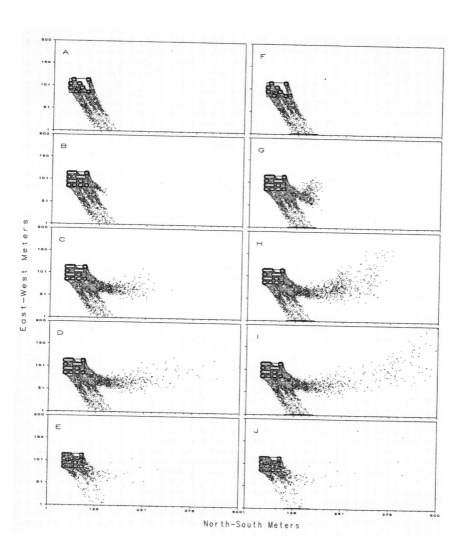

ATMOSPHERIC MICROBIAL AEROSOLS

Bruce Lighthart
U. S. Environmental Protection Agency
Corvallis, OR

A. J. Mohr
U.S. Army, Dugway Proving Ground
Dugway, UT

ATMOSPHERIC MICROBIAL AEROSOLS

Theory and Applications

Edited by

BRUCE LIGHTHART
ALAN JEFF MOHR

CHAPMAN & HALL
New York • London

First published in 1994 by
Chapman & Hall
One Penn Plaza
New York, NY 10119

Published in Great Britain by
Chapman & Hall
2-6 Boundary Row
London SE1 8HN

Color illustration: Dispersion maps of 50,000 (or 10,000 in bottom row) simulated viable bacterial droplets deposited (left column) or deposited and airborne (right column) 15 (A,F), 30 (B,G), 45 (C,H), and 60 min (D,E,I,J) after spray initiation at Tulelake, CA on 28 May 1988.

Library of Congress Cataloging-in-Publication Data
Atmospheric microbial aerosols—theoretical
  and applied aspects / edited by Bruce Lighthart and A.J. Mohr.
      p.   cm.
   Includes bibliographical references and index.
   1. Air—Microbiology.   2. Aerosols.   I. Lighthart, B.   II. Mohr.
A. J. (Alan Jeff)
QR101.F9 1994
576′. 190961—dc20   93-2375
                                                                           CIP

ISBN 0-412-03181-7

British Library Cataloguing in Publication Data available.

Please send your order for this or any other Chapman & Hall book to **Chapman & Hall, 29 West 35th Street, New York, NY 10001, Attn: Customer Service Department.** You may also call our Order Department at 1-212-244-3336 or fax your purchase order to 1-800-248-4724.

For a complete listing of Chapman & Hall's titles, send your request to **Chapman & Hall, Dept. BC, One Penn Plaza, New York, NY 10119.**

Dedication

This book is dedicated with pride to my wonderful family Anne, Ian and Allie Mohr, and to my dear wife Gloria Lighthart.

Contents

Contributors

Dr. John Burckle
U.S. Environmental Protection Agency
Risk Reduction Engineering Laboratory
Cinncinati, OH

Dr. Donald E. Gardner
U.S. Army
Aberdeen Proving Ground, MD 21010

Dr. Janina Gitelman
Institute for Biological Research
Ness-Ziona, Isreal

Dr. Eitan Israeli
Institute for Biological Research
Ness-Ziona, Isreal

Dr. Paul A. Jensen
National Institute of Occupation Safety and
    Health
4676 Columbia Rd.
Cinncinati, OH 45226

Ms. Barbara Johnson
U.S. Army
Dugway Proving Ground
Dugway, UT 84022

Dr. Jinwon Kim
Lawrence Livermore Laboratory
Livermore, CA

Dr. Gerry LaVeck
U.S. Environmental Protection Agency
Office of Pesticides and Toxic
    Substances
401 M St.
Washington, DC 20460

Dr. Bruce Lighthart
U.S. Environmental Protection Agency
200 SW 35th St.
Corvallis OR 97333

Dr. G. Macek
U.S. Environmental Protection Agency
Office of Pesticides and Toxic
    Substances
401 M St.
Washington, DC 20460

Dr. Balkumar Marthi
Hindustan Lever Ltd.
Microbiology Department
Research Ctr, Chakala
Andheri (East)
Bombay, India 400099

Dr. Alan J. Mohr
U.S. Army
Dugway Proving Ground
Dugway, UT 84022

Dr. I. Gary Resnick
U.S. Army
Dugway Proving Ground
Dugway, UT 84022

Dr. Harry Salem
U.S. Army
Aberdeen Proving Ground, MD 21010

Dr. Philip Sayre
U.S. Environmental Protection Agency
Office of Pesticides and Toxic
  Substances
401 M St.
Washington, DC 20460

Ms. Brenda T. Shaffer
ManTech Environmental, Inc.
200 SW 35th St.
Corvallis, OR 97333

Dr. Linda Stetzenbach
Environmental Research Center
University of Nevada, Las Vegas
Las Vegas, NV 89154

Dr. Hollis H. Wickman
National Science Foundation
Washington, DC 20550

1

Introduction

An aerosol is a colloidal suspension of liquid droplets or solid particles in air. A **bioaerosol** is then defined as an aerosol whose components contain, or have attached to them, one or more microorganisms. Consequently, a microbial bioaerosol contains microorganisms such as viruses, bacteria, fungi, protozoa, or algae that are usually viable, i.e., alive. Bioaerosols may be relatively solid particles or liquid droplets and range in size from a single microorganism to large droplets. These droplets may contain many microorganisms, pollen grains, and agglomerations and/or rafts of microorganisms attached to particulate plant debris, skin flakes, and/or soil particles (Fig. 1.1). Liquid droplets larger than one organism may change in size upon evaporation (or condensation), leaving an aeroplanktonic residue of nonvolatile solute, particulate matter and/or viable or non-viable microorganisms. (To avoid the confusion of the changing condition of a liquid droplet evaporating to a dry particle, bioaerosols in this state will be termed droplet/particles or D/Ps.)

The atmospheric load of bioaerosols comes from many natural and anthropogenic sources. Natural bioaerosols are generated when wind conditions entrain individual droplets and particles into the surrounding bulk atmosphere (Fig. 1.2). Wind-driven soil dust, plant debris, wave spray, bubble bursting, and rainsplash also contribute to the atmospheric load. Bioaerosols generated by human activities may be due to extramural and intramural processes. Extramural bioaerosols are produced as sprays in liquid droplets from nozzles in agricultural applications (Fig. 1.3), spray from manufacturing processes, bursting bubbles from wastewater treatment plants, and spray drift from nuclear cooling towers, or as dry particles from urban vehicular activity and rural agricultural practices, and perhaps even shedding from buildings. Intramural bioaerosols may be generated during walking, sneezing, coughing, talking, shedding of skin flakes, surgical and dental processes, and household activities including ventilation with or without air conditioning, household cleaning, toilet flushing, and so on.

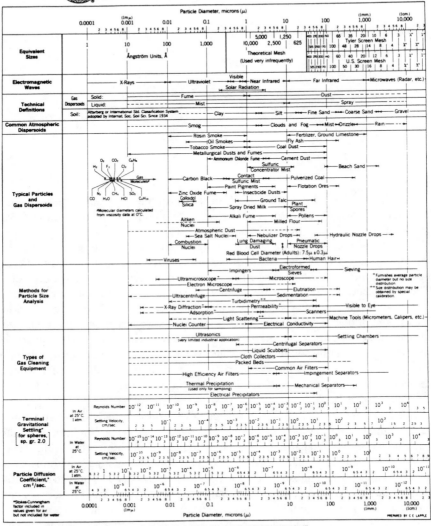

SRI International • 333 RAVENSWOOD AVE. • MENLO PARK, CALIFORNIA 94025 • (415) 326-6200

Figure 1.1. Particle size ranges for aerosols (from C.E. Lapple, *SRI Journal* 5:94).

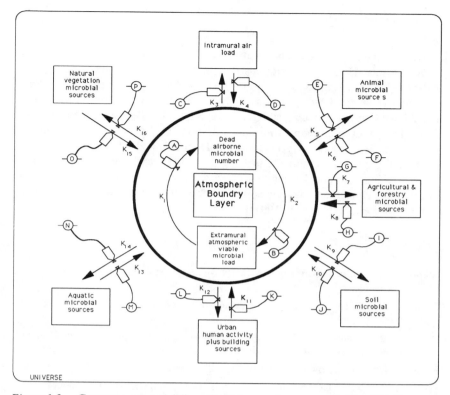

Figure 1.2. Compartment model diagram of bioaerosol sources and sinks. [See Forsester (1961) for definition of symbols, and Atkins (1969) for description of compartment models).

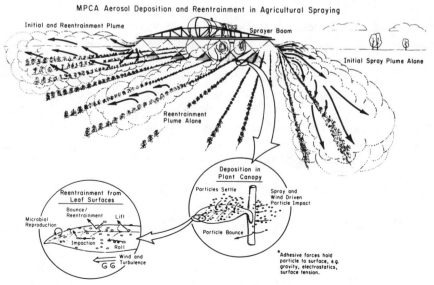

Figure 1.3. MPCA aerosol deposition and reentrainment in agricultural spraying.

3

Many of the physical and chemical processes that describe aerosols also apply to bioaerosols. Detailed descriptions may be found in books by Hinds (1982), Reist (1984), and Willeke and Baron (1993), which contain many useful examples, and Hesketh (1986), whereas Hidy (1984) and Fuchs (1964), the classic aerosol texts, are good technical references. Shaw (1978) contains detailed topic areas such as droplet evaporation and filtration. Descriptions of experimental aerobiological methods may be found in Dimmick and Akers (1969), results of aerobiological measurements in Gregory (1973), and a good discription of laboratory and theoretical aerobiology in Cox (1987). A list of useful references and journals concerned with aerosols is given in the *Journal of Aerosol Science and Technology* [14(1): 1–4 (1991)].

References

Atkins, G. L., Multicompartment Models for Biological Systems. Methuen and Co. Ltd. London, p. 153.

Cox, C. S. 1987. The aerobiological pathways of microorganisms. John Wiley & Sons, New York, p. 293.

Dimmick, R. L., and A. B. Akers. 1969. An introduction to experimental aerobiology. Wiley–Interscience, New York. p. 494.

Forrester, J. W. 1961. Industrial dynamics. MIT Press, Cambridge, MA.

Fuchs, N. A. 1964. The mechanics of aerosols. Dover Publications, Inc. New York, p. 408.

Gregory, P. H. 1973. The microbiology of the atmosphere, 2nd ed. John Wiley & Sons. New York, p. 377.

Hidy, G. M. 1984. Aerosols. Academic Press, Inc., New York, p. 774.

Hesketh, H. E. 1986. Fine particles in gaseous media. Lewis Publishers, Inc. Chelsea, MI.

Hinds, W. C. 1982. Aerosol technology. John Wiley & Sons, New York.

Reist, P. C. 1984. Introduction to aerosol science. Macmillan Publishing Co., New York, p. 299.

Shaw, D. T. (ed.). 1978. Fundamentals of aerosol science. John Wiley & Sons, New York, p. 372.

Willeke, K. and P. A. Baron (Eds.). 1993. Aerosol Measurement. Van Nostrand Reinhold. New York, p. 876.

2

Physics of Microbial Bioaerosols

Bruce Lighthart

2.1. Bioaerosol Physics

Bioaerosol droplets may be generated from a suspension of microorganisms sprayed into the atmosphere as a polydipersed aerosol, that is, a dispersion made up of many different sized droplets. There, each droplet follows its unique trajectory all the while evaporating to a packed residue of particles. The shape and size of the residue particle depends on the quality and quantity of microorganisms in the droplet. The rate of evaporation (or condensation) of solvent in the droplet, usually water, depends on chemical parameters that control droplet evaporation (condensation) such as the droplet solvent and solute, and ambient relative humidity (RH) and temperature. Physical forces which are acting on a droplet/particle in motion are atmospheric drag which usually tends to slow a particle from the bulk air velocity and changes with droplet size and shape, and gravity, which usually accelerates a particle to its terminal settling velocity. On a larger scale, droplets are dispersed by the atmospheric motion and carried downwind where they are impacted or settle onto surfaces. Electrostatic forces may be important in attracting particles to surfaces, including the surface of other particles or droplets.

The physical and chemical factors described in the following sections can be integrated to produce dynamic simulation models of bioaerosol trajectories and dispersal patterns (as explained in Chapter 8). Although biological decay of the aerosol is included in the model, it is described in more detail in a later section (Section 7.2).

2.1.1. Trajectory of Uniform Bioaerosol Particles

To give the reader a more realistic understanding of a bioaerosol droplet's sojourn through the atmosphere, calculations of its path location and physico-chemical processes affecting the droplet will be combined in a description of its trajectory. The trajectory description will start by defining the forces acting on the droplet/particle[1] (Sec. 2.1.1) and then how the trajectory may be predicted in time and space (Sec. 2.1.1.2), through velocity adjustment after ejection from a nozzle (Sec. 2.1.1.3), and, finally, to its terminal velocity (Sec. 2.1.1.5). The flow regime will be initially assumed laminar. (Chapter 8 will give some turbulent flow functions that may be incorporated into a simulation model.) It is assumed in this discussion that the bioaerosol droplets are liquid water, spherically shaped, and contain spherical bacteria 1 μm in diameter. Further discussion of an application of the trajectory equations are given in Lighthart et al. (1991) for various size droplets in a laminar flow regime.

2.1.1.1. Forces Acting on Bioaerosol Droplet/Particle

When a bioaerosol droplet/particle is in motion in the atmosphere, it is subject to two effects: its inertia as defined by the Newton Resistance Law [Eq. (2.1)] and viscous or drag effects of the atmosphere as defined by Stokes Law [Eq. (2.2)]. The ratio of the inertial effects to viscous effects is useful to determine which effect is dominant in a fluid regime (and, therefore, which descriptive equations apply) and is expressed by a dimensionless number, the Reynolds number [Eq. (2.3)]. For Re > 1000, inertial forces dominate viscous forces, whereas for RE < 1, viscous forces dominate, and between the two a transition region occurs (Fig. 2.1).

$$F_D = \frac{C_D \pi \rho_g D_p^2 V^2}{8}, \tag{2.1}$$

where C_D is the drag coefficient defined in Eq. (2.1), ρ_g is the density of ambient air (g/ml), D_p is the diameter of the particle (m), and V is the velocity of particle (m/s).

$$F_D = \pi \eta V D_p, \tag{2.2}$$

where η is the coefficient of viscosity of ambient air (dynes × s/cm^2).

$$Re = \frac{\text{inertial force}}{\text{frictional force}} = \frac{\rho_g U d}{\eta}, \tag{2.3}$$

where U is the relative velocity between air and bioaerosol particle (m/s) and, d is the diameter of the pipe (m).

---

[1]Because of the continuum of property changes on transition from a liquid droplet to a dry residue particle, this notion is expressed in this book as a droplet/particle or in abbreviated form as D/P.

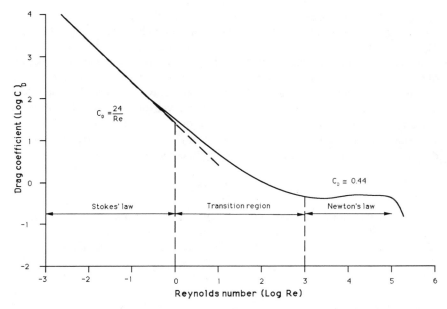

Figure 2.1. Drag coefficient versus Reynolds number for spherical droplet/particles.

The physical constants for ρ_g and η may be found in *The Handbook of Chemistry and Physics* (Chemical Rubber Co., 1953) and Table 2.1).

2.1.1.2. Trajectory Calculation

The trajectory of a bioaerosol droplet through the atmosphere on ejection from a source such as a spray nozzle or sneeze is described mathematically by physical and chemical relationships. The simplest trajectory would be followed in a laminar flowing atmospheric regime, that is, where the flow of air is in parallel, straight lines; in the present case, parallel to the ground. The horizontal (or x-plane) and vertical (or y-plane) locations of a D/P may be located independently

Table 2.1. Some properties of air at 20°C, and constants.

Property	Value	Units
Viscosity (η)	1.81×10^{-4}	g/cm s
Density (ρ)	1.205×10^{-3}	g/cm^3
Diffusion coefficient (D_0)	0.19	cm^2/s
Mean free path (γ)	0.066	μm
Molecular weight of air	28.9	g/mole
Gas constant (R)	8.31×10^{-7}	dyn cm/K mole
Gravity (g)	981	cm/s^2
Boltzman constant (k)	1.38×10^{-16}	dyn cm/K

in either the x- or y-direction by sequentially calculating the distance traveled by each D/P over each of many known short time intervals or steps, and velocities. The horizontal or vertical velocity of the D/P at each time step (V_t) may be calculated by using the final x,y-values of the previous step as the initial values of the present step. Equation (2.4) describes either the vertical velocity component which is influenced by both atmospheric drag and gravity (where V_0 is the initial spray or nozzle velocity) until it attains its terminal velocity (V_{TS}), or the horizontal velocity component where only atmospheric drag is considered (i.e., V_0 is 0). The summation of all distances traveled through all time steps determines the droplets location at any time interval in its trajectory (Fig. 2.2).

$$V_t = V_f - (V_f - V_0)e^{-t/\tau}, \qquad (2.4)$$

where V_t is the vertical or horizontal velocity at time t (m s^{-1}), V_0 is the initial spray velocity for the horizontal axis and zero for the vertical axis (m s^{-1}), V_f is the final velocity which is the bulk airstream velocity for the horizontal axis

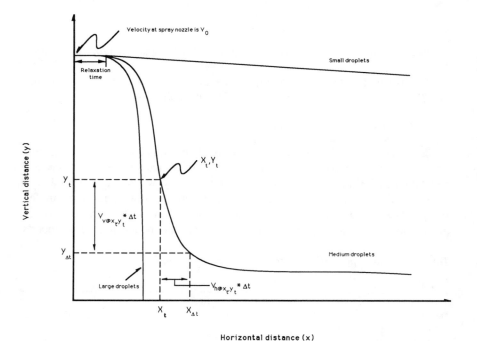

Figure 2.2. Graph showing the idea of a stepwise droplet trajectory for a bioaerosol droplet, where V_{ht} and V_{vt} are the vertical and horizontal components and Δt is the time step for idealized trajectories in a laminar flow atmospheric regime for small ($\leq 6 \mu$m), medium ($\geq$ droplet $\leq 80 \mu$m), and large ($\geq 80 \mu$m) droplets at 50% RH and 20°C.

and the terminal settling velocity (V_{TS}) for vertical axis calculations (m s^{-1}), for example, Hind [1982, Eq. (5.15)], t is the time step (s), and τ is the relaxation time (s) (see Sec. 2.1.1.3).

All the time the droplet/particle is on its trajectory path the water is evaporating and the microorganisms are dying. Droplet evaporation will be discussed later in the chapter, and microbial death in Chapter 7.

2.1.1.3. Initial Velocity

The **initial velocity** (V_0) of a forcibly emitted droplet at the beginning of the first time step in a trajectory can be calculated from Eq. (2.5) if the pressure at the source is known. The initial velocity at the source is simply that which would occur if the droplet fell from a height equal to the height of a water column which would give that source pressure.

$$V_0 = \left(\frac{2P}{\rho_{\bar{p}}}\right)^{1/2},$$
(2.5)

where V_0 is the velocity at the source nozzle (m s^{-1}), P is the pressure at the source (= $2\rho_p g h$) (g m^{-1} s^{-2}), h (m) is the height of water column that would give the pressure at the nozzle, and $\rho_{\bar{p}}$ is the density of the spray fluid plus particle suspension (gm^{-3}).

2.1.1.4. Relaxation Time

The relaxation time (τ) or adjustment time is an important descriptor of how much time it will take an aerosol droplet to go from its initial velocity to a final velocity, for example, from the velocity coming out of a nozzle and slowing to the local atmospheric velocity, or the initial velocity (V_0) of a drop falling to its terminal settling velocity. The relaxation time is a function of the droplet size and density, and atmospheric viscosity and density [Eq. (2.6), Fig. 2.3, and Table 2.2]. For bioaerosol droplets, the relaxation time may occur in a fraction of a second, even for larger droplets, so it only needs to be accounted for in the initial time steps of the droplets trajectory (Fig. 2.2). (This, of course, depends on the length of the time step used.)

$$\text{Relaxation time} = \tau = \frac{\rho_p D_p^2 C_c}{18\eta},$$
(2.6)

where C_c is the Cunningham slip correction factor (defined in Sec. 2.1.1.5).

Table 2.2. Unit density spheres.

Particle Diameter Micrometers (μm)	Slip Correction Factor	Sedimentation Velocity (cm/s)	Corrected Sedimentation Velocity (cm/s) v (S. C.)	Reynolds Number Terminal Velocity	Diffusion Coefficient (cm²/s)	Corrected Diffusion Coefficient (cm²/s)	Mobility (s/g)	Corrected Mobility (s/g)	Relaxation Time (s)	Corrected Relaxation Time (s)
0.001	2.1697+02	3.0110−09	6.5330−07	2.0046−15	2.3545−04	5.1084−02	5.8620+09	1.2719+12	3.0694−12	6.6595−10
0.002	1.0877+02	1.2044−08	1.3100−06	1.6037−14	1.1772−04	1.2804−02	2.9310+09	3.1880+11	1.2277−11	1.3354−09
0.003	7.2700+01	2.7099−08	1.9701−06	5.4124−14	7.8482−05	5.7057−03	1.9540+09	1.4206+11	2.7624−11	2.0083−09
0.004	5.4666+01	4.8177−08	2.6337−06	1.2829−13	5.8861−05	3.2179−03	1.4655+09	8.0117+10	4.9110−11	2.6848−09
0.005	4.3850+01	7.5276−08	3.3009−06	2.5057−13	4.7089−05	2.0649−03	1.1724+09	5.1410+10	7.6734−11	3.3648−09
0.006	3.6638+01	1.0840−07	3.9715−06	4.3299−13	3.9241−05	1.4377−03	9.7701+08	3.5796+10	1.1050−10	4.0484−09
0.007	3.1488+01	1.4764−07	4.6457−06	6.8758−13	3.3635−05	1.0591−03	8.3744+08	2.6369+10	1.5040−10	4.7357−09
0.008	2.7625+01	1.9271−07	5.3235−06	1.0264−12	2.9431−05	8.1302−04	7.3276+08	2.0242+10	1.9644−10	5.4266−09
0.009	2.4621+01	2.4390−07	6.0050−06	1.4613−12	2.6161−05	6.4411−04	6.5134+08	1.6037+10	2.4862−10	6.1213−09
0.01	2.2218+01	3.0110−07	6.6901−06	2.0046−12	2.3545−05	5.2312−04	5.8620+08	1.3025+10	3.0694−10	6.8197−09
0.02	1.1415+01	1.2044−06	1.3749−05	1.6037−11	1.1772−05	1.3438−04	2.9310+08	3.3458+09	1.2277−09	1.4015−08
0.03	7.8247+00	2.7099−06	2.1204−05	5.4124−11	7.8482−06	6.1409−05	1.9540+08	1.5290+09	2.7624−09	2.1615−08
0.04	6.0366+00	4.8177−06	2.9082−05	1.2829−10	5.8861−06	3.5532−05	1.4655+08	8.8466+08	4.9110−09	2.9645−08
0.05	4.9690+00	7.5276−06	3.7405−05	2.5057−10	4.7089−06	2.3399−05	1.1724+08	6.8257+08	7.6734−09	3.8129−08
0.06	4.2613+00	1.0840−05	4.6192−05	4.3299−10	3.9241−06	1.6722−05	9.7701+07	4.1634+08	1.1050−08	4.7087−08
0.07	3.7591+00	1.4754−05	5.5462−05	6.8758−10	3.3635−06	1.2644−05	8.3744+07	3.1480+08	1.5040−08	5.6538−08
0.08	3.3849+00	1.9271−05	6.5230−05	1.0264−09	2.9431−06	9.9621−06	7.3276+07	2.4803+08	1.9644−08	6.6493−08
0.09	3.0960+00	2.4390−05	7.5511−05	1.4613−09	2.6161−06	8.0994−06	6.5134+07	2.0166+08	2.4862−08	7.6973−08
0.1	2.8667+00	3.0110−05	8.6316−05	2.0046−09	2.3545−06	6.7494−06	5.8620+07	1.6804+08	3.0694−08	8.7988−08
0.2	1.8693+00	1.2044−04	2.2514−04	1.6037−08	1.1772−06	2.2006−06	2.9310+07	5.4789+07	1.2277−07	2.2950−07
0.3	1.5611+00	2.7099−04	4.2306−04	5.4124−08	7.8482−07	1.2252−06	1.9540+07	3.0505+07	2.7624−07	4.3125−07
0.4	1.4149+00	4.8177−04	6.8166−04	1.2829−07	5.8861−07	8.3283−07	1.4655+07	2.0736+07	4.9110−07	6.9486−07
0.5	1.3299+00	7.5276−04	1.0011−03	2.5057−07	4.7089−07	6.2623−07	1.1724+07	1.5592+07	7.6734−07	1.0205−06
0.6	1.2742+00	1.0840−03	1.3812−03	4.3299−07	3.9241−07	4.9999−07	9.7701+06	1.2449+07	1.1050−06	1.4079−06
0.7	1.2347+00	1.4754−03	1.8217−03	6.8758−07	3.3635−07	4.1631−07	8.3744+06	1.0340+07	1.5040−06	1.8570−06
0.8	1.2053+00	1.9271−03	2.3227−03	1.0264−06	2.9431−07	3.5472−07	7.3276+06	8.8318+06	1.9644−06	2.3677−06
0.9	1.1824+00	2.4390−03	2.8839−03	1.4613−06	2.6161−07	3.0933−07	6.5134+06	7.7017+06	2.4862−06	2.9398−06

1	1.1642+00	3.0010−03	3.5054−03	2.0046−06	2.3545−07	2.7410−07	5.8620+06	6.8245+06	3.0694−06	3.5733−06
2	1.0821+00	1.2044−02	1.3033−02	1.6037−05	1.1772−07	1.2739−07	2.9310+06	3.1716+06	1.2277−05	1.3285−05
3	1.0547+00	2.7099−02	2.8582−02	5.4124−05	7.8482−08	8.2777−08	1.9540+06	2.0609+06	2.7624−05	2.9136−05
4	1.0410+00	4.8177−02	5.0154−02	1.2829−04	5.8861−08	6.1277−08	1.4655+06	1.5257+06	4.9110−05	5.1125−05
5	1.0328+00	7.5276−02	7.7748−02	2.5057−04	4.7089−08	4.8635−08	1.1724+06	1.2109+06	7.6734−05	7.9254−05
6	1.0274+00	1.0840−01	1.1136−01	4.3299−04	3.9241−08	4.0315−08	9.7701+05	1.0037+06	1.1050−04	1.1352−04
7	1.0235+00	1.4754−01	1.5100−01	6.8758−04	3.3635−08	3.4424−08	8.3744+05	8.5707+05	1.5040−04	1.5393−04
8	1.0205+00	1.9271−01	1.9666−01	1.0264−03	2.9431−08	3.0035−08	7.3276+05	7.4779+05	1.9644−04	2.0047−04
9	1.0182+00	2.4390−01	2.4834−01	1.4613−03	2.6161−08	2.6638−08	8.5134+05	6.6322+05	2.4882−04	2.5315−04
10	1.0164+00	3.0110−01	3.0605−01	2.0046−03	2.3545−08	2.3931−08	5.8620+05	5.9583+05	3.0694−04	3.1198−04
20	1.0082+00	1.2018+00	1.2117+00	1.6002−02	1.1772−08	1.1869−08	2.9310+05	2.9551+05	1.2277−03	1.2378−03
30	1.0055+00	2.6904+00	2.7051+00	5.3734−02	7.8482−09	7.8911−09	1.9540+05	1.9647+05	2.7624−03	2.7775−03
40	1.0041+00	4.7367+00	4.7561+00	1.2614−01	5.8861−09	5.9103−09	1.4655+05	1.4715+05	4.9110−03	4.9311−03
50	1.0033+00	7.2867+00	7.3106+00	2.4255−01	4.7089−09	4.7244−09	1.1724+05	1.1763+05	7.6734−03	7.6988−03
60	1.0027+00	1.0263+01	1.0291+01	4.0994−01	3.9241−09	3.9348−09	9.7701+04	9.7966+04	1.1050−02	1.1080−02
70	1.0023+00	1.3572+01	1.3604+01	6.3250−01	3.3635−09	3.3714−09	8.3744+04	8.3940+04	1.5040−02	1.5075−02
80	1.0021+00	1.7127+01	1.7162+01	9.1217−01	2.9431−09	2.9491−09	7.3276+04	7.3426+04	1.9644−02	1.9684−02
90	1.0018+00	2.0870+01	2.0908+01	1.2505+00	2.6161−09	2.6208−09	6.5134+04	6.5253+04	2.4862−02	2.4907−02
100	1.0016+00	2.4803+01	2.4844+01	1.6512+00	2.3545−09	2.3583−09	5.8820+04	5.8717+04	3.0694−02	3.0744−02

2.1.1.5. Terminal Velocity

When bioaerosol droplets/particles fall or are drawn (or propelled) through air by gravitational force (F_g) or an electric field, their velocity increases or decreases from some initial value until the atmospheric drag force (F_D) is equal to the force field, causing it to move. In the case of gravitational force, this is the terminal settling velocity of the droplet/particle (V_{TS}). Thus, when F_D equals F_g, then V equals V_{TS}, that is,

$$F_D = F_g,$$

$$3\pi\eta V_{TS} = \frac{(\rho_p - \rho_g)\pi D_p^2 g}{6},$$

and, by rearrangement,

$$V_{TS} = \frac{(\rho_p - \rho_g)g D_p^2}{18\eta} \text{ for } D_p > 1 \text{ }\mu m \text{ and Re} < 1.0, \tag{2.7}$$

where ρ_p is the density of the particle (g/ml); ρ_g is the density of the gas (g/ml), and g is the acceleration of gravity (9.801 m s^{-2}).

The terminal settling velocity for virus size particles may be nil ($< 10^{-9}$ cm s^{-1}), whereas 100-μm agglomerations or rafts may settle at 25 cm s^{-1} (Fig. 2.3).

2.1.1.6. Cunningham Slip Correction (C_c)

The estimation of velocity for small droplets (or particles) less than or equal to 1.0 μm must be corrected for molecular slip at the droplet's surface that causes particles to fall faster than expected by Stokes law, that is,

$$V_{TS} = \frac{(\rho_p - \rho_g)g D_p^2 C_c}{18\eta}, \tag{2.8}$$

where

$$C_c = 1 + \frac{2\lambda}{D_p}\left[1.257 + 0.400 \exp\left(-1.1\frac{D_p}{2\lambda}\right)\right], \tag{2.9}$$

where λ is the mean free path which is defined as the average distance traveled by a molecule between successive collisions,

$$\lambda = 1/(2n\pi D_m^2)^{1/2}, \tag{2.10}$$

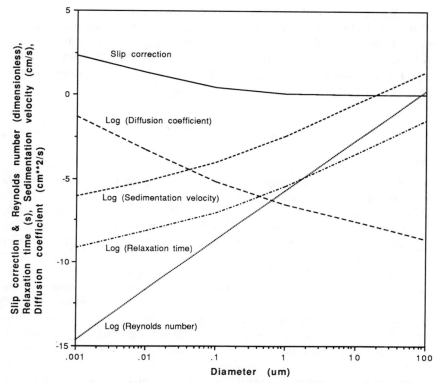

Figure 2.3. Graph of some values used in aerosol physics as a function of particle size (unit density spheres).

and is equal to 0.066 μm for air at 1 atm and 20°C, where n is the number of molecules/unit volume of air at standard condition or $2.5 \times 10^{19}/cm^3$ and D_m is the collision diameter of air molecules $(3.7 \times 10^{-8}$ cm$)$.

For droplets larger than 1 μm, the slip correction approaches unity and is often neglected (Fig. 2.3).

Droplets (or particles) greater than or equal to 1000 μm must be corrected for drag (also known as Newton's drag):

$$C_D = \frac{24n}{r_g V D_p} = \frac{24}{Re} = 1 + \frac{Re^{2/3}}{6}, \qquad (2.11)$$

where Re is the Reynolds number (see the definition in the next section).

2.1.1.7. Reynolds Number

The aerodynamic fluid flow of a bioaerosol, in a pipe for example, is characterized by a dimensionless number, the Reynolds number (Re). The Reynolds number expresses whether a flow regime is turbulent or laminar by evaluation of the inertial ($F_{inertia}$) to frictional ($F_{friction}$) forces acting on bioaerosol particles in the flow regime [Eq. (2.12), Fig. 2.3)]. It depends on the relative velocity between a movable particle such as a bioaerosol particle and the surrounding gas. It is aerodynamically equivalent for the gas to flow past a slower moving or stationary particle:

$$\text{Re} = \frac{F_{inertia}}{F_{friction}} = \frac{\rho \mu_g U D}{\eta}, \tag{2.12}$$

where U is the relative velocity between air and the bioaerosol particle and D is the diameter of the pipe.

In pipes, the flow is laminar for $\text{Re} < 2000$ and turbulent for $\text{Re} > 4000$. For a Reynolds number greater than 3 and less than 400, an error of 2% in the C_D can be expected from Eq. (2.12) or for a Reynolds number of 1000, an error less than 10%.

In the bulk atmosphere, the aerodynamic fluid flow cannot be adequately characterized by the Reynolds number as can be seen by trying to determine the "pipe size" or D in Eq. (2.12). What is that value in the bulk atmosphere situation? It is proposed that droplet/particle aerodynamic characteristics are laminar if the vertical airflow velocity component of the velocity regime is less than 10% of the variance of the horizontal velocity component. Greater than 10%, the droplet/particle would have turbulent aerodynamic characteristics. This means that in an atmospheric regime that is vertically stable, and probably with weak winds, laminar droplet/particle aerodynamics could be expected.

Example 1.1. Where will a 5-μm-diameter water droplet 2 s after emission be located in space with respect to the originating nozzle which is operating at 20 psi and pointed downwind in a 20°C, $1-\text{m s}^{-1}$ wind?

Using Eq. (2.4) to calculate the vertical and horizontal positions of the droplet with respect to the droplet emission nozzle,

$$P = 20 \text{ psi} \times \text{gravity [from Eq. (2.5)]}$$
$$= 1.41 \times 10^4 \text{ kg m}^{-2} \times 9.81 \text{ m s}^{-1}$$
$$= 138{,}232 \text{ kg m}^{-1} \text{s}^{-2},$$

$$V_0 = \left(\frac{2 \times 132{,}232}{10^3}\right)^{1/2}$$
$$= 16.6 \text{ m s}^{-1},$$

$$C_c = 1 + \frac{2 \times 0.066}{5 \times 10^{-4}} \left[1.257 + 0.400 \exp\left(\frac{1.1 \times 5 \times 10^{-4}}{2 \times 0.066}\right) \right] \text{[from Eq. (2.9)]}$$

$$= 1.03;$$

$$\tau = 1 \times (0.00005)^2 \times (1.03/18) \times (1.79 \times 10-4) \text{ [from Eq. (2.6)]}$$

$$= 8.0 \times 10^{-5} \text{ s.}$$

The velocity of the droplet in the horizontal axis (V_x) is

$$V_x = 1 - (1 - 16.6)e^{-1/8.0 \times 10E-5} \text{ [from Eq. (2.4)]}$$

$$= 1.00 \text{ m s}^{-1}.$$

Note that the relaxation time did not significantly affect the particle velocity; therefore, the distance traveled downwind from the nozzle in 2 s is

$$\text{distance} = V_x \times \text{travel time}$$

$$= 1 \text{ m s}^{-1} \times 2\text{s}$$

$$= 2 \text{ m.}$$

The velocity of the droplet in the vertical axis (V_y) is

$$V_{TS} = (1 - 1.2 \times 10^{-3}\pi \times (5 \times 10^{-6}) \times (9.81/18) \times (1.79 \times 10^{-4})$$

$$= 1.52 \text{ m s}^{-1}$$

$$V_y = 1.52 \times 10^{-2} - (1.52 \times 10^{-2} - 0)e^{-1/8.0E-5}$$

$$= 1.52 \times 10^{-2} \text{ m s}^{-1}.$$

Again the relaxation time does not significantly affect the settling velocity; therefore, the particle settles below the plane of the spray nozzle:

$$\text{distance} = V_y \times \text{travel time}$$

$$= 1.52 \times 10^{-2} \text{ m s}^{-1} \times 2 \text{ s}$$

$$= 3.04 \times 10^{-2} \text{ m.}$$

2.1.2. Nonuniform Bioaerosol Particles Trajectory

In nature, bioaerosol particles are not all spherical. The sources of these particles may be as rafts of microorganisms attached to plant debris dislodged by wind action or as the residue after droplet evaporation. In the former, the debris might be plant surface components or fungal pieces with associated microbes, both of which are growing on the plant surface (Shaffer and Lighthart, 1991). The residue from bioaerosol droplets may be formed during raindrop splash (Butterworth and

McCartney, 1991), bubble bursting during wave action, or aeration at a sewage treatment plant. Their shape affects their trajectory and can be quasi-accounted for using a dynamic shape factor (Sec. 2.1.2.1). The residue from the evaporation of droplets might pack into rhombohedrallike agglomerations of spherical microbes is estimated in Sec. 2.1.2.2.

2.1.2.1. Dynamic Shape Factor (χ)

The previous discussion assumed that the bioaerosol particles were spherical. The dynamic shape factor is defined as the ratio of the actual resistance force of the nonspherical particle to the resistance force of a spherical particle having the equivalent volume and density. In terms of the particle terminal setting velocity, the equivalent diameter, D_e, may be used:

$$V_{TS} = \frac{\rho_p D_e^2 g}{18 \eta \chi},$$
(2.13)

where D_e is the equivalent volume diameter of particle [see Hinds (1982, Section 19.1) for methods of evaluating d_e) and χ is the dynamic shape factor (see Table 2.3).

2.1.2.2. Microbial Packing

As the water evaporates from a microbial bioaerosol, the microbes will agglomerate. It is assumed that they will form the most compact arrangement energetically possible (Dallavalle, 1948). For spherical microbes in a droplet, the agglomerate would have the diameter described by Eq. (2.14).

$$D_{pack} = \left(\frac{v}{0.71}\right)^{1/3},$$
(2.14)

where v is the rhombohedral-type packed volume ($n[4/3]\pi(D_p/2)^3]$) and n is the number of spherical bacteria suspended in the initial spray droplet.

2.1.2.3. Bioaerosol Trajectory Fractionation

Simulated droplet trajectories of a polydispersed microbial aerosol in a laminar airflow regimen (Fig. 2.4) have been compared with observed dispersal patterns of aerosolized *Bacillus subtilis* subsp. *niger* spores in a quasi-laminar airflow. Simulated dispersal patterns could be explained in terms of initial droplet sizes and whether the droplets evaporated to residual aeroplanktonic size before settling to the ground. For droplets that evaporated prior to settling out, a vertical downwind size fractionation is predicted in which the microbial residue of the smallest

Basis of the Table 2.3.

Physical constants used in preparation of table:

$\lambda = 6.53 \times 10^{-4}$ cm = mean free path of gas molecules in air
$\eta = 1.810 \times 10^{-4}$ Poise = viscosity of air
$k = 1.3708 \times 10^{-14}$ erg/°C = Boltzmann's constant
$T = 29\ 3°K$ = absolute temperature (20°C)
$\rho = 1$ g cm^{-3} = particle density
$\rho' = 1.205 \times 10^{-3}$ g cm^{-3} = density of air
$g = 981$ cm/s^2 = acceleration due to gravity

Slip Correction Factor[a]

$$\text{S.C.} = 1 + \frac{2A\lambda}{d}A = 1.257 + 0.400 \exp(-1.10d/2\lambda).$$

Mobility[b]

$$B = (3\pi\eta d)^{-1}$$

Diffusion Coefficient[b]

$$D = kBT$$

Relaxation Time[c]

$$\tau = d^2\rho/18\eta$$

Reynolds number (terminal velocity)[b]

$$\text{Re} = C_D(\text{Re})^2/24 = \rho\rho'g\,d^3/18\eta^2; \qquad \text{note } C_D = \text{drag coefficient}$$

Sedimentation Velocity[b]

1. For values of Re up to 0.05:

$$v = (\rho-\rho')gd^1/18\eta$$

2. For values of Re from 0.05 to 4:

$$v = [C_D(\text{Re})^2/24 - 2.3368 \times 10^{-4}\,(C_D(\text{Re})^2)^2$$
$$+ 2.0154 \times 10^{-4}\,(C_D(\text{Re})^2)^3 - 6.9105 \times 10^{-9}\,(C_D(\text{Re})^2)^4]\eta/\rho'd$$

---

[a]Davies, C. N. 1945. Definitive equations for the fluid resistance of spheres. *Proc.* Phys. Soc. 57 (322).

[b]Green, H. L., and Lane, W. R. 1964. Particulate clouds: Dusts, smokes and mists. E. & F. N. Spon Ltd., London.

[c]Davies, C. N. (ed.) 1966. Aerosol science. Academic Press, London.

Compiled by R. A. Gussman, 1971; revised, 1980.

droplets settles the least and is found in the airstream at about sprayer height, and progressively larger droplet residues settle to progressively lower heights. Observations of spore particle size distributions from the particular spray nozzle used had three size fractions: one containing large, presumably nonevaporated droplets of ≥ 7 μm in aerodynamic diameter, and two smaller fractions, one with diameters of 2–3 μm (probably the residue of droplets containing more

Table 2.3. Tabulation of some dynamic shape factors.

Geometric Shape	Dynamic Shape Factor (χ)
Sphere	1.0
Multiple sphere	
2-sphere chain	1.12
3-sphere chain	1.27
3-compact spheres	1.18
4-sphere chain	1.32
4-compact spheres	1.17
Cylinder ($L/D=4$)	
Horizontal axis	1.32
Vertical axis	1.07

Source: Davies (1979).

than one spore) and 1–2 μm in diameter (probably the residue from single-spore droplets). As predicted by the simulation, an observed bioaerosol spray settled and progressed downwind, with the number of small droplets and particles increasing in proportion to the height and distance downwind (Fig. 2.4A, B).

2.1.3. Bioaerosol Electrostatic Charge

Gases are poor conductors of electricity; therefore, airborne droplet/particles may hold, for extended periods of time, the charge they may have gained by various mechanisms. A droplet, upon separation from an atomizing source, will retain the charge of the source or may gain charge on collision with migrating electrons or ions. This is called diffusion charging. Another charging mechanism is by collision with electrons when passing through an electric field. This is called field charging. The larger the object, the greater the charge it can attain. The estimated maximum charge an object can have in dry air is a function of size [Hesketh (1986, Fig. 4.5)]. For droplet/particles greater than 0.8 μm in diameter, field charging is the dominant charging mechanism, whereas for droplets less than 0.2 μm in diameter, diffusion charging is dominant. Upon collision with another object, the charges on the two objects tend to equalize, and because like charges repel each other, there a reduction in the probability of further collisions.

The movement of a droplet/particle through a nonconductive gas is related to the electric force field (E) by Coulomb's Law [Eq. 2.(15)]. E is a measured value.

$$F_E = qE = n_p eE, \qquad (2.15)$$

where q is the electrical charge on an object, n_p is the number of electron charges on an object, and e is the charge on one electron ($= 1.605 \times 10^{-19}$ Cs).

The terminal migration velocity (V_{TE}) of a charged spherical object in an electric force field may be calculated using a form of Stokes' Law [Eq. (2.16)]:

$$V_{TE} = \frac{\text{electric force field}}{\text{drag}} = \frac{\eta_g eEC_c}{3\pi\eta_g D_p},$$ (2.16)

where C_c is the Cunningham slip correction, D_p is the object's aerodynamic diameter, and η_g is the viscosity of gas.

2.1.4. Bioaerosol Coagulation/Agglomeration

Thermal force causing Brownian motion or gravity and eclectically caused kinematic force result in *interdroplet coagulation* and *interparticle agglomeration*. For monodispersed spherical droplet/particles, it is assumed that they adhere to each other at every collision and that the accumulated particle/droplet changes size slowly. The rate of change of the droplet concentration in a monodispersed aerosol is shown in Eq. (2.17):

$$\frac{dn}{dt} = -K_s n^2,$$ (2.17)

where

$$K_s = 4kTC_c/3\eta \approx 3.0 \times 10^{-10} C_c \,(\text{cm}^3/\text{s}),$$ (2.18)

where n is the concentration of particles/droplets, k is Boltzmann's constant (1.38×10^{-16} dyn cm/°K), η is the viscosity of air, and T is the absolute temperature.

The change in an aerosol's particle/droplet concentration in a unit volume over a given period of time due to accumulation is given by integrating Eq. (2.17) (assuming K_s is constant) as shown in Eq. (2.19):

$$N_0 = \frac{N_0}{1+N_0 K_s t},$$ (2.19)

where N_t is the number of particles at time t in a unit volume, N_0 is the initial number of particles at time $t = 0$, K_s is the coagulation coefficient [see Eq. (2.18) or Tab. 2.4), and t is the time interval.

As coagulation proceeds, the size of the particle increases as the cube root of time [Eq. (2.20)], assuming agglomeration proceeds uniformly over the surface of the particle.

$$d_t = d_0(1+N_0 K_s t)^{1/3},$$ (2.20)

where d_t is the diameter of a particle at time t and d_0 is the initial particle diameter. This function applies to relatively spherical-shaped objects.

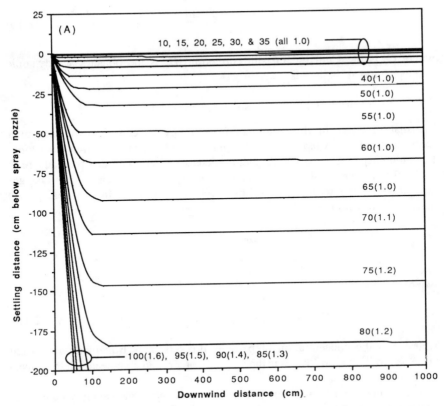

Figure 2.4. Calculated deposition trajectory of various-sized, viable, evaporating bio-aerosol droplets sprayed (at 20 psi) from a 1-m height in a quasi-laminar flow (at 0.163 m s^{-1}) airstream at 22°C and 50% RH showing (A) small droplet and particle trajectories (numbers in the body of the figure represent particle diameters in micrometers with packed residue diameters in parentheses), and (B) large droplet (200–500 μm in diameter) trajectories. [From Lighthart et al. (1991).]

For polydispersed bioaerosols, the coagulation coefficients become a function of the two interacting particle/droplet sizes, for example,

$$K_{1,2} = \pi(d_1 + d_2)(D_1 + D_2), \tag{2.21}$$

where d_1 and d_2 are two different particles and D_1 and D_2 are the diffusion coefficients of d_1 and d_2, respectively, where

$$D_{1 \text{ or } 2} = kTC_c/3\pi\eta d_{p1 \text{ or } 2}, \tag{2.22}$$

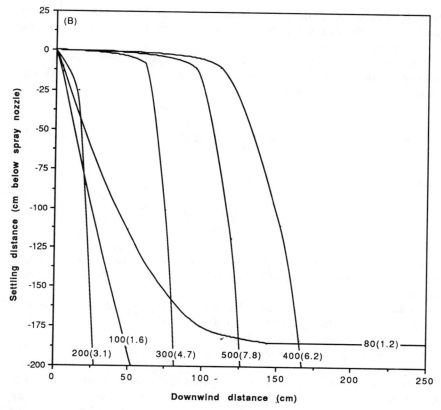

Figure 2.4. (*Continued*).

where k is Boltzmann's constant, T is the absolute temperature, d_p is the particle diameter, and η is the viscosity of air. Equation (2.22) is known as the Stokes–Einstein equation.

Note that coagulation will proceed more rapidly with coagulation coefficients for polydispersed [e.g., Eq. (2.21)] than monodispersed [e.g., Eq. (2.18)] bioaerosols. Further consideration shows that when large particles adhere to small

Table 2.4. *Coagulation coefficients* (K_s) *at standard conditions.*

Diameter (μm)	Coagulation Coefficient (cm³ s⁻¹)
0.01	6.7×10^{-10}
0.1	8.6×10^{-10}
1.0	3.5×10^{-10}
10.0	3.0×10^{-10}

ones, the size and mass of their combination alter the size or mass of the large particle relatively little and leads to the notion that large particles "consume" or "mop up" small bioaerosol particles.

The capture efficiency (E) for a relatively large particle with a diameter d_{pL} by another with smaller diameter, d_{pS}, moving at a relative velocity of δV with Stokes' number (Stk) of ≥ 0.1 is

$$E = \left(\frac{\text{Stk}}{\text{Stk}+0.25}\right)^2,$$ (2.23)

where

$$\text{Stk} = \frac{\rho_p D_{pS} C_c d \delta V}{18 n \eta D_{pL}}.$$ (2.24)

Thus, a 1-μm-diameter bacterium falling at its terminal velocity (V_{TS}), colliding with an 80-μm-diameter droplet falling at its V_{TS}, would have about a 1% capture efficiency.

Additional consideration of coagulation of polydispersed aerosols is given in Hinds (1982) and Mercer [in Shaw (1978)].

2.1.5. Bioaerosol D/P Size Distribution

The distribution of bioaerosol droplets/particles (D/Ps) in the atmosphere varies from a single virus, bacterium, or fungal spore, to agglomerations of microorganisms which may or may not be "rafted" on plant debris or soil particles. The spectrum of D/P sizes have been arbitrarily differentiated into intersection size classes based on sampler design. Each of the size classes could be (but have not been) further categorized into subsets of microbial species in various catagorizable physiological states; each state more or less able to resist the rigors of the aerial environment (e.g., solar irradiation, drying and temperature sensitivity or resistance).

Aerodynamic size will greatly affect the potential for a D/P on a surface to be entrained into the atmosphere and its subsequent time aloft. For example, larger D/Ps will have a greater drag force exerted on them by the wind to overcome surface adhesion or other attachment mechanisms. Large-size D/Ps will also tend to settle out of the atmosphere faster (Lighthart et al., 1991). Additionally, preliminary investigations have indicated that microbial survival in bioaerosols is a function of particle size; larger particles remain viable longer (Sorber et al., 1976; Lighthart et al., 1991). Figure 2.5 shows the particle size distribution of bacteria and fungi found in the urban atmosphere (Lighthart et al., 1979). Note that more than the majority of the bacteria were greater than 8.2

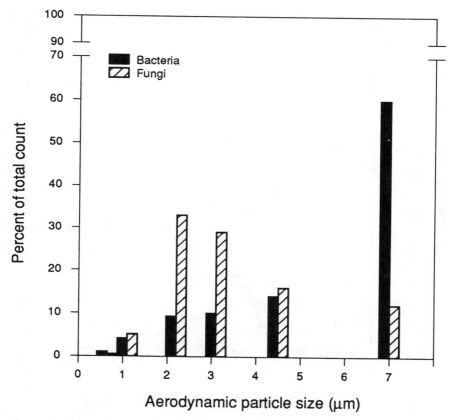

Figure 2.5. Mean percent bacterial and fungal particle size distribution from samplers collected almost daily for 9 months at an outdoor urban location.

μm aerodynamic diameter, whereas most of the fungi were 2–5 μm aerodynamic diameter. This suggests that the bacteria which are about 1 μm in diameter are agglomerated or rafted on something, and the fungi are quite often individual.

Particle size distributions can be displayed on log-probability paper to yield the quantitative distribution information for comparison purposes in terms of the count median diameter (CMD is defined as the 50% cumulative size) of the particles and the slope of their regression line. [See Hinds (1982) for a excellent description of this process.] For example, the upwind and downwind D/P size distributions of airborne bacteria from a large activated sludge wastewater treatment plant (WWTP) are shown in Fig. 2.6. The two distributions are characterized by a relatively small (CMD of approximately 3 μm aerodynamic diameter), broad size range (shallow slope) of upwind D/Ps, and large (CMD of approximately 10 μm aerodynamic diameter), narrow size range (high slope) downwind D/Ps. It

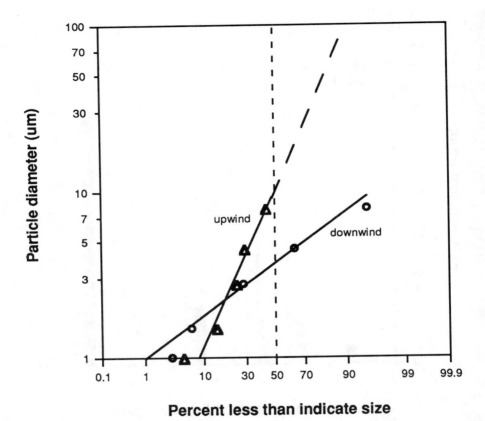

Figure 2.6. Lognormal microbial bioaerosol D/P size distribution of upwind and down-wind air at an urban wastewater treatment plant.

would appear that the small number of D/Ps in the upwind air relative the large number being generated by the WWTP did not significantly contribute to the downwind population. This suggests that the differences in the particle size distributions could be used to trace the bioaerosol emitted by the WWTP.

Measurements of aerodynamic diameters of viable bioaerosol D/Ps are made with the Anderson cascade impactor (see Chapter 6) and are conveniently displayed on log-probability paper as described in Hinds (1982, p. 89). Note that extrapolation of the fitted line through the data points shows that less than 50% of the viable D/Ps upwind of the WWTP are greater than or equal to the 50% cutoff size of the last stage of the Anderson sampler used to make these measurements. This means that more than 50% of the viable D/Ps are greater than 8.2 μm aerodynamic diameter. Extramural samples taken at other locations seem to indicate that this is generally true.

2.2. Bioaerosol Chemistry (Salt Effect)

The rate of droplet size change due to evaporation or condensation of water from and to the droplet is dependent on the bulk air temperature and relative humidity, the temperature at the surface of the droplet, the droplet salt concentration, the size of the droplet, and how far from the droplet the kinetic motion of the water vapor molecules are restricted. The last two factors are known as the Kelvin effect and Fuchs effect, respectively. The Kelvin effect is due to the decreased energy necessary for water molecules to escape from a droplet surface as the curvature of the surface increases, that is, the droplet gets smaller due to evaporation. The Kelvin effect is important for droplets less than 0.1 μm in diameter such as virus-containing bioaerosols.

The rate of change in droplet size as a result of evaporation at each time step may be calculated with Eq. (2.25), where the change in droplet diameter (dD_p) with respect to the simulation time step in our droplets trajectory is the product of the first two terms on the right of the equation describing evaporative effects, including the Kelvin effect on droplet size, and the last term, which describes the Fuchs effect on the mass transfer rate of water vapor away from the droplet (Davies, 1978).

$$\frac{dD_p}{dt} = \frac{4D_v M}{R \rho_p D_p} \left(\frac{p_\infty}{T_\infty} - \frac{p_d}{T_d} \right) \left(\frac{2\lambda + D}{D_p + 5.33 \left(\frac{\lambda^2}{D_p} \right) + 3.42\lambda} \right), \qquad (2.25)$$

where D_v is the diffusion coefficient of water in air at the ambient temperature, M is the molecular weight of water, R is the gas constant, p_∞ is the ambient water vapor pressure (i.e., the relative humidity), p_d is the saturation water vapor pressure at droplet surface temperature, T_∞ is the ambient temperature, T_d is the droplet surface temperature, λ is the mean free path of air molecules (0.066 μm).

For each simulated time step, the change in droplet surface temperature, the critical site for droplet evaporation, is given in Eq. (2.26). The surface cooling of the droplet at each time step, δT, in the simulation of a droplet's trajectory, is subtracted from the present droplet temperature, D_{d-1}, and the new temperature, T_d, used for the next simulated time step, that is, $T_d = T_{d-1} - \delta T$. The equation expresses the heat loss by droplet evaporation in the left-hand term and heat gain by the conduction from the air to the droplet in the right-hand term.

$$dT = \frac{3t}{r^2 \rho_p C_p} \left[\frac{D_v MH}{R} \left(\frac{p_d}{T_d} - \frac{P_\infty}{T_\infty} \right) - k(T_\infty - T_d) \right], \qquad (2.26)$$

where δT is the temperature change increment at each time step, t is time, r is the droplet radius, C_p is the specific heat of water vaporization as a function of ambient temperature, H is the latent heat of fusion for water as a function of ambient temperature, and k is the thermal conductivity of air.

Droplets of solution cease to evaporate when the partial pressure of water vapor escaping the droplet surface is the same as the ambient relative humidity. Thus, solutes in the droplet solution will affect the vapor pressure at the droplet surface and, consequently, its evaporative rate. (Imagine what the concentration of salt would be in a 100-μm-diameter 0.01% (W/V) aqueous solute droplet that evaporated to one-hundredth its original diameter. How might impurities in the solution change?)

For droplets with a solute such as sodium chloride, there is a change in the solvent vapor pressure of the droplet as evaporation of the solvent proceeds. The change in vapor pressure may be calculated with Eq. (2.27) and the value for P_d used in Eq. (2.25).

$$P_d = P_s \left(1 + \frac{6imM_w}{M_s \rho \pi D_p^3} \right)^{-1} \exp\left(\frac{4\gamma M_w}{\rho RTD_p} \right), \qquad (2.27)$$

where P_s is the saturation vapor pressure of solvent at ambient conditions, i is the number of ions formed when solute dissolves, for example, 2 for NaCl, m is the mass of dissolved solute with weight M_s, M_w is the molecular weight of the solvent, M_s is the molecular weight of the solute, ρ is the density of the solvent, R is the gas constant, and T is the absolute temperature.

Note that the first term in parentheses in Eq. (2.27) accounts for the reduced vapor pressure of the solvent due to the solute in the droplet and the second term accounts for the droplet curvature (i.e., the Kelvin effect).

During each time step of our simulated droplet trajectory, the ion concentration changes until the droplet solvent is saturated with solute and then the solute crystallization occurs; therefore, it must be recalculated at each time step until solvent saturation occurs.

Other physical and chemical topics that are beyond the scope of this book include some factors that affect droplet evaporation such as ambient wind conditions, atmospheric pressure, and surface films on the droplet. Relatively nonvolatile particles like microorganisms suspended in the droplet solution may be accounted for by assuming they randomly get trapped in the droplets surface film, causing a reduction in the evaporative surface equal to their projected area.

References

Adamson, A. W. 1982. Physical chemistry of surfaces. 4th ed. John Wiley & Sons, New York.

Atkins, G. L. 1969. Multicompartment models for biological systems. Methuen & C. Ltd., London. 152 pp.

Butterworth, J. and H. A. McCartney. 1991. The dispersal of bacteria from plant surfaces by water splash. J. Appl. Bacteriol. 71:484–495.

Chemical Rubber Publishing Co. 1953. Handbook of chemistry and physics. Chemical Rubber Publishing Co., Cleveland, OH.

Dallavalle, J. M. 1948. Micrmeritics. The technology of fine particles. Pitman Publishing Corp., New York.

Davies, C. N. 1978. Evaporation of airborne droplets, pp. 135–164. In D. T. Shaw (ed.), Fundamentals of aerosol science. John Wiley & Sons, New York.

Davies, C. N. 1979. Particle fluid interaction. J. Aerosol Sci. 10:477–513.

Hesketh, H. E. 1986. Fine particles in gaseous media. Lewis Publishers, Inc. Chelsea, MI.

Hinds, W. C. 1982. Aerosol technology. John Wiley & Sons, New York.

Lighthart, B., J. C. Spondlove, & T. G. Akers. 1979. Bacteria & viruses, pp. 11–2. *In* Aerobiology—The ecological systems approach. R. L. Edmonds (Ed.), Dondon, Hutchinson & Ross, Inc., Stroudsburg, PA.

Lighthart, B., B. T. Shaffer, B. Marthi, and L. Ganio. 1991. Trajectory of aerosol droplets from a sprayed bacterial suspension. Appl. Environ. Microbiol. 57(4):1006–1012.

Reist, P. C. 1984. Introduction to aerosol science. Macmillan Publishing Co., New York.

Shaffer, B. R., and B. Lighthart. 1991. Proc. Amer. Soc. Microbiol. Los Angeles, CA.

Shaw, D. T. (ed.). 1978. Fundamentals of aerosol science. John Wiley & Sons, New York.

Sorber, C. A., H. T. Bausum, S. A. Shaub, and M. J. Small. 1976. A study of bacterial aerosols at a wastewater irrigation site. J. Water Pollut. Control Fed. 48:2367–2379.

3

Atmospheric Environment of Bioaerosols

Jinwon Kim

3.1. Introduction

Bioaerosols are influenced by the atmosphere from the first moment they are exposed to the environment. Therefore, knowledge of local weather and climate is an important factor in planning the time and location of aerosol release and for assessing the area and period influenced by bioaerosol dispersal and survival. The most important influences of the atmospheric environment on bioaerosols are the dispersion and evaporation of droplets and the survival of organisms contained in the droplets. Combined effects of the dispersion, removal, and survival of microorganisms in sprayed droplets determine the area and period affected by sprayed aerosols.

Aerosols such as sprays are carried away from the generation area by the wind. The size and mass of individual aerosols are so small that individual droplets rapidly adjust to the movement of the air (i.e., wind). Therefore, the movements of sprayed aerosols can be described by the local wind of the spray site.

During their sojourn, aerosols are removed from the air to the Earth's surface. Gravitational settling and scavenging are the most important mechanisms to remove airborne particles. Terminal velocity of a particle or droplet usually increases with size. Therefore, atmospheric temperature and relative humidity (RH) play a major role in aerosol dispersion and settling by affecting the evaporation of droplets. That is, upon evaporation to smaller droplets, aerosols reach their terminal velocity faster than larger droplets (Fig. 3.1). Airborne aerosols are also removed from the air by scavenging. During precipitation, rainwater or snow collects airborne particles including aerosols and brings them to the ground (wet scavenging). When particles carried by the air hit the surface of an object (ground, plant, water, etc.), some of them stick to the surface (dry scavenging).

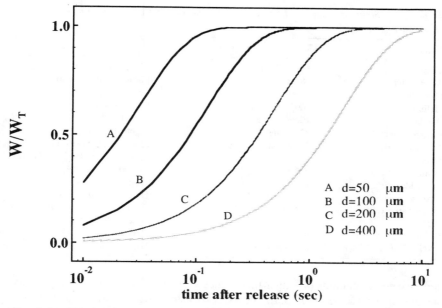

Fig. 3.1. The ratio between the instantaneous speed (w) to terminal velocity (w_T) of particles of different sizes as a function of time after release

Survival of microorganisms contained in sprayed droplets is also influenced by various atmospheric conditions such as solar radiation, temperature, RH, and chemical composition of air. Some factors affecting airborne bacterial survival are shown in Table 3.1.

A comprehensive understanding of short- and long-term local atmospheric environment requires the knowledge of atmospheric motions of various time and length scales ranging from turbulent eddies to planetary-scale circulations. Local atmospheric conditions such as wind speed, temperature, and RH are strongly influenced by the features of large-scale flow field, geographical locations, and local topography. The average temperature and wind field on the globe shows characteristic dependence on the season, latitude, and relative location of the major continents and oceans. Atmospheric turbulence responsible for diffusion during transport by the mean wind is strongly influenced by the local atmospheric conditions and diurnal variation of solar radiation reaching the ground. The local precipitation is mostly determined by the large-scale flow; however, surface characteristics such as topography and land–sea contrast can have a significant impact in the detailed pattern of local precipitation. Usual concentration of precipitation on the upslopes of mountain ranges and winter storms near the Great Lakes are good examples of such effects.

Mountain ranges on the Earth's surface also affect local climate by shaping the

Table 3.1. Some atmospheric factors affecting the survival of microorganisms (Lighthart et al. 1979)

Factor	Level at which effects seen	Reference
Ultraviolet	1×0.75 (mW cm^{-2}) $\times$ s	Riley & Kaufman (1972)
Temperature	-40° C	Ehrlich *et al.* (1970a)
Oxygen	0–30% of 1 atm	Cox *et al.* (1974)
Helium	100% of 1 atm	Cox (1968b)
Argon	100% of 1 atm	Cox (1968b)
Carbon monoxide	85 ml/liter	Lighthart (1973)
Sulfur dioxide	2.5 mg/m^3	Lighthart *et al.* (1971)
Ozone		Elford & Van den Ende (1942)
Nitrogen dioxide	2.5 ppm	Chatigny et al. (1973)
Formaldehyde	1 ml/liter	Won & Ross (1969)
Open air factor	not known	Druett (1973b), Harper (1973)
Water vapor (RH)	0–100%	Dimmick & Akers (1969)
Pressure		Druett (1973a,b)
Oxygen	0–100%	Hess (1965)
PAN	not available	Hacumin *et al.* (1964)

large-scale circulations of longer time scales as well as inducing topographically generated mesoscale (Fig. 3.10) flows. Downslope wind storms and the orographic convection (convection induced by lifting the low-level flow by topographic slopes) are examples of highly localized topographic influence on the local flow field. Thermal contrast due to differential heating for different types of the Earth's surface following the distribution of mountain–valley, soil moisture, vegetation, land, and water induces characteristic local circulations. The examples of thermally induced circulations are the mountain (valley) winds and land (sea) breezes during the nighttime (daytime).

3.2. Structure of the Mean Atmosphere

The mean state of the Earth's atmosphere may be viewed as the consequence of the spatial temperature gradient due to differential radiative heating of the Earth–atmosphere system and the subsequent motion of the atmosphere, which transports heat to alleviate such temperature gradient. The solar radiation per unit area arriving on the Earth's surface varies significantly depending upon the season and latitude. The absorption of the incident solar radiation by the Earth–atmosphere system varies following the chemical composition of the air and the characteristics of the underlying Earth's surfaces. As a consequence, the mean state of the Earth's atmosphere exhibits characteristic structure in both the vertical and the horizontal planes for each season owing to the variation of solar heating, differential responses of water and land surfaces to solar heating, and subsequent atmospheric motions.

When the global and time mean structure is considered, the Earth's atmosphere is stably stratified with warm, light air located above cold, dense air. Note that even though the atmospheric mean temperature decreases with height, if an air parcel at higher altitude is brought down to the lower level without heat exchange with its surroundings during the descent (adiabatic process), its temperature at the lower level becomes warmer than the new environment owing to compressibility. As a result, the vertical movement of air is strongly restricted by stratification. The horizontal variation of the atmospheric mean temperature is largest in the meridional (north–south) direction with a maximum at the middle latitude of the winter hemisphere showing the importance of latitudinal variation of incoming solar radiation in determining the atmospheric time and spatial mean state.

3.2.1. Vertical Structure of the Atmosphere

The atmospheric wind and temperature vary most significantly in the vertical direction. For example, the wind speed varies by an order of 100 ms^{-1} over a few tens of kilometers in the vertical direction, while it changes by an order of 10 ms^{-1} over a few thousand kilometers in the horizontal direction. The variation of the atmospheric temperature is hundreds of times larger in the vertical direction than in the horizontal direction. For example, in the lowest 10 km of the Earth's atmosphere, the mean temperature decreases in the vertical direction by about 60 K/10 km while the mean temperature changes in the horizontal direction by about $10 \text{ K}/10^3$ km.

The vertical structure of the mean atmosphere is often divided into four distinct layers following the vertical distribution of thermal structure (Fig. 3.2). These layers are called the troposphere, stratosphere, mesosphere, and thermosphere in order of increasing height. The boundaries of these layers are called the tropopause, stratopause, and mesopause.

The troposphere contains approximately 80% of the total mass of the atmosphere and virtually all of the atmospheric water, i.e., water vapor, liquid water in the forms of cloud water and rain, and ice such as cloud ice, snow, graupel, and hail. The depth of the troposphere varies from less than 8 km in the polar regions to 15 km in the equatorial region. The mean temperature in the middle latitude troposphere decreases vertically at approximately 6.5 K/km. This is a relatively weak stratification and sometimes allows the vertical movement of the air to cover the entire depth of the troposphere as often seen from the development of deep convective systems such as thunderstorms. The lowest part of the troposphere is usually called the atmospheric boundary layer (ABL) where the direct exchange of momentum, heat, moisture, and other pollutants between the atmosphere and the underlying surface occurs.

The stratosphere is located immediately above the troposphere, and is separated from it by the tropopause. The lowest 10 km of the stratosphere is approximately isothermal (constant temperature); then the temperature increases gradually up to the stratopause. As a result, the stratosphere is so strongly stratified that

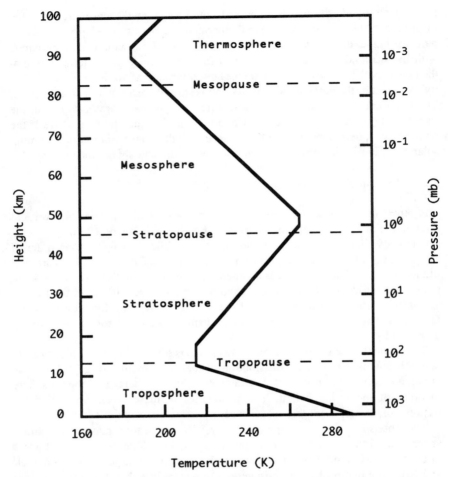

Fig. 3.2. Vertical temperature structure of the U.S. Standard Atmosphere.

the vertical movement of air is severely restricted. Owing to this very stable stratification, pollutants ejected into the stratosphere have very long residence times, especially if they are small particulates or neutrally buoyant and chemically stable gases. Across the tropopause, there is a sudden decrease of the water vapor concentration while the ozone concentration increases sharply within the lowest few kilometers of the stratosphere. Absorption of solar radiation by ozone plays a crucial role in determining the thermal structure and atmospheric motions in the stratosphere. The stratosphere and troposphere combine about 99% of the total atmospheric mass.

The mesosphere is about 40 km deep. The temperature in the mesosphere decreases with height as in the troposphere, but at a rate somewhat less than that in the troposphere. The weak stratification in the mesosphere allows less restricted vertical movement of air compared to that in the stratosphere. The mesosphere contains very little water vapor; however, a thin cloud is sometimes formed in the mesosphere by strong vertical motion. Ozone concentration in the mesosphere decreases with height primarily because of the reduced availability of oxygen.

The thermosphere extends up to several hundred kilometers from the mesopause. The temperature increases with height in the thermosphere. The temperature in the thermosphere is strongly influenced by the solar activity, especially in the upper part of the thermosphere, with a higher temperature for the period of stronger solar activity. Owing to high temperature and low particle concentration, the molecular mean free path, which is the average distance a molecule can travel without colliding with another molecule, is larger in the thermosphere. As a result, the composition of the atmosphere becomes heterogeneous so that lighter gases such as the hydrogen and helium become more abundant than heavier gases such as the oxygen and nitrogen. On the other hand, the composition of air is almost uniform below the mesopause (about 90 km above the ground). For that reason, the layer above the mesopause is often called the heterosphere, while the layer below the mesopause is called the homosphere or turbosphere.

3.2.2. Meridional Structure of the Atmosphere

Variation of the atmospheric mean structure in the north–south direction is the consequence of the local imbalance in the radiative heat budget and the heat transport by the atmospheric and oceanic circulations to alleviate such local radiative imbalance. Meridional variation of the zonal mean (averaged along the latitude circle) temperature is shown in Fig. 3.3a. When local radiative heat budget of the Earth–atmosphere system is concerned, there is a significant imbalance between the absorption of solar radiation and the emission of terrestrial long-wave radiation in the meridional (north–south) direction. When averaged over the depth of the atmosphere, absorption of the solar radiation (heat gain) exceeds the emission of the long-wave radiation (heat loss) in the equatorial regions, and the polar regions emit more long-wave radiation (heat loss) than the incoming solar radiation (heat gain). Observations indicate that the actual difference of atmospheric temperature between the poles and the equator is about one-half that expected from the radiative equilibrium temperature distribution (the latitudinal distribution of the mean temperature in the absence of atmospheric motions so that there is an local equilibrium between the solar heating and infrared cooling for a column of the atmosphere). In other words, polar regions lose more heat than they receive through the radiative processes, and vice versa in the tropics. The excess radiative heat input in the equatorial regions is balanced by the amount of the heat transported from the equator to the poles by the

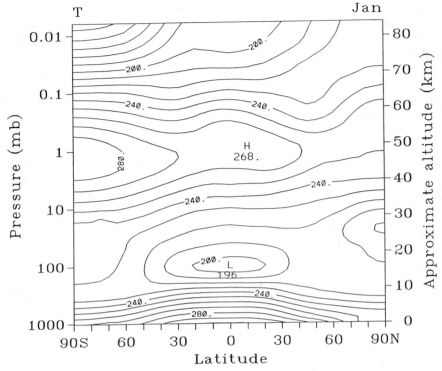

Fig. 3.3. (a) Meridional cross section of zonal mean temperature during January (courtesy of W.-K. Choi). The unit is in degrees Celsius. (b) Meridional cross section of zonal mean wind during January (courtesy of W.-K. Choi). The unit is in ms^{-1}.

atmospheric and oceanic meridional motions. About half of the required heat transport to compensate for the imbalance due to radiative transfer is carried out by atmospheric motions and the other half of the required heat is transported by oceanic circulations.

In the lower troposphere, the temperature decreases toward the pole while the tropopause over the tropics is higher and colder than that over the polar regions. The largest meridional variation of the zonal mean tropospheric temperature appear over the middle latitude between 30° and 50° latitude in the winter hemisphere. In the summer hemisphere, meridional variation of tropospheric temperature is also largest in the middle latitude, but its magnitude is smaller than that in the winter hemisphere. This meridional temperature variation is closely related to the vertical variation of zonal mean wind.

The lower stratosphere is characterized by the warmer pole and colder equator in the summer hemisphere. In the winter hemisphere, the lower stratospheric temperature is distinctly warmer over the middle latitude with cold pole and

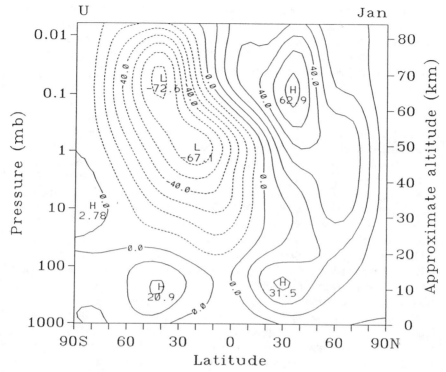

Fig. 3.3. (*Continued*).

equator. Sometimes the lower stratospheric temperature over the winter pole increases suddenly by as much as 70 K over approximately one week. This phenomenon is often called the sudden warming. When this occurs, the strato-spheric circulation around the winter pole is sometimes completely reversed. The temperature decreases from summer pole to winter pole in the upper strato-sphere near the stratopause. Mesospheric temperature has a minimum over the summer pole and increases monotonically toward the winter pole.

This structure and strength of zonal mean wind is closely related to the longitu-dinal gradient of the mean temperature via the thermal wind relationship explained in Subsection 3.3.2.3. In most of the troposphere, the mean wind is westerly except in the tropics where easterly wind (trade wind) prevails. Tropospheric easterly wind also appears in the shallow layer at high latitudes (Fig. 3.3b). The maximum westerly jet (the westerly wind of maximum intensity) in the tropo-sphere appears near the subtropical tropopause. The westerly jet in the summer hemisphere also appears near the tropopause level but with weaker intensity than in the winter hemisphere.

The zonal mean wind in the stratosphere and mesosphere slows significant difference following each season. In the winter hemisphere, the westerly flow extends up to the mesopause with maximum jet speed in the middle of the mesosphere. In the summer hemisphere, westerly wind extends only up to the lower stratosphere. Above this level easterly winds appear with maximum intensity near the middle of the mesosphere.

3.2.3. Atmospheric Boundary Layer

Aerosols are usually sprayed near the ground within the atmospheric boundary layer (ABL). Because they evaporate and settle to the underlying surfaces in a relatively short period under typical atmospheric conditions, dispersion of aerosols is greatly influenced by the local atmospheric environment in the vicinity of the spray site. Owing to the relatively small mass exchange between the ABL and the atmosphere above the ABL, except during the periods when strong convective systems develop, most aerosols sprayed near the ground will stay within the atmospheric boundary layer.

The ABL may be defined as "the part of the troposphere that is directly influenced by the presence of the Earth's surface, and responds to surface forcings with a time scale of about an hour or less" (Stull 1990). In contrast to the ABL, the portion of the atmosphere lying above the ABL is often called the free atmosphere because the underlying surfaces have less direct influence on the atmosphere above the ABL than within the ABL. The ABL is usually turbulent, occupying the lowest several tens of meters to a few kilometers of the troposphere. Convection induced by heated surface or mechanical stirring by wind shear (spatial variation of wind speed and direction) and surface roughness (unevenness of the surface that disturbs the wind near the surface) are the most important causes of turbulence in the atmospheric boundary layer.

The structure of the ABL is closely related to the temperature of the underlying surfaces and the characteristics of the low-level flow, such as wind speed, stratification, and humidity. The temperature of the land surface is determined by the partitioning of the available radiative energy (incoming solar radiation and the downward atmospheric long-wave radiation) into various components of heat fluxes, such as sensible heat flux, latent heat flux, soil heat flux, upward long-wave radiation, from the land surface. Note that the signs of the sensible, latent, and soil heat fluxes indicate the direction of the heat flux. A schematic of the surface energy budget is provided in Fig. 3.4. The most important factors that determine the surface energy budget are the soil moisture content and vegetation. Snow cover is also important in determining the amount of reflected solar radiation at the surface. The most immediate response of the land surface to the radiative heating (or cooling) occurs over a shallow layer at the land surface. On the other hand, solar radiation can penetrate for some depth into water and is absorbed over the depth. Therefore, combined with the large heat capacity of water, the

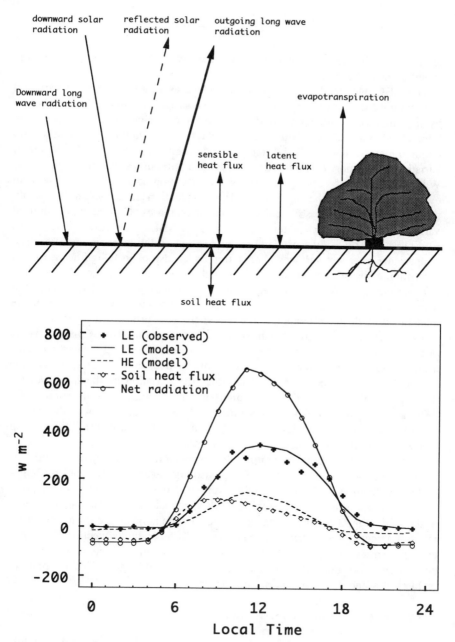

Fig. 3.4. (a) A schematic illustration of energy budget over the land surface. (b) An example of energy balance over the land surface: Latent heat flux [LE: solid cross is the observed values over the forest site during the Hydrological Atmospheric Pilot Experiment (Gash *et al*. 1989) and solid line is a model result], sensible heat flux (H: dashed line), soil heat flux (dashed line with open diamond), and net radiation (solid line with open circles).

temperature change is much slower over the ocean than over the land for the same radiative heating.

The ABL over land surfaces usually undergoes significant diurnal variation following the daytime solar heating of the land surface and the nighttime cooling by the emission of long-wave radiation from the ground surface; deep (a few kilometers) and well mixed during a sunny daytime and shallow (less than several hundred meters, in general) and stably stratified at night (Fig. 3.5a). Such diurnal variation is the most pronounced over dry land without significant cloud cover. When the land surface is wet, most of the incoming solar radiation is used to evaporate water from the land surface. Heat conduction and the soil heat capacity tend to increase as the soil moisture content increases. Consequently, diurnal variation of the land surface temperature and the ABL over wet land (Fig. 3.5b) is not as dramatic as over dry land owing to smaller diurnal variation of surface temperature over wet surfaces. Vegetation cover also plays a significant role in determining the surface energy budget at the land surface (hence, the surface temperature). Plants bring up water from the root zone, which is usually too deep for direct interaction with the atmosphere. Therefore, vegetation over the land can provide some evaporation even when near surface soil layer is dry.

The structure of the ABL over the ocean is also determined by the characteristics of the low-level air flow and the sea surface temperature (SST). Over the ocean, the magnitude of SST variation is much smaller than over the land surface for a period of several days or so. Therefore, the variation of the low-level air moving over the ocean is the most important factor, causing the variation of the ABL structure over the ocean. Warmer (colder) air moving over the colder (warmer) sea surface will develop a shallow and stably stratified (relatively deep and well mixed) atmospheric boundary layer.

In addition to the diurnal variation following heating and cooling of the underlying surface, the structure of the ABL is also significantly influenced by large-scale-flow features such as mean wind, vertical motion, and temperature advection. Large-scale subsidence (sinking motion) can substantially reduce the depth of the ABL (Fig. 3.6) while a strong low-level wind is favorable for the development of a deeper atmospheric boundary layer. Temperature advection also influence the structure of the ABL by changing the thermal structure within. The characteristics of the ABL are also greatly influenced by the existence of low-level clouds and development of convective systems.

3.2.3.1. Convective Boundary Layer

When the underlying surface is warmer than the overlying air, the ABL exhibits quite uniform distributions of the potential temperature (temperature that a hypothetical air parcel will obtain when it is moved from its original pressure level to the 1000 mb level without heat exchange with its environment) and other pollutants in the vertical direction (Fig. 3.7). Such an ABL is sometimes called

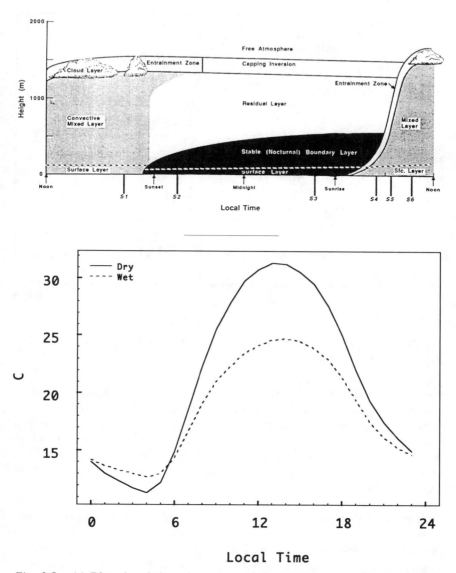

Fig. 3.5. (a) Diurnal variation of the atmospheric boundary layer (after Stull 1989, reprinted by the permission of Kluwer Academic Publishers). (b) Diurnal variation of the surface temperature during a sunny day over dry and wet land surfaces.

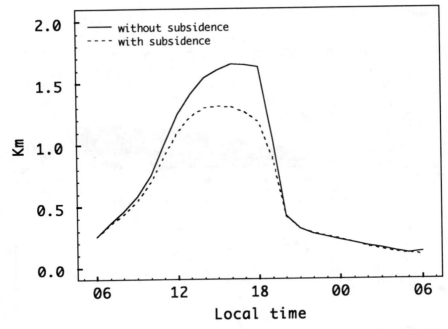

Fig. 3.6. Diurnal variation of the depth of the ABL with (solid line) and without (dashed line) large scale subsidence.

the mixed layer. The convective boundary layer develops when the surface heating is strong owing to daytime solar heating of the ground surface or when cold air moves over warm surfaces. A well-mixed layer is usually capped by a thin inversion layer which marks the top of the mixed layer (e.g., Θ in Fig. 3.7). This inversion layer is an effective barrier for the exchange of the atmospheric mass and other properties between the mixed layer and the overlying free atmosphere.

Owing to strong vertical mixing in the convective boundary layer, pollutants ejected within the convective boundary layer are mixed rapidly owing to strong turbulence, soon contaminating the entire boundary layer. Such a phenomenon is often called fumigation. Strong vertical movement of air in the convective boundary layer increases the chance for the pollutants to be removed through the deposition process in the region relatively close to the source of the pollutants.

3.2.3.2. Stable Boundary Layer

During the night, the ground surface is cooled by the emission of long-wave radiation and the atmosphere near the ground surface becomes stably stratified. For a stably stratified boundary layer, mechanical stirring of air flow due to

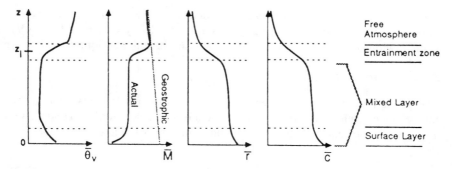

Fig. 3.7. Profiles of the mean virtual potential temperature (θ_v), wind speed (M), water vapor mixing ratio (r), and pollutants concentration (c) in the mixed layer (after Stull 1990, reprinted by the permission of Kluwer Academic Publishers).

surface friction and wind shear become major sources of turbulence. A stable boundary layer also develops when warm air advects over cold surfaces. In some cases, a well-mixed layer appears on top of the stable boundary layer, which is, in many occasions, a remnant of the well-mixed daytime boundary layer (residual layer).

Stable stratification suppresses vertical motion in the stable boundary layer so that turbulence in the stable boundary layer is usually weak and intermittent (Mahrt 1985). Strong stability also enhances the horizontal component of wind fluctuations compared to vertical movement so that dispersion is more lateral than vertical. Pollutants, including microbial air pollution (MAP), ejected in the stable boundary layer are not well dispersed in the vertical but are mostly carried away from the source of pollutants by the mean wind.

3.3. Atmospheric Motions

The atmosphere is a mixture of gases. The motion and the state of the atmosphere may be represented by the basic governing equations of fluid mechanics, derived from the conservation principles of mass, momentum, and thermodynamic energy. The equation of state is also used to relate the atmospheric pressure, density, and temperature.

3.3.1. Conservation of Mass

Conservation of mass states that the net change of mass in a fixed volume element is the same as the net mass flux across the boundary of the volume element. In a differential form, the conservation of mass is written:

$$\frac{\partial \rho}{\partial t} + \nabla \cdot (\rho \mathbf{V}) = 0, \tag{3.1}$$

where $\partial\rho/\partial t$ is the change of atmospheric density with time at a fixed location in space, ρ the density, $\mathbf{V}$ is the three-dimensional velocity vector, and $\nabla \cdot (\rho\mathbf{V})$ is the divergence of the vector $\rho\mathbf{V}$ (mass flux) in three dimensions.

3.3.2. Conservation of Momentum

The acceleration of a fluid element is the same as the net force exerted on the fluid element divided by the mass of the fluid element (Newton's Second Law). This principle of conservation of momentum may be written as

$$\rho\frac{\partial\mathbf{V}}{\partial t} = \mathbf{F}, \tag{3.2}$$

where $\partial\mathbf{V}/\partial t$ is the local (relative to a fixed point in space) acceleration of a fluid element and $\mathbf{F}$ is the net force (vector) acting on the fluid element.

The force $\mathbf{F}$ for atmospheric motions may be written as

$$\mathbf{F} = -\rho\mathbf{V}\cdot\nabla\mathbf{V} - \nabla p - \rho f\hat{k}\times\mathbf{V} - \rho g\hat{k} + \tau, \tag{3.3}$$
$$\text{(a)} \quad \text{(b)} \quad \text{(c)} \quad \text{(d)(e)}$$

where ∇ is the three-dimensional gradient operator, $\cdot$ is the inner (or scalar) product of two vectors, $\times$ denotes the vector product of two vectors, $\hat{k}$ is the unit vector normal to the Earth's local surface, $f = 2\Omega \sin \phi$ is the Coriolis parameter ($\Omega = 2\pi/86400 \text{ s}^{-1}$ is the Earth's angular speed and ϕ is the latitude), g is the local gravity, which is usually taken as a constant, and τ is the friction. The terms on the right-hand side of Eq. (3.3) represent the (a) advective force, (b) pressure gradient force, (c) Coriolis force due to the rotation of the Earth, (d) gravitational force, and (e) friction within the atmosphere and at the interface between the atmosphere and the Earth's surface, respectively. The pressure gradient force is directed toward the lower pressure, the opposite direction of the gradient of pressure; the Coriolis force to the right (left) of the motion in the Northern (Southern) Hemisphere (note: the sign of the Coriolis force is negative in the Southern hemisphere); and the friction in the opposite direction of the motion.

Since most of the atmospheric observations are made on constant-pressure surfaces, it is often convenient to express the equation of motion in a coordinate system where the pressure is the vertical coordinate. This coordinate system is called the pressure coordinate system. In the pressure coordinate system, the force $\mathbf{F}$: is represented as:

$$\frac{\mathbf{F}}{\rho} + -\mathbf{V}\cdot\nabla\mathbf{V} - \nabla\Phi - f\hat{k}\times\mathbf{V} - g\hat{k} + \frac{\tau}{\rho}, \tag{3.4}$$
$$\text{(a)} \quad \text{(b)} \quad \text{(c)} \quad \text{(d)} \quad \text{(e)}$$

where Φ is the geopotential height ($\Phi \equiv gz$: z is the geometrical height) and the gradient of variables are calculated on constant-pressure levels. The terms (a–e) in Eq. (3.4) have the same meaning as those in Eq. (3.3).

For many atmospheric motions, there exists near equilibrium states between the most dominant driving forces in Eq. (3.3) or in Eq. (3.4). In that case the local acceleration term ($\partial \mathbf{V} / \partial t$) in Eq. (3.2) becomes negligible and the resulting motion can be expressed by the most dominant forces in Eq. (3.3) or in Eq. (3.4.) Important examples of such balanced states are shown in the following.

3.3.2.1. Hydrostaticity

For a static atmosphere (without any motion), the vertical pressure gradient force is balanced by the gravitational force (Fig. 3.8). For many atmospheric motions, vertical acceleration of an air parcel ($dw / dt = \partial w / \partial t + \mathbf{V} \cdot \nabla w$) is negligible compared to the vertical pressure gradient force and gravity so that the balance of forces in the direction of the local vertical can be closely approximated by hydrostatic relationships. For a hydrostatic atmosphere the vertical component of equation of motion [Eq. (3.2)] and [Eq. (3.3)] can be simplified to

$$O = \frac{\partial p}{\partial z} + \rho g \cdot \qquad (3.5)$$

The hydrostaticity is a valid assumption for motions whose horizontal length scale is much larger than the vertical length scale. Many atmospheric motions

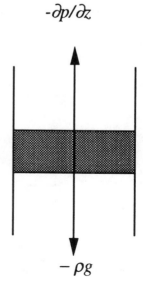

Fig. 3.8. The force balance for the hydrostatic atmosphere.

fall into this category. Important exceptions from the hydrostatic balance are the convective plumes and turbulent eddies where vertical acceleration due to the latent heat release and shear is strong, respectively.

3.3.2.2. Geostrophy

Atmospheric motions whose characteristic time scales for the horizontal and vertical motions are larger than one pendulum day and the period of buoyancy oscillation, respectively, may be represented by the approximate balance between the horizontal pressure gradient force and the Coriolis force (Lindzen, 1990). Such a balanced state is called the geostrophic balance.

For motions in the geostrophic balance the horizontal components of the momentum equation [Eq. (3.2) or Eq. 3.3)] can be reduced to

$$\rho f U_g = -\frac{\partial p}{\partial y}, \tag{3.6}$$

$$\rho f V_g = \frac{\partial p}{\partial x}, \tag{3.7}$$

where U_g and V_g are the east–west and north–south components of the geostrophic wind. The geostrophic wind is parallel to the isobar (line of equal pressure) with high pressure at the right- (left-) hand side of the wind direction (Fig. 3.9) in the Northern (Southern) Hemisphere so that it does not contribute to the net mass transport across isobars.

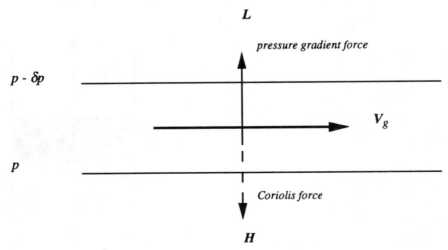

Fig. 3.9. The force balance for the geostrophic flow.

The geostrophic balance is established for much of the atmospheric motions whose horizontal length scales exceed a few thousand kilometers except in equatorial regions where the Coriolis parameter (f) becomes vanishingly small. Note that $f = 0$ at the equator. Atmospheric motions of such scales play important roles ranging from affecting the day-to-day weather systems to the global transport of the atmospheric mass and trace gases.

3.3.2.3. Thermal Wind

Combining the hydrostatic relationship with the geostrophy, the vertical shear of the geostrophic wind is related to the horizontal temperature gradient. This relationship between the vertical shear of geostrophic wind and the horizontal temperature gradient is called the thermal wind relationship. The geostrophic wind given by Eqs. (3.6) and (3.7) may be rewritten in the pressure coordinate using Eq. (3.4) as

$$fU_g = -\frac{\partial \Phi}{\partial y}, \tag{3.8}$$

$$fV_g = \frac{\partial \Phi}{\partial x}, \tag{3.9}$$

where the horizontal derivatives $\partial/\partial x$ and $\partial/\partial y$ are taken on an isobaric surface. The hydrostatic equation [Eq. (3.5)] can also be written in the pressure coordinate using the relationship:

$$g\partial z = \partial \Phi \tag{3.10}$$

together with the equation of state as

$$\frac{\partial p}{\partial \Phi} = -\frac{1}{\rho} = -\frac{R_a T}{p}, \tag{3.11}$$

where R_a is the gas constant for air ($R_a = 287$ JKg^{-1}K^{-1} for dry air). Differentiating Eqs. (3.8) and (3.9) with respect to p and using the hydrostatic equation in the pressure coordinate system [Eq. (3.11)], the thermal wind relationship on a pressure coordinate is obtained:

$$\frac{\partial U_g}{\partial p} = \frac{R_a}{fp}\frac{\partial T}{\partial y}, \tag{3.12}$$

$$\frac{\partial V_g}{\partial p} = -\frac{R_a}{fp}\frac{\partial T}{\partial x}. \tag{3.13}$$

Equations (3.12) and (3.13) state that U_g (V_g) increases with height when the temperature decreases toward the north (east) in the Northern Hemisphere. The opposite is true in the Southern Hemisphere where the Coriolis parameter f changes sign. The vertical structure of the zonal mean wind can be closely related to the meridional variation of the zonal mean temperature (Section 3.2) via the thermal wind relationship. Note that the pressure decreases with height, therefore decreasing U_g with increasing pressure ($\partial U_g / \partial p < 0$) implies that U_g increases with height ($\partial U_g / \partial z > 0$).

3.3.3. Conservation of Thermodynamic Energy

The first law of thermodynamics states that the change of internal energy of a system is the sum of the heat and work added (or subtracted) to (or from) the system. In a differential form, it may be written as:

$$\frac{\partial \theta}{\partial t} = V \cdot \nabla \theta + Q, \tag{3.14}$$

where $\partial \theta / \partial t$ is the local (fixed in space) variation of the potential temperature θ per unit time, Q is the local heating (or cooling) rate given in the unit of K s^{-1}. The potential temperature θ is the temperature of an air parcel when it is moved from the present pressure level (p) to the reference pressure level (p_0) and is defined as

$$\theta \equiv T \left(\frac{p_0}{p} \right)^{R_d/C_p}, \tag{3.15}$$

where R_a is again the gas constant for dry air, C_p is the specific heat of the dry air at constant pressure ($C_p = 1004$ J kg^{-1} K^{-1}), and the reference pressure p_0 is usually taken to be 10^3 mb.

The main cause of the local heating (cooling) of the atmosphere is the gain (loss) of heat by the radiative processes, phase change (evaporation, condensation, melting, freezing, and sublimation) of water substances, and the heat exchange between the atmosphere and the Earth's surface. The imbalance between the absorption and emission of radiative energy by the air results in net heating or cooling; latent heat is released (absorbed) when water vapor is condensed into liquid water causing net heating of the air in the region. Evaporation has an opposite role in the local temperature.

3.3.4. The Equation of State

The relationship between the atmospheric pressure (p), and density (ρ), and temperature (T) can be expressed by the equation of state as

$$p = \rho R_a T \qquad (3.16)$$

Equations (3.1), (3.2), (3.3), or (3.4), (3.14) or (3.16) are often called the governing equations for atmospheric motions. Depending on the scale of atmospheric motions, the governing equations can be further simplified to highlight the most dominant forcing terms.

3.3.5. Atmospheric motions of various scales

The characteristics of atmospheric motions largely depend on the horizontal scale that ranges from the planetary scale ($\sim 10^7$ m) to the molecular scale ($\sim 10^{-7}$ m). An example of the classification of atmospheric motions following the characteristic horizontal length scale and time scale is shown in Fig. 3.10.

Depending on the length and time scales of a motion, relative importance of the various terms in the governing equation changes significantly leading to approximate equations of motions. For motions whose length scales are of the order of 10^4 km (global or planetary scale motions), the variation of the Coriolis

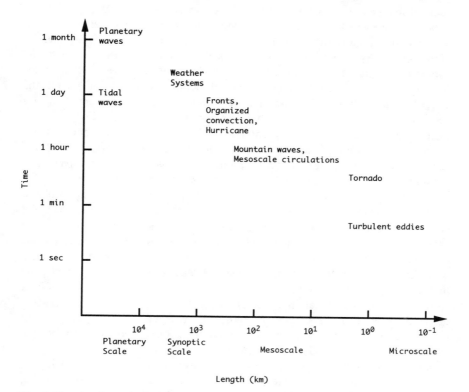

Fig. 3.10. A schematic description of scale definitions following time and length scales.

force following the latitude ($\beta = \partial f / \partial \phi$) becomes important. Motions of this scale are important in the global transport of pollutants and trace gases such as ozone. For example, the observed global distribution of ozone is the consequence of global-scale transport from the equatorial region to the polar regions.

The motions most closely related to daily weather are of time scales from a few days to a week with length scales of the order of a few thousand kilometers (synoptic scale). Synoptic-scale motions include the high- and low-pressure systems and fronts that appear frequently in weather maps. For the motions of this scale, the Earth's rotation is one of the most important factor, but its latitudinal variation can be neglected. As a result, the Coriolis force may be assumed to be constant within an interested region. The pressure-gradient force and the Coriolis force are two of the most important components of forcing, but friction becomes important for longer periods.

Mesoscale motions extend for the horizontal and time scales of 10^1–10^3 km and 1 min–1 day, respectively. Mesoscale motions contain a wide variety of phenomena: squall lines, convective systems, mountain waves, land and sea breeze circulations, to name a few. Mesoscale motions govern much of the detailed features of local flows. They also drive smaller scale motions that are important for short-term dispersion of pollutants near their sources. Mesoscale motions are primarily driven by large-scale flow and are strongly influenced by local topography, temperature variation, and moisture sources. Owing to the wide range of time and length scales, various terms in the governing equations become important depending on the scales of motion considered. In general, the effects of the Earth's rotation becomes less important as time and length scales decrease. For the motions of length scales less than 10^2 km, the Coriolis force is frequently neglected from the force terms in Eq. (3.3) or (3.4). For such scales of motions, geostrophy is not established. For smaller-scale motions with strong vertical motion, such as convective plumes, small-scale gravity waves, and turbulent eddies, hydrostaticity ceases to be a valid assumption.

Turbulence is the smallest scale of motion considered in the atmospheric sciences; however, it plays an important role in the dislodgment of bioaerosol particles from plant and other sources, in dispersion of airborne pollutants, in exchange of momentum and thermodynamic energy between the lower part of the atmosphere and the Earth's surface, as well as in the long-term energy balance of the atmospheric motions. Development and decay of turbulence following the thermal structure of the ground surface and the low-level atmosphere is the direct cause of the diurnal variation of the atmospheric boundary layer over most of the land surfaces. Some studies of atmospheric energetics even suggest that as much as 25% of total atmospheric kinetic energy is dissipated by the turbulence in the free atmosphere (Heck et al. 1977).

Atmospheric turbulence is generated by various causes such as surface friction, convection, wind shear, and breaking gravity waves. In the well-mixed ABL, turbulence is generated mainly by convection, and surface friction assumes secondary importance as a source of turbulence. On the other hand, turbulence is usually weak in the stably stratified ABL, and is generated primarily by surface roughness and wind shear. Outside the ABL, strong turbulence is most frequent in synoptic-scale frontal zones, within convective systems, and over mountainous regions where the atmospheric flow is disturbed by strong wind shear and temperature gradient (synoptic-scale frontal zones), strong vertical movement of air due to unstable stratification associated with the release of latent heat (convective systems), and the onset of hydrodynamic instability induced by orographic gravity waves.

3.4. Atmospheric Radiation

Radiative transfer is the main mechanism for energy exchange between the Earth and outer space. A monochromatic radiation is an electromagnetic wave of a wavelength λ, which is made up of mutually perpendicular in-phase electric and magnetic fields. Electromagnetic waves travel at the speed of light in the medium. The frequency ν is related to its wavelength via the relationship $\nu = c/\lambda$. Electromagnetic waves transport energy as well as momentum. The energy (E) and momentum (p) carried by a photon (a light quantum at wavelength λ) is

$$E = h\nu = \frac{hc}{\lambda}, \quad p = \frac{h\nu}{c} = \frac{h}{\lambda}, \quad (3.17)$$

where h is Plank's constant. As the frequency of an electromagnetic wave increases (decreasing wavelength), photon energy and momentum carried by the wave increase.

The electromagnetic waves involved in the energy transfer between the sun and Earth–atmosphere system and within the Earth–atmosphere system are usually categorized as the short-and long-wave radiations depending on the wavelength as shown in Fig. 3.11. In general, the electromagnetic waves of wavelengths

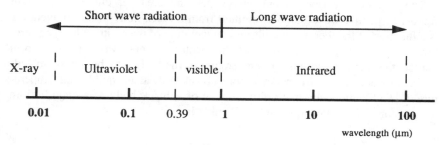

Fig. 3.11. The spectrum of the electromagnetic waves important for the Earth's atmosphere.

shorter than 1 μm (10^{-6} m) are called the short-wave radiation, while those of wavelengths longer than 1 μm are called the long-wave radiation within the context of the atmospheric radiative transfer.

Long-wave radiative transfer plays an important role in energy exchange between various atmospheric layers and the Earth's surface. Absorption of the long-wave radiation from the Earth's surface by the atmosphere and re-emission from the atmosphere maintains the Earth's surface temperature much higher than the radiative equilibrium temperature. This effect is often called the atmospheric effect (or greenhouse effect or blanket effect). The constituents of the atmosphere that are active absorbers and emitters of long-wave radiation are sometimes called the greenhouse gases. Water vapor is the most important greenhouse gas in the atmosphere; carbon dioxide (CO_2) and methane (CH_4) are also important greenhouse gases.

3.4.1. Black Body Radiation

The radiative transfer process between the sun–Earth–atmosphere system can be explained by physical laws governing the emission and absorption of electromagnetic waves by matter. The details of radiative properties of individual matter are usually very complex and need much experimentation to explain. However, with respect to the bulk of the atmosphere, sun, and the Earth's surface, their radiative properties can be approximated by the black body radiation.

A black body is defined as the ideal emitting and absorbing body in thermal equilibrium with respect to the radiative transfer process, that is, a black body absorbs all radiation incident on it and it reradiates at a rate corresponding to the radiative equilibrium temperature. In reality, most materials are not perfect absorbers. Such materials are called gray bodies. Some of the useful principles governing the black body radiation are presented in the following subsections.

3.4.1.1. Plank's Law

The flux density (the radiant energy per unit area per unit time per unit frequency or unit wave number) emitted from a black body at the frequency v is a function of the thermal equilibrium temperature of the emitting body. Therefore, by measuring the spectrum of the electromagnetic waves emitted from matter, the thermal equilibrium temperature of the body may be approximated.

The flux density at a frequency v from a black body at an equilibrium temperature T [$E_v(T)$; monochromatic flux density of a black body at an equilibrium temperature T] can be expressed by Plank's law as

$$E_v(T) = \frac{2hv^3}{c^2(\exp[hv/kT] - 1)}, \tag{3.18}$$

where h is Plank's constant ($h = 6.6262 \times 10^{-27}$ erg s), k is the Boltzmann constant ($k = 1.3806 \times 10^{-16}$ erg K^{-1}), and c is the speed of light ($c = 2.998 \times 10^{10}$ cm s^{-1}); Eq. (3.18) is called the Plank function. The Plank function can be written in terms of the wavelength using the relationship between the frequency (v) and the wavelength (λ) and their derivatives [$\lambda = c / v$ and $d\lambda = -(c / v^2)dv$] as (note that $E_\lambda d\lambda = E_v dv$):

$$E_\lambda(T) = \frac{2hc^2}{\lambda^5(\exp[hc / k\lambda T] - 1)}. \qquad (3.19)$$

Total radiant energy per unit time per unit area emitted by a black body at a temperature T is obtained by integrating Eq. (3.18) or Eq. (3.19) over the interval of entire frequency or wavelength.

3.4.1.2. Wien's Displacement Law

Wien's displacement law states that the wavelength of the maximum intensity (λ_m) for black body radiation is inversely proportional to the temperature:

$$\lambda_m = \frac{a}{T}, \qquad (3.20)$$

where $a = 0.2897$ cm K. Wien's displacement law can be derived from Plank's law of black body radiation [Eq. (3.19)] by differentiating it with respect to the wavelength and is a useful tool for finding the approximate temperature of a matter without analyzing the entire details of the spectrum.

3.4.1.3. Stefan–Boltzmann Law

The flux density emitted by a black body at a temperature T can be obtained by integrating the Plank function with respect to the frequency (or wavelength) in all directions, which yields

$$E(T) = \sigma T^4, \qquad (3.21)$$

where $\sigma = 5.67 \times 10^{-5}$ erg cm^{-2} s^{-1} K^{-4} is the Stefan–Boltzmann constant. This law can be obtained by integrating Plank's function with respect to the entire frequency (wavelength) and over a closed surface containing the black body.

3.4.2. Solar Radiation

The sun is the major source of the radiative energy incident on the Earth. The total solar energy incident on the Earth–atmosphere system can be computed by measuring the solar spectrum at the top of the atmosphere. The solar constant (S) is defined as the flux of solar energy (energy/time) across a surface of unit area normal to the solar beam at the average distance between the sun and the Earth (R_m). Recent measurement suggests the value of the solar contrast to be 1353 W m^{-2}.

The total solar energy incident on the Earth (E) can be computed by multiplying the solar constant with the area of the cross section of the Earth ($A_c = \pi r^2$, so that

$$E = S \times A_c = S \times \pi r^2 \left(\frac{R}{R_m}\right)^2, \qquad (3.22)$$

where r is the radius of the Earth and R and R_m are the instantaneous and average distance between the sun and the Earth, respectively. If the variation of the distance between the Earth and the sun is neglected, the average solar energy on the Earth's surface is then

$$\frac{E}{A_{\text{Earth}}} = \frac{E}{4\pi r^2} = \frac{S}{4}, \qquad (3.23)$$

where $A_{\text{Earth}} = 4\pi r^2$ is the surface area of the Earth (Fig. 3.12).

The solar spectrum in the visible and infrared range can be well fit with a Plank curve at $T = 6000$ K (Fig. 3.13) with the maximum intensity of the solar radiation occurring at the wavelengths in the vicinity of 0.45 μm (the short-wave end of visible light). About 50% of solar energy is contained in the range

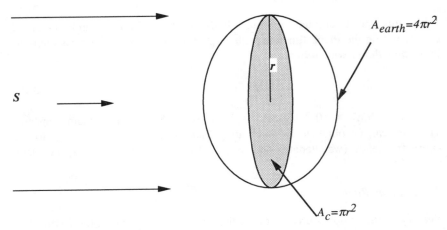

Fig. 3.12. A schematic illustration for solar energy intercepted by the Earth.

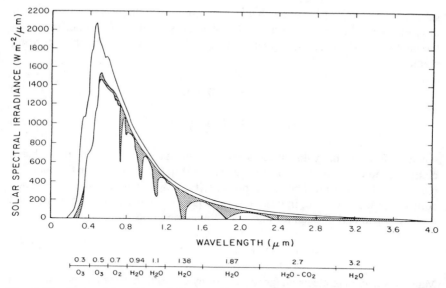

Fig. 3.13. Spectral irradience distribution curves related to the sun: Outer curve is the observed solar irradience at the top of the atmosphere and the inner curve is the one at the Earth's surface. Shaded areas represent absorption by various gases in the clear atmosphere. The difference between the two envelops is due to scattering (after Liou 1980).

longer than visible wavelength ($\lambda < 0.7$ μm); about 40% of solar energy is contained within the visible range (0.4 μm $< \lambda < 0.7$ μm); and about 10% of solar energy is contained in the wavelength shorter than 0.4 μm.

The difference between the solar spectrum observed at the top of the atmosphere and at sea level (Fig. 3.13) indicates the occurrence of absorption and scattering of the solar radiation by atmospheric gases and other pollutants. The absorption of solar radiation is almost complete in the ultraviolet range (see also Section 3.4.3), while the atmosphere is relatively transparent to radiation in the visible range.

3.4.3. Photochemical Reactions in the Upper Atmosphere: Ozone Layer

Absorption of photon energy by a molecule causes the transition of the molecule to a higher electronic energy level. When the absorption of photon energy is large enough, it may cause reactions such as photodissociation of the molecule or interaction with other molecules, leading to the overall photochemical process.

Photochemical processes are crucial for the generation and maintenance of the ozone layer in the stratosphere. Ozone (O_3) prevents harmful high-energy electromagnetic waves from reaching the Earth's surface. In the upper atmosphere, O_3 is created by photochemical reactions which combine a molecular oxygen (O_2) with an oxygen atom (O). The chief source of atomic oxygen is the photolysis of molecular oxygen. An O_3 molecule is created by the following reaction:

$$O_2 + h\nu \rightarrow 2O, \tag{3.24}$$

$$O + O_2 + M \rightarrow O_3 + M, \tag{3.25}$$

where M is a third body that absorbs extra energy in the reaction. The photodissociation of molecular oxygen into oxygen atoms [Eq. (3.24)] occurs in the radiation band $\lambda < 0.286$ μm. The latter process [Eq. (3.25)] is called a three-body collision process. Ozone, in turn, absorbs a photon energy contained in the band $\lambda < 1.18$ μm and dissociates into one O and one O_2 as

$$O_3 + h\nu \rightarrow O_2 + O. \tag{3.26}$$

Major absorption of electromagnetic waves by absorption occurs in the ultraviolet range ($\lambda < 0.334$ μm) with maximum absorption at $\lambda = 0.254$ μm, filtering out most of the high-energy electromagnetic waves incident on the top of the Earth's atmosphere. Ozone also absorbs in the visible range, even though small compared to the ultraviolet range, with maximum absorption of visible light occurring at $\lambda = 0.6$ μm.

Destruction of O_3 also occurs via reaction with other gâses. As man-made pollutants in the atmosphere increase, the destruction of the ozone layer also increases. The most damaging process in the destruction of O_3 by trace gases emitted by human activity may be the reaction of ozone with nitric oxide (NO) [Eqs. (3.27)–(3.29)], methane (CH_4) and the chlorine atom (Cl) [Eqs. (3.30)–(3.32)]. The chemical process leading to destruction of ozone by the reaction with NO is

$$NO + O_3 \rightarrow NO_2 + O_2, \tag{3.27}$$

$$NO_2 + O \rightarrow NO + O_2. \tag{3.28}$$

The net chemical reaction of this process is

$$O_3 + O + NO \rightarrow 2O_2 + NO. \tag{3.29}$$

The chlorine atom destroys ozone in a similar process:

$$Cl + O_3 \rightarrow ClO + O_2, \tag{3.30}$$

$$ClO + O \rightarrow Cl + O_2. \qquad (3.31)$$

The net chemical reaction is

$$O_3 + O + Cl \rightarrow 2O_2 + Cl. \qquad (3.32)$$

The importance of the preceding O_3 = destroying process by NO and Cl is that NO and Cl are regenerated as a product of the net reaction, leading to chain reactions. Nitric oxides and chlorine atoms in the upper atmosphere are generated from the photochemical reaction of nitrogen oxide (N_2O) emitted by high-flying aircraft and chlorofluorocarbon ($CFCl_3$ and CF_2Cl_2) gases (e.g., Freon), widely used as a refrigerant and spray propellant.

3.4.4. Terrestrial Radiation

As a whole, the Earth–atmosphere system is in close thermal balance with respect to the incoming solar radiation. In other words, when averaged over a long period over the entire globe, the absorption of solar radiation by the atmosphere and underlying surface is almost exactly balanced by the emission of the long-wave radiation from the Earth–atmosphere system.

The Earth–atmosphere system emits radiation corresponding to its radiative equilibrium temperature. The radiative equilibrium temperature of the Earth is about 250 K. However, the spectrum of the radiation emitted by the Earth is closely approximated by the spectrum emitted by a black body of 290 K, which is close to the average temperature of the Earth's surface (Fig. 3.14). This difference between the Earth's radiative equilibrium temperature and the actual emitting temperature indicates that some of the infrared radiation emitted by the Earth's surface is trapped by the atmospheric greenhouse gases (Fig. 3.14).

The most important absorbers of the infrared radiation are water vapor (H_2O), CO_2, and ozone. Water vapor absorbs infrared radiation at 6.3 μm and in the range between 12 μm < 18 μm. Carbon dioxide absorbs at 4.3 μm, 15 μm, and in the range 6 μm < λ < 8 μm. Ozone absorbs strongly at 9.6 μm. Atmospheric trace gases such as CH_4 and carbon monoxide (CO) are also efficient absorbers of infrared radiation, but the total amount absorbed by such trace gases is thought to be smaller than that absorbed by major atmospheric greenhouse gases. Recent numerical studies suggest that by doubling the concentration of atmospheric CO_2, the global mean surface temperature may increase by a few degrees Celsius. Although the assessment of the influence of the increase of greenhouse gases on global and regional climate is not complete, even a slight change of the global mean temperature can cause a large unnatural variation of climate.

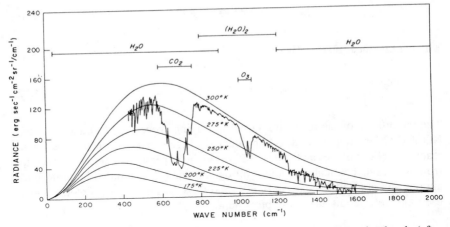

Fig. 3.14. The infrared spectra from the Earth and the various absorption bands (after Liou 1980).

3.5. Air Pollution

Air pollutants are usually defined as substances contained in the atmosphere that may cause adverse effects under certain conditions to the environment of human, animal, and plant life. The most important factor that determines the severity of air pollution is the concentration of pollutants. Local concentration of air pollutants are determined by the amount of emission at the source; transport and dispersion by atmospheric motions; transformation of pollutants by chemical reaction; and removal from the atmosphere by settling, scavenging, absorption, and impaction (Oke 1978).

3.5.1. Emission of Pollutants

The emission at the source determines the total amount of pollutants released into the atmosphere. Pollutants in the atmosphere originate from natural processes and human activity. A list of the most important types of air pollutants are given in Table 3.2. Increasing emission of man-made pollutants into the atmosphere becomes a serious environmental problem not only at local level but at global level as well.

The majority of particulates in the atmosphere are emitted from natural sources, such as sea spray, forest fires, and dust carried by wind. The moist important anthropogenic sources of particulates are combustion of fossil fuel, industrial processing, and perhaps farming activities. In general, settling speed increases with increasing size for particles of the same density so that smaller particles (less than 10 μm in diameter) are suspended longer in the atmosphere than larger

Table 3.2.   Types and sources of atmospheric pollutants (after Oke, 1978)

Type of pollutants	Source	
	Natural	Man made
Particulates	Volcanos Wind action Sea spray Fires	Combustion Industrial processing Farming
Sulfur compounds	Bacteria Volcanos Sea spray	Burning fossil fuel Industrial processing
Carbon monoxide/dioxide	Volcanos Fires Plants Phytoplanktons	Combustion Industrial processing
Hydrocarbons	Bacteria Plants	Internal combustion
Nitrogen compounds	Bacteria	Combustion

particles. For liquid droplets, evaporation, by changing the size of droplets, also affects the settling speed. As a consequence, dispersion of small particulates (especially liquid droplets) are strongly affected by atmospheric conditions such as temperature, wind, and relative humidity.

Oxides of sulfur and nitrogen are emitted into the atmosphere in many forms. The most important pollutants are sulfur dioxide (SO_2), NO_1 and NO_2 which are responsible for various chemical reactions that lead to the formation of acid rain and destruction of the Earth's ozone layer. Most man-made sources of sulfur and nitrogen oxides are associated with the burning of fossil fuel such as coal and petroleum.

Most atmospheric hydrocarbons (H_c) due to human activities are emitted by internal combustion engines, especially from automobiles. Hydrocarbons are chemically reactive so that under suitable atmospheric conditions with sufficient sunlight, a high concentration of hydrocarbons in an urban area can cause photochemical smog and the formation of secondary pollutants.

Carbon monoxide and CO_2 are generated by burning carbonaceous material. The main sources of man-made CO are internal combustion engines. Carbon dioxide is one of the important constituents of the Earth's atmosphere and is not regarded as a pollutant, in general. However, rapid increases in the concentration of atmospheric CO_2 due to the burning of fossil fuels raises concerns about the possible impact on the Earth's climate by changing the radiative property of the

Earth's atmosphere. Recent studies on the impact of the doubling the atmospheric CO_2 concentration suggests a 1–4 C increase of the global mean surface temperature, which can significantly alter the Earth's climate.

3.5.2. Atmospheric Transport and Mixing

Movement of pollutants in the atmosphere is controlled by the movement of air so that wind speed, direction, and atmospheric stability have an important role in determining the local concentration of pollutants released into the atmosphere. Transport and dispersion of pollutants are generally enhanced under strong wind and weak atmospheric stability (or strong instability). Pollutants emitted into the mixed layer (see Section 3.2.3) are rapidly dispersed vertically and laterally, soon contaminating a large area near the source of the emission. In air pollution terms this is called fumigation. Active movement of air in the vertical direction enhances removal of pollutants by impaction with the ground in the vicinity of the source. In the stable layer, vertical and lateral dispersion is suppressed owing to weak turbulent intensity and pollutants emitted into the stable layer being carried away from the source by the mean wind. This is called fanning. A schematic of fumigation and fanning near the ground surface is shown in Fig. 3.15.

The temperature inversion layer is very strongly stratified, so that the vertical movement of air is strongly restricted near the inversion layer. The temperature inversion near the top of the boundary layer suppresses mass exchange across the inversion layer so that it works as an effective lid for the pollutants released

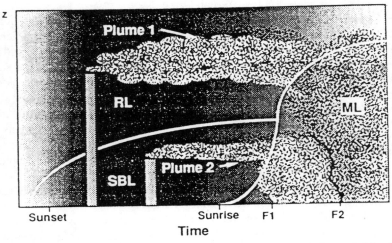

Fig. 3.15.   A schematic illustration of the fanning (plumes 1 and 2 before reaching ML) and the fumigation (plumes 1 and 2 after reaching ML) (after Stull 1990, reprinted by the permission of Kluwer Academic Publishers).

into the atmospheric boundary layer. Local concentration of pollutants may become high under stable conditions with weak wind. In mountainous regions nocturnal drainage flow, formed during clear nights due to the surface radiative cooling, carries pollutants from the surrounding hills toward the valley. Thus, very high concentration of pollutants can occur in the valley. Topographic blocking of the low-level air can significantly enhance the concentration of pollutants at the upstream side of a hilly terrain.

3.5.3. Atmospheric Flux Methods for Estimating the Dry Deposition Rate

Airborne particles and gases are deposited on the ground surface by gravity settling, wet scavenging by precipitation, and by dry deposition. Without wet scavenging, most large airborne particles are removed by gravitational settling, while small particles and gases are removed largely by contacting the underlying surface. Dry deposition can be defined as the aerodynamic transfer of tracer gases and aerosol from the air to the surface and the gravitational settling of particles.

The rate of dry deposition of airborne particles and gases may be measured by observing the fluxes of such constituents in the atmosphere. This is sometimes called the atmospheric flux methods for estimating the rate of dry deposition. Estimating the rate of dry deposition from the observed atmospheric fluxes of interested constituents may be understood easily by considering the conservation of the airborne constituents, which will be presented in the next section.

3.5.3.1. Conservation Equation for Atmospheric Constituents

Let C be the spatially averaged concentration of an atmospheric constituent. Then, the conservation principle for the constituent may be expressed as

$$\frac{\partial C}{\partial t} = -\nabla \cdot (\mathbf{V}C) - \frac{\partial}{\partial z}WC - \nabla \cdot \langle \mathbf{v}'c' \rangle - \frac{\partial}{\partial z}\langle w'c' \rangle + P - S, \qquad (3.33)$$

where C, $\mathbf{V}$, W, c', $\mathbf{v}$, and w' are the spatially averaged values and the deviation from the mean values of the concentration of any atmospheric constituents, horizontal, and vertical velocities, respectively, and $\langle \# \rangle$ is the averaging operator in the horizontal. Therefore, for any variable ψ, the following relationship is established:

$$\psi = \Psi + \psi', \quad \text{where } \Psi = \langle \psi \rangle.$$

The term on the left-hand side of Eq. (3.33) denotes the local change of mean concentration C and those terms on the right-hand side of Eq. (3.33) denote the horizontal $[\nabla \cdot (\mathbf{V}C), \nabla \cdot \langle \mathbf{v}'c' \rangle]$ and vertical $[\partial(WC)/\partial z, \partial \langle w'c' \rangle/\partial z]$ flux divergence of C (c') by the mean wind (fluctuating wind), respectively, and the source (P) and sink (S) of the constituent. In general, W and $\partial(WC)/\partial z$ are small compared to the other terms on the right-hand side of Eq. (3.33). Neglecting this term, the budget equation for an atmospheric constituent C, [Eq. (3.33)] is rewritten as

$$\frac{\partial C}{\partial t} \approx -\nabla\cdot(\mathbf{V}C) - \nabla\cdot\langle \mathbf{v}'c'\rangle - \frac{\partial}{\partial z}\langle w'c'\rangle + P - S. \qquad (3.34)$$

It is clear from Eqs. (3.33) and (3.34) that, except for the sources and sinks of the constituents, the change of the mean concentration C at a fixed location is determined by the divergence of the fluxes due to the mean and fluctuating components of the concentration and wind [first three terms on the right-hand side of Eq. (3.34)]. Depending on the measurement techniques (and instruments) and associated assumptions for each terms on the right-hand side of Eq. (3.34), the methods for estimating the rate of dry deposition on the ground may be categorized as follows.

3.5.3.2. Eddy Correlation Method

Assume that, for a layer between the ground surface and a level of measurement at a height z above the ground surface (Fig. 3.16), the rate of the change of local concentration $\partial C/\partial t$ and the horizontal divergence terms [$\nabla\cdot(\mathbf{V}C)$ and $\nabla\cdot\langle \mathbf{v}'c'\rangle$] in Eq. (5.2) are known (as a special case, if the flow is steady and horizontally homogeneous, $\partial C/\partial t$ and the divergence terms vanish). Then approximating the vertical flux convergence as

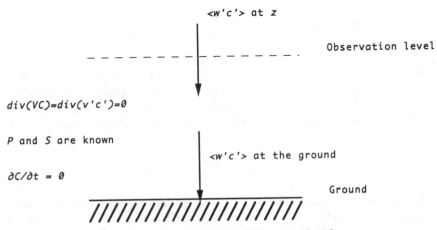

Fig. 3.16. The vertical flux of a constituent near the ground surface.

$$-\frac{\partial}{\partial t}\langle w'c'\rangle \approx -\frac{\langle w'c'\rangle_z - \langle w'c'\rangle_s}{z},$$ (3.35)

the conservation equation [Eq. (3.34)] can be written as

$$\langle w'c'\rangle_s = \langle w'c'\rangle_z - z\overline{Q},$$ (3.36)

where the subscripts s and z denote the values at the ground and the observation level, respectively, and $\overline{Q}$ is the vertically averaged sum of the known rate of change, horizontal flux divergence, source, and sink terms defined by

$$\overline{Q} = -\frac{1}{z}\int_0^z \left(\partial C/\partial t + \nabla\cdot\left(\mathbf{V}C\right) + \nabla\cdot\langle\mathbf{v}'c'\rangle - P + S \right) dz.$$

Equation (3.36) states that the rate of deposition on the ground ($\langle w'c'\rangle_s$) is obtained as the difference between the net source (or sink) of the constituents in the layer between the ground surface and the observational level and the flux across the observational level in Fig. 3.16. The flux $\langle w'c'\rangle_s$ is in the form of the covariance between the fluctuations of the vertical velocity and the concentration of the constituents, and for this reason, this method is called the eddy correlation method.

In order to use this method, fast response sensors for the measurement of vertical velocity and the concentration of the interested constituent must be available. It is also necessary to compile a large sample to reduce the error in estimating the correlation. Such measurements are typically made by an aircraft or an observational tower. Measurements by aircraft can cover wide areas in a relatively short period, so that horizontal and vertical variations of the flow field may be measured directly. However, the rate of change of local concentration is somewhat difficult to measure unless the aircraft flies the same path over a long period of time. On the other hand, observations using the sensors mounted on a tower cannot capture the horizontal flow structure and the vertical structure can be obtained only near the ground, limited by the height of the tower. However, the rate of local change may be measured quite accurately from tower data ranging over a sufficient period.

3.5.3.3. Vertical Gradient Method

The eddy correlation method discussed in the preceding subsection is the most direct method to measure fluxes in the atmosphere. However, it requires fast response sensors that are capable of sampling the concentration of the interested constituents at a very high rate. Such sensors may not be always available for every species all the time. Moreover, eddy correlation method requires large amount of data to compute fluxes accurately.

In case high-frequency data for computing the eddy correlation are not available, empirical formulations derived from the statistics of flux observations may be used. One indirect method is to estimate the fluxes (or eddy correlations) of constituents using the vertical profiles of the interested constituents and the mean state of the lower atmosphere. The assumptions for using this method are the same as those for the eddy correlation method.

Assuming that turbulent transport resembles the molecular transport that depends on the molecular viscosity (or conductivity) and the spatial gradient of velocity (or temperature), the eddy correlation between the vertical velocity fluctuation w' and a constituent ϕ' can be expressed using the eddy diffusivity K as

$$\langle w'\phi' \rangle = -K\frac{\partial \Phi}{\partial z}, \tag{3.37}$$

where Φ is the horizontal mean value of the constituent assumed to vary only in the vertical. The eddy diffusivity K depends on the characteristics of underlying surfaces and atmospheric conditions such as the roughness and temperature of the underlying surface, wind shear, and stratification. A variety of formulations have been used to represent the eddy diffusivities using the vertical profiles of mean atmospheric variables. One of the most widely used formulations for eddy diffusivities used the mixing length and the mixing velocity as

$$K = \ell^2 \left| \frac{d\mathbf{V}}{dz} \right| \tag{3.38}$$

where ℓ is the mixing length and the mixing velocity is represented by the mixing length and the vertical shear. The mixing length near the ground surface is usually represented as

$$\ell = \frac{kz}{\psi} \tag{3.39}$$

where k is the von Karman constant (0.3–0.4), z is the distance from the ground surface, and ψ is a dimensionless function depending on the static stability of the low atmosphere.

3.5.3.4. Variance Method

Surface flux $\langle w'c' \rangle_s$ is sometimes expressed in terms of the characteristic scales of the velocity (u_*: friction velocity) and concentration (c_*) as

$$\langle w'c' \rangle_s = u_* c_*. \tag{3.40}$$

The surface flux is sometimes expressed in terms of a stability- and friction-velocity-dependent coefficient (g) and the standard deviation of c' (σ_c) as

$$u_* c_* = g u_* \sigma_c. \tag{3.41}$$

This method also requires a fast response sensor that can sample the vertical velocity and c at a high rate sufficient to compute the standard deviation. One advantage of this method is that computing the standard deviation is less vulnerable to the error associated with sampling problems than computing covariances. There are several methods to compute u_* [note that $u_*^2 = \langle u'w' \rangle_s$] from the observed vertical profiles of the wind and temperature near the ground. However, a functional form of the coefficient g must be determined before this method is applied. The coefficient g depends on the surface friction and the static stability near the ground as for $\langle u'w' \rangle$, but the functional forms for the momentum and other properties are generally different.

3.5.3.5. Atmospheric Mass Balance Method

Dry deposition over a wide area can be estimated by measuring the horizontal and vertical fluxes at the lateral and top boundaries of a box covering the region of interest (Fig. 3.17). By integrating the conservation equation [Eq. (3.34)] over the box shown in Fig. 3.17, we can obtain the rate of dry deposition integrated over the horizontal distance L as

$$\{\langle w'\, c' \rangle_s\}_L = \{VC\}_z \big|_x^{x+L} - \{\langle w'c' \rangle_z\}_L + \left\{ P - S - \frac{\partial C}{\partial t} \right\}_{Box} \tag{3.42}$$

where $\{\ \}_L$, $\{\ \}_z$, and $\{\ \}_{Box}$ denote the integration over the horizontal distance L, vertical depth z, and a box consisting of L and z, respectively. If the source, sink, and the rate of the change of concentration in the box are known, the deposition rate can be obtained by measuring the fluxes across the boundaries of the designated box.

This method is applied for a region of about 100 km in the horizontal and sometimes vertically extending over almost the entire troposphere. (It is often assumed that the flux at the top of the troposphere is negligible owing to strong stability at the tropopause.) The flux measurement is required to cover a very wide region, so that aircraft observation is usually used to evaluate the flux terms. In reality, the aircraft cannot cover a wide area completely, so that this method is useful when the concentration of the target pollutants varies slowly in the domain. In other words, if the pollutants to be measured are concentrated

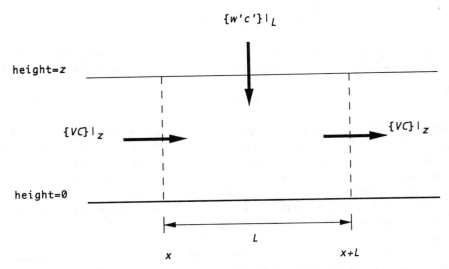

Fig. 3.17. A schematic illustration for the budget of a constituent for a box.

in a small area compared to the total area of observation, there are more chances to miss such localized high concentration owing to incomplete coverage of the total area by instruments, so that the budget calculated using such data can be misleading.

3.5.4. Chemical Reaction

Polluted urban atmospheres contain large amounts of NO_2, NO, hydrocarbons, and byproducts of partial oxidation of hydrocarbons emitted primarily by burning coal and internal combustion processes. Under suitable atmospheric conditions (strong solar radiation, concentration of chemical species, etc.) chemical reactions in the atmosphere can transform such pollutants into secondary pollutants.

Sulfur dioxide, mainly generated by burning coal, can be oxidized to form sulfur trioxide (SO_3) by the following reactions (Oke 1978):

$$SO_2 + SO_2 \rightarrow SO_3 + SO,$$
$$SO + SO_2 \rightarrow SO_3 + S, \tag{3.43}$$

$$SO_2 + O_2 \rightarrow SO_3 + S, \tag{3.44}$$

$$SO_2 + SO_2 + NO \rightarrow SO_3 + SO + NO, \tag{3.45}$$

$$SO_3 + HO \rightarrow H_2SO_4. \tag{3.46}$$

Oxidation of SO_2 into sulfur trioxide (SO_3) is enhanced by the presence of strong sunlight. Sulfur trioxide can react with atmospheric water vapor in the presence of catalysts to form sulfuric acid mist (H_2SO_4) (Eq. 3.46)]. Sulfuric (London-type) smog is formed by such a process.

The Los Angeles-type smog is formed by photochemical reaction of NO_2. Such reactions of nitrogen oxides are enhanced by the presence of hydrocarbons and generate a wide variety of secondary pollutants such as O_3 and peroxyacetyl nitrates [PAN, Eq. (3.55)]. The photochemical smog process is centered around the NO_2 photolytic cycle shown:

$$NO_2 + h\nu \rightarrow NO + O, \qquad (3.47)$$

$$O + O_2 \rightarrow O_3, \qquad (3.48)$$

$$O_3 + NO \rightarrow NO_2 + O_2. \qquad (3.49)$$

In the presence of reactive hydrocarbons (H_c) released by vehicle exhaust, the oxidation of NO can be performed without consuming O_3 in the reaction [Eq. (3.48)], giving a higher concentration of ozone. This reaction process is (where an asterisk denotes the reactive species):

$$O + H_c \rightarrow H_cO^*, \qquad (3.50)$$

$$H_cO^* + O_2 \rightarrow H_cO_c^*, \qquad (3.51)$$

$$H_cO_3 + NO \rightarrow H_cO_2^* + NO_2. \qquad (3.52)$$

Active hydrocarbon is a source of other secondary photochemical pollutants through the reactions

$$H_cO_3^* + H_c \rightarrow \text{aldehydes, ketones,} \qquad (3.53)$$

$$H_cO_3^* + H_c \rightarrow O_3 + H_3O_2^*, \qquad (3.54)$$

$$H_cO_x^* + NO_2 \rightarrow \text{PAN.} \qquad (3.55)$$

An example of the diurnal cycle of the pollutant concentration in Los Angeles smog is shown in Fig. 3.18. Observed daytime variation of NO, NO_2, and O_3 concentrations agree well with smog-chamber results (Bailey et al., 1978). Smog-chamber simulation shows the important role of hydrocarbons in the formation of Los-Angeles-type photochemical smog. Ozone, aldehyde, and peroxyacetyl nitrate cause eye and throat irritation and plant damage. Photochemical smog

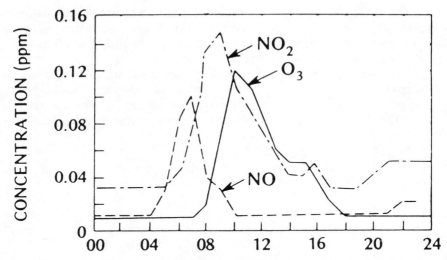

Fig. 3.18. The observed daytime variation of NO, NO_2, and O_3 concentrations (after Oke 1978).

is also accompanied by a characteristic odor partially due to aldehydes and formaldehyde and a brown haze due to NO_2 and light-scattering particles.

Acknowledgment

The author wants to express sincere appreciation to Dr. B. Lighthart for providing an opportunity to write this article and for many helpful comments. Discussions with Dr. Woo-Kap Choi were useful. Special thanks are due to the publishers and authors who supplied or granted permission for the tables and figures to be used in this chapter.

References

Bailey, R., H. Clarke, J. Ferris, S. Krause, and R. Strong, 1978: *Chemistry of the environment*, Academic Press, New York, 575 pp.

Gash J., W. Shuttleworth, and C. Lloyd, 1989: Micrometerological measurements in Les Landes Forest during HAPEX-MOBILTHY., *Agr. For. Met.*, **46**, 131–147.

Heck, W. J., H. A. Panofski, and M. A. Bender, 1977: The effect of clear-air turbulence on a model of the general circulation of the atmosphere. *Beitrage zur Physik der Atm.*, **50**, 89–97.

Lighthart, B., J. C. Spondlove, and T. G. Akers, 1979: Factors in the production, release, and viability of biological particles. p 11–22 in R. H. Edhounds (Ed.), *Aerobiology:*

The ecological systems and approach. US/IBP synthesis series 10. Dowdon, Hutchinson, and Ross Inc., Stroudsburg, PA.

Lindzen, R. S., 1991: *Dynamics in atmospheric physics*. Cambridge U.P., New York.

Liou, K. -N., 1980: *An introduction to atmospheric radiation.*, International geophysics series, **v. 26**, Academic Press, New York.

Mahrt, L., 1985: Vertical structure and turbulence in the very stable boundary layer. *J. Atmos., Sci.*, 44, 1106–1121.

Oke, T. R., 1978: *Boundary layer climates*. Methuen & Co Ltd., London.

Stull, R., 1990: *An introduction to boundary layer meteorology*, Kluwer Academic Publishers Boston. 666 pp.

Wallace, J. M. and P. V. Hobbs, 1977: *Atmospheric Science*. Academic Press, New York.

4

Distribution of Microbial Bioaerosol

Bruce Lighthart and Linda D. Stetzenbach

4.1. Introduction

Although the atmosphere forms a continuous bioaerosol transport medium be-tween indoor and outdoor air, barriers occur that hinder airflow. Because of these hinderances, it has been assumed that it is adequate to study these two environments separately. As more is learned about both populations, this assump-tion may need to be changed.

In the past, the distribution of outdoor viable droplet/particles (D/P) in urban and rural atmospheric environments seemed chaotic. Presumably because they could not be seen, and when sampled, their concentration variation within and among collection sites was very large. With the development of viable D/P samplers such as the slit and cascade impactors (see Chapter 6), and airplane-borne devices (Timmons et al., 1966), a much better understanding of the micros-cale and mesoscale distribution patterns of viable bioaerosols is emerging. Pres-ently, the dynamic processes that contribute to the airborne concentration or standing stock (i.e., the concentrations during a sampling time period of viable bioaerosols) are being elucidated.

A notion used to help understand the viable D/P distribution patterns in the atmosphere is their colony count balance in an imaginary static unit air volume through which an airstream passes. Figure 4.1 shows such a unit air volume where D/Ps are entrained into the atmosphere from some source(s) at some influent flux (number of D/Ps m^{-2} s^{-1}) to an air volume, and lost from that volume at some effluent flux. The concentration at any instant is the summation of those two fluxes. The effluent flux may be due to many factors, including meteorological processes such as advective and convective turbulence, physical process such as gravitational settling and agglomeration, and/or biological pro-cesses such as death and repair damage.

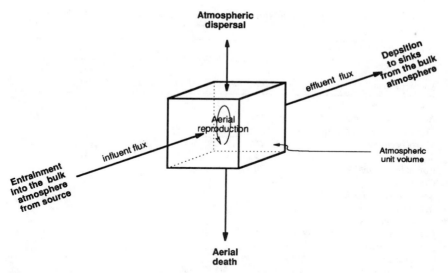

Figure 4.1. Generalized diagram of a viable bioaerosol D/P number balance in a unit volume of air.

Because of the continuous motion of the atmosphere, the sources and sinks of bioaerosols may be great distances and times apart. A bioaerosol cloud would be thought of as D/Ps added and removed by the aforementioned process, while at the same time being turbulantly mixed on various scales as well as advectively (i.e., movement of the air mass within which turbulent eddies exist) transported by the atmosphere.

4.2. Extramural (Outdoor) Bioaerosol Distributions

4.2.1. Quality of Microbial Bioaerosols

The species of an extramural airborne microorganism that might be found in an air sample at any particular location depends on (a) the species that can and do escape from a source, (b) their survival characteristics in the prevailing meteorological conditions, and (c) the travel time (or distance, which is the product of the wind speed and travel time) from the source. Air samples taken downwind near a source will tend to have all of the kinds of microorganisms that are found on the source. Farther downwind, a more "hardy" population will be selected. Those that remain, are able to survive the prevailing atmospheric conditions. Tables 4.1 and 4.2 list the surviving bacteria and fungi found in the air at several locations. The most prevalent bacteria (found in 50–95% of the samples) were Gram-positive cocci in the genera *Micrococcus, Sarcina,* and *Staphlococcus,* and occasionally *Coryneforms.* Among the Gram-negative bacte-

Table 4.1. Percentage of some kinds of bacteria that have been cultured from the atmosphere.

	Location					
Microorganism	Arctic[a]	Arctic[b]	Arctic[b]	Boreal[c]	Ocean, N. Atlantic[d]	Ocean, N. Atlantic[d]
Gram(+)pleomorphicrods		23	46	24	20	37
Aerobic spore-forming rods	5	51	18	38	33	29
Cocci in clusters and tetrads		23	15			
Micrococcus sp.	28			19 }	41 }	13
Sarcina sp.	3			5 }		
Corynebacterium-like	30	0	20			
Gram(−)rods					4	21
Achromobacter (probably)	5					
Achromobacter or *Flavobacterium*	16					
Flavobacterium, Achromobacter, Pseudomonas					5	
Miscellaneous		7	1			
Total Samples	145	45	1085	1173	367	189

Sources: (a) Pady et al. (1948); (b) Pady and Kelly (1953a); (c) Kelly and Pady (1953b); (d) Pady and Kelly (1954).

ria, *Flavobacterium, Achromobacter,* and, rarely, *Pseudomonas* are found. The most prevalent genus of fungus is often *Cladosporium* although *Alternaria* is frequently observed. Many samples contain fungi that do not sporulate and, therefore, cannot be identified.

4.2.2. Sources

The sources of microbial bioaerosols include just about all surfaces and porous materials. The question is not if it is a source, but how much, and what kind of source. The "much" is in terms of emission rate per area per unit time (or flux), and "what kinds" in terms of species. The sources will be discussed as (a) animal and human sources, (b) terrestrial including rural (including plants) and urban sources, (c) aquatic sources, and (d) atmospheric sources. Table 4.3 lists some of the sources, concentration found at the sources, emission rate, fluxes, and death rates of airborne bactiera. For further details on sources of airborne microorganisms, see Lighthart, Spendlove, and Akers (1979).

4.2.2.1. Animal and Human Sources

Talk, particularly with consonants, coughing, and sneezing (Wells, 1934; Wells and Wells, 1942) results in the injection of microorganisms in the atmosphere. Millions of droplet nuclei may be generated from sneezing. Constant

Table 4.2. Percentage of some of kinds of fungi that have been cultured from the atmosphere.

Microorganism	Arctic[a]	Arctic[b]	Arctic[b]	Boreal[c]	Ocean, N. Atlantic[d]
Cladosporium sp.	21[e]	26.5	26.5	73	82
Alternaria sp.		9.5	9.5	7	3
Pullularia sp.					2
Penicillium sp.	10	8	8	3	2
Botrytis sp.					1.5
Stemphylium sp.		3	3	1.5	1
Aspergillus sp.				1	
Streptomyces sp.	14			3	
Phyllosticta sp.	0.6				
Pestallozia sp.	0.6				
Leptosphoeria sp.	6				
Choetomium sp.	0.6				
Papularia sp.		5	5		
Yeast	10	3	3	1	2
Non-conidia formers	48	32	32	8	
Other fungi	13	3	3	3	3
Total Samples	154	125	125	3622	4991

Sources: (a) Pady et al. (1948); (b) Pady and Kelly (1953a); (c) Kelly and Pady (1953); (d) Pady and Kelly (1954); (e) Formerly classified as *Hormodendrum*.

shedding of skin flakes and hair may be sources of airborne viruses, bacteria, and fungi (Clark and Shirley, 1973; Riemensnider, 1967). Although some of these sources may be of epidemiological importance, and perhaps of significance in local atmospheric loadings, they are probably negligible contributors to large-scale extramural loading of the atmosphere. Some pathogenic bacteria of veterinary importance are discussed by Constantine (1969) and Gordon (1965).

4.2.2.2. Terrestrial Sources

Plants including both farm crops and forests, exposed soil in rural areas, and, to a much lesser extent, urban sources are thought to be the major contributors to the extramural atmospheric load over land. Loading from these sources is increased by turbulent weather frontals and convective turbulence (Fulton, 1966c) and rainsplash (Butterworth and McCartney, 1991), whereas sedimentation and raindrop washout are thought to reduce their numbers.

4.2.2.2.1. Rural Sources

Plants and exposed soil are thought to be major sources of airborne microorganisms although at least 30% of the load in a large-scale farming area could be

Table 4.3. Tabulation of some sources, concentrations, emission rates, fluxes, and death rates of bacteria cultured from atmospheric samples.

Microorganism	Aerial Source	Airborne Microorganisms				Reference
		Concentration (# m^{-3})	Emission Rate (# s^{-1})	Death Rate (k min^{-1})	Flux (# m^{-2} s^{-1})	
		Rural				
Mixed bacteria		4.2–1640				Jones and Cookson (1983)
Mixed bacteria/fungi	690 m altitude	6–810				Fulton (1966a)
Mixed bacteria/fungi	1600 m altitude	5–260				Fulton (1966a)
Mixed bacteria/fungi	3127 m altitude	1–110				Fulton (1966a)
Mixed bacteria/fungi	Frontal at 1200 m altitude	700–3802				Fulton (1966b)
Mixed bacteria/fungi	Land to sea at 152 m altitude	20–600				Fulton (1966c)
Mixed bacteria	Rural air	360				Camann et al. 1988
Mixed bacteria	Bare soil	663 (123)[a]			124	Lindemann, et al. (1982)
Mixed bacteria	Bare soil, dry	738 (198)			43 (9.2)	Lindemann and Upper (unpublished data)
Mixed bacteria	Bare soil, wet	1,545.5 (439)			154.5 (57)	Lindemann and Upper (unpublished data)
Mixed bacteria	Extramural air	450				Camann et al. (1988)
Viruses	Extramural air	<0.005				Camann et al. (1988)
Mixed bacteria	Compost pile	55,500				Millner, et al. (1980)
		Agriculture				
Peronospora tabacina (fungus)	Tobacco field	384.6–0.9			3.9(2.3)	Aylor and Taylor (1983)
Streptomycetes	Fallow field	0–6715				Lloyd (1969)
Uromyces phaseoli (fungus)	Bean field					Aylor and Ferrandino (1985)
Single spores	Bean field				19.4(8.7)	Aylor and Ferrandino (1985)
Double spores	Bean field				5.6(3.4)	Aylor and Ferrandino (1985)
Triple spores	Bean field				2.1(1.4)	Aylor and Ferrandino (1985)
Spore range	Bean field				6 to 31	Aylor and Ferrandino (1985)
Mixed bacteria	Corn	46–141				Lindemann et al. (1982)
Mixed bacteria	Winter wheat	2408–6500			57	Lindemann et al. (1982)
Mixed bacteria	Bean field	2240			449	Lindemann et al. (1982)
Mixed bacteria	Alfalfa	2690 (167)			543	Lindemann et al. (1982)

Organism	Source	Value	Reference
Mixed bacteria	Pea field	92	Lindemann et al. (1982)
Mixed bacteria	Upwind backround	51 (5.3)	Lindemann et al. (1982)
Mixed streptomycetes	Cultivated field	0–6715	Lloyd (1969)
Mixed bacteria	Open farmland	165	Lighthart (1984)
Mixed fungi	Open farmland	237	Lighthart (1984)
Mixed bacteria	Grass seed combine source	6.4×10^8	Lighthart (1984)
Mixed fungi	Grass seed combine source	4.7×10^6	Lighthart (1984)
Mixed bacteria	Downwind grass seed combine	3,923–18,520	Lighthart (1984)
Mixed fungi	Downwind grass seed combine	3,310–33,740	Lighthart (1984)
Mixed bacteria	Downwind grass straw baler	2,150–6,660	Lighthart (1984)
Mixed fungi	Downwind grass straw baler	2,340–3,310	Lighthart (1984)
Mixed bacteria	Loading grass seed into truck	3,450	Lighthart (1984)
Mixed fungi	Loading grass seed into truck	10,530	Lighthart (1984)

Urban

Organism	Source	Value	Reference
Mixed bacteria/fungi	Land air mass at 670m	200–3,500	Fulton and Mitchell (1966)
Mixed bacteria/fungi	marine air mass at 690 m	5–1,400	Fulton and Mitchell (1966)
Mixed bacteria	Urban air	610	Camann et al. (1988)

Wastewater

Organism	Source	Value	Reference
Escherichia coli phage C3000	Sewage treatment plant	0–0.6	Fannin et al. (1976)
Escherichia coli phage K-12 Hfrl	Sewage treatment plant	0–0.6	Fannin et al. (1976)
Mixed coliform bacteria	Sewage trickling filter	3–19,737	
Mixed coliform bacteria	Dewatered sludge in forest	0–15,000	
Mixed coliform bacteria	Cooling tower makeup water	299–1,220	
Mixed bacteria		6.7×10^4– 2.7×10^5	Adams et al. (1980)
Aspergillus fumigatus (fungus)	Sewage sludge compost	4,600,000	Millner et al. (1980)
Mixed fungi	Sewage sludge compost	0–55,000	Millner et al. (1980)
Mixed bacteria	Sewage sludge compost	0–55,500	Millner et al. (1980)
Escherichia coli	Trickling filter	0–15,300	Millner et al. (1980)
		70–1,130	Teltsch and Katzenelson (1978)

[a]Standard error.

accounted for by certain harvesting activities (Lighthart, 1984). How tillage contributes to atmospheric loading in rural areas is unknown, but it is thought to be significant. It is anticipated that spray applications and drift from microbiological pest control agents (MPCAs) from use in large-scale forested and agricultural areas will make a significant contribution to the downwind airborne loading in the future. Presently, deleterious effects of MPCA applications are anticipated and seen on susceptable, nontarget insects (James et al., 1993; Miller, 1990). The use of wastewater in sprinkler irrigation also contributes to the rural microbial load (Adams and Spendlove, 1970). However, the contribution of forestry and mining, including ore-processing operations on airborne microbial loading, is unknown.

4.2.2.2.2. Plant Sources

Plants are the natural habitat for many epiphytic microorganism, including phytopathogenic and saprophytic viruses, bacteria, and fungi (Preece and Dickinson, 1971; Bernstein et al., 1973). These microorganisms have a lognormal distribution (Hirano et al., 1982) on leaves and, therefore, some leaves could contribute more significantly than others to the atmospheric microbial load as the result of wind turbulance than others. Many of the microorganisms released from plants are rafted on plant and fungal debris, or soil particles (Lighthart et al., 1992a, 1992b). Because the release mechanism is thought to be largely the result of a wind-associated (i.e., turbulance and abrasion) process, one might imagine that there is a plume of microorganisms dispersed downwind from them. Plants are probably a major, large-scale source contributor to the atmospheric loading.

4.2.2.2.3. Urban Sources

Airborne microorganisms enter the atmosphere from many point sources that are usually categorized as general urban activities. Vehicular activities may generate large numbers of microbes rafted in dust particles (Roberts, 1973). Other point sources contributing to the urban loads include sewage treatment plants (Lighthart, 1967; Randell and Ledbetter, 1966; Fannin et al., 1976), power-generating cooling towers where it was estimated that 10^{10} bacteria/s may be emitted (Lighthart and Frisch, 1976; Adams et al., 1979; Adams et al., 1980), and industrial sources of textile mills, abattoirs, rendering plants [e.g., *Bacillus anthrasis* (Spendlove, 1957)], automobile demolition, and agricultural processing (Spendlove, 1974).

4.2.2.3. Aquatic (Including Marine) Sources

The atmosphere over marine areas has been observed to have fewer airborne microorganisms than over land, although nearshore areas with spray generated by wave action has higher numbers (Zobell, 1949: Fulton, 1966b). Blanchard and Syzdek (1972) have shown that bubbles rising to the surface concentrate bacteria and, when they burst at the surface, inject many bacteria-laden micro-droplets as high as 20 cm into the atmosphere. Lighthart (unpublished data) has found an increase of up to 1000 times (i.e., 10^5 cfu/ml) in the bacterial concentration in the first 250 μm of the surface film of a lake as compared to the bulk water, suggesting that the surface film may be a significant contributor to the downwind atmospheric load when it has been disturbed by wind action and so forth.

4.2.2.4. Atmospheric Sources

Airborne bacteria have been shown to metabolize, repair themselves, and reproduce (Dimmick et al., 1979a, 1979b).

4.2.3. Spatial Distribution

The quantitative atmospheric distribution of bacterial and fungal bioaerosols has been found, on average, to vary with altitude, diurnal and seasonal periodicity, proximity to land or sea, urban or rural locations, and, perhaps, latitude.

4.2.3.1. Altitude

Usually the concentration of viable D/Ps varies inversally with altitude (Fulton, 1966a). This is logical because the source of the D/Ps is at the surface of the Earth (e.g., Fulton, 1966a; Wright, et al., 1969). The concentration at the surface varies from up to several thousand per cubic meter to nil at an altitude of 60–70 km (Imshenetsky et al., 1978). However, within the atmosphere's mixed layer near the Earth's surface, which may be thousands of meters thick, continuous vertical mixing tends to distribute evenly the D/Ps (Fulton, 1966c). The greatest reduction in verticle concentration occurs in the first 50 m above the ground (e.g., Wright et al., 1969). Atmospheric inversion may trap viable D/Ps below this layer, increasing their concentration with time.

The exception to the general decrease in D/P concentration with altitude occurs during atmospheric inversions where higher concentrations may be isolated in a layer above the inversion away from particle-removing conditions than in the layer below the inversion.

4.2.3.2. Frontal Activity

Changing weather conditions between two different air masses is called a front and is often the site of great turbulence. As a result of the turbulence, large numbers of microorganisms may be swept into the atmosphere. For example, where cold air replaces warm air in a fast moving dry, cold front, increased winds at the ground surface may sweep viable D/Ps into the air where the turbulence will keep them suspended and transport them to higher altitudes. A wet frontal with similar turbulence will sweep fewer viable D/Ps into the atmosphere because of the moist ground surface and raindrop washout from the atmosphere. After passage of the frontal, turbulence subsides and the viable D/P concentration returns to that characteristic of the new air mass (Fig. 4.3).

4.2.3.3. Air Masses

On passage of an air mass over land, microorganisms are entrained into the bulk atmosphere, increasing their airborne concentration; then when the land loaded air masses pass over the ocean, they lose their terrigenous D/P load from their lower altitudes (Figs. 4.2 and 4.3; Fulton, 1966c). The increase in concentration over land is thought to be a function of the duration of the air mass over the land, the land area, the meteorological conditions, and the source of microorganisms (Fig. 4.2). This is the inverse of the usual case and is thought to occur because the lower-altitude populations are removed by gravitational settling, and/or perhaps death, faster than they were replaced from the ocean source. Or in terms of an advecting air volume similar to that in Fig. 4.1, the effluent flux from the air mass over the ocean was greater than the influent flux from the ocean source, resulting in a decreasing concentration in the overlying air volume.

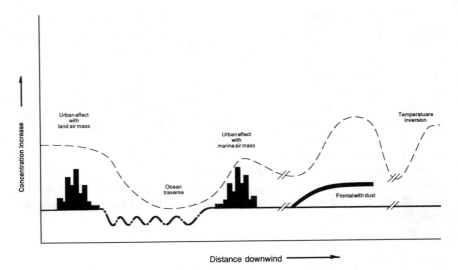

Figure 4.2. Diagram showing relative change in bacterial concentration isopleth (dashed) line for land and marine air masses traversing a city, dust-generating weather frontal, and temperature inversion [adapted from Mitchel and Fulton (1966); Fulton, 1966a, 1966b, 1966c].

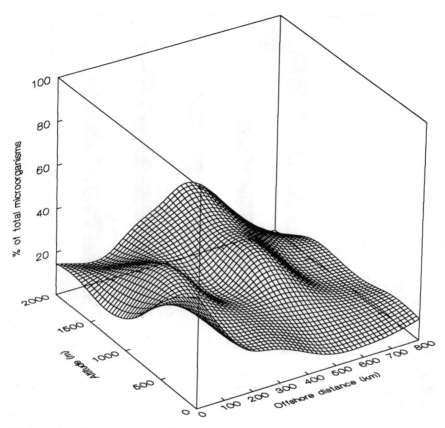

Figure 4.3. Percentage response surface of total microorganisms in the air column (altitude) of an air mass traversing from land to sea [after Fulton and Mitchell (1966)].

The influence of large local bacterial sources on the loading in an air mass was shown by Fulton and Mitchell (1966) to be substantial for a relatively clean marine air mass as it pass over a city, and insignificant for a relatively unclean land air mass passing over the same city (Fig. 4.4; Fulton and Mitchell, 1966). The influence may be seen as an increase in the population concentration isopleth (i.e., line of constant concentration) in Fig. 4.2 as the marine air mass goes from upwind to downwind of a city.

4.2.3.4. Long-Distance Transport

Microorganisms had been transported in the atmosphere before contaminating Pasteur's famous experiment showing air was not sterile. Later, it was found that microorganisms were in the air far out at sea (Certes, 1884; Fisher, 1886);

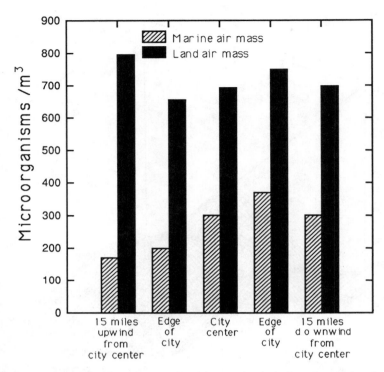

Figure 4.4. Microorganisms in two air masses, one relatively unloaded (marine) and one loaded (land) with airborne microorganisms on passage over a city [after Fulton and Mitchell (1966)].

it was not known whether they were aerosolized in situ or transported there from distant lands. Later observations were made from Lindberg's airplane crossing the north Atlantic in 1933 (Meier, 1935a, 1935b) and from a commercial aircraft over the Caribbean Sea (Meier, 1936) that microorganisms were in the marine air, but, again, their transport distance was unknown. In the early 1950s, Kelly and Pady (1953, 1954) at McGill University in Montreal carried out a series of experiments that culminated by relating the bacteria to their transporting air masses whose trajectories were determined from meteorological observations. From their observations, bacteria and fungi, both alive and ungrowable and presumed dead, where found associated with continental, polar, tropical and marine air (Pady and Kelly, 1953a, 1953b, Kelly and Pady, 1953, 1954, Pady, Kelly, and Polunin, 1948). Again however, when the microorganisms found in the air masses were entrained and how far they were transported is problematical. Long-distance transport of bacteria has been observed. Red sand and associated bacterial spores of genus *Bacillus* (e.g., *B. megaterium, B. pumilis, B. licheniformis,* and *B. firmuslentus*) were deposited on white snow in Scandinavia (Bovallius et al., 1978). The transport path of the spores was determined by calculating horizontal geostrophic trajectores and showed the red sand and spores that fell on Sweden came from a sand storm in Turkey 2 days earlier. In the vertical direction, both bacteria and fungi have been transported at least 77 km above the surface of the earth (Imshenetsky et al., 1978).

Although it has been shown that bacteria and fungi can be transported in the air for long distances, there is accumulating evidence that most survive for relatively short times. This means that to maintain a dynamic equilibrium, the upward flux of bacteria over a wind fetch must equal the loss rate in the bulk air due to death and deposition. For example, if the downwind concentration, upward flux of bacteria, in a constant wind, and fetch over a large field crop were approximately 5000 colony-forming units (cfu) m^{-3}, 500 cfu m^{-2} s^{-1}, 1 m s^{-1}, and 10 m, respectively (e.g., Lindemann et al., 1982), then the deposition rate ($<<$ 500 cfu m^{-2} s^{-1}) due to a slow sedimentation velocity for these small particles and the bulk concentration, which is thought to be relatively constant, must have a relatively high death rate ($\leq$500 cfu m^{-3} s^{-1}) to maintain the equilibrium.

4.2.3.5. Urban/Rural

The urban atmosphere appears to have a more uniformly high concentration of microorganisms in the atmosphere than the rural atmosphere. Bovallius and colleagues (1978) showed in their comparison of rural and urban collection sites in Sweden that the latter may have up to 2000 times more bacteria in the air than the former. The bacteria were also associated with larger particles in the urban locations. The concentration of urban bacteria in the air may be significantly affected by air pollutants such as carbon monoxide (CO) and sulfur dioxide (SO_2) which correlate negatively with viable bacterial loads and nitric oxide (NO), nitrogen dioxide (NO_2), and hydrocarbons which correlate positively. The bacterial particle size was negatively correlated with relative humidity and CO (Lee, 1973).

That is not to say that rural locations do not have high airborne concentrations of microorganisms. Continuous sources of high concentrations may be associated with local environments such as animal rearing facilities like swine houses (Sellers and Parker, 1969; Elliot et al., 1976), or high intermittant concentrations associated with dust storms and frontals (Fulton, 1966a, 1966b, 1966c) and crops (e.g., Lindemann et al., 1982).

4.2.4. *Temporal Distributions*

Temporal changes in the environment causes changes in the density of microorganisms from soil and vegetation sources of airborne microorganisms. The environmental changes have been found to produce both diurnal and annual cyclic changes in the quantity of airborne bacteria and fungi.

4.2.4.1. Diurnal Cycle

Although a diurnal periodic cycle is known to occur in the aerial release of some fungi (Gregory, 1973) and may be readily observed in some restricted environments, it is difficult to observe a general diurnal periodiciy of airborne microorganisms into the bulk atmosphere. This is primarily due to the large concentration differences in the atmosphere from day to day for the same place and time. A diurnal periodicity has been detected by Pady and Kramer (1967) in the atmosphere over the Kansas plains. They found the hourly mean concentrations for 127 days over 4 months in the Manhatten, Kansas wheat country had two peaks, one at about 0800 h in the morning and the other just after nightfall as illustrated in Fig. 4.5. The morning peak is presumably related to sunrise, whereas the large, late afternoon/evening peak is related to the usual afternoon convective warming and increased winds that would tend to entrain plant debris

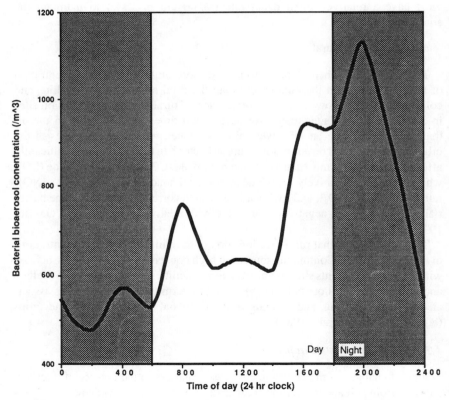

Figure 4.5. Diurnal cycle of mean hourly values of bacteria near the ground for 127 days over a 4-month period at Manhatten, KA [after Pady and Kramer (1967)].

and soil particles with attached microbes into the atmosphere. To illustrate the likelihood of this mechanism, small radar-reflective chaff sprinkled over chaparral in the Colorado plains was observed to be lifted several kilometers into the air by convective updrafts (A.S. Frisch, NOAA Laboratory, Boulder, CO, personal communication).

4.2.4.2. Annual Cycle

The annual cycle of bacterial and fungal population concentrations in the atmosphere in Montreal were observed over a 1.5-year period by Pady and Kelly (1954; Fig. 4.6). Both populations had winter minimums, which increased in spring. The bacteria reached a peak in late spring, whereas the fungi did not peak until midsummer; at the same time, the bacteria reached a midsummer secondary minimum. In the fall, the bacteria again increased to a primary maximum, then decreased; both populations then decreased to their winter minima. The winter primary minima for both populations are presumed to be the result of the source flux restrictions of ice and snow cover, the cold weather inhibition of microbial growth, and the lack of leaves on trees for microbial growth. The spring increase in both populations is thought to be due to the development of conditions conducive to microbial growth and dissemination (i.e., warmer, wet weather, leaves on trees, and wind). The summer decrease in bacteria is thought to be due to the development of drier conditions, higher temperatures, and/or increased solar radiation, in which bacteria cannot survive, but through which fungal sporulation and dissemination is well adapted. Again, the wetter fall conditions are thought to reduce fungal sporulation and stimulate bacterial growth, resulting in the fall bacterial increase and fungal decrease to their winter minima. When the weather conditions of winter arrive, the bacterial populations decline to their winter minima. This is the annual concentration where distinct seasonal variation occurs. It might be expected that where seasonal variations are not as dramatic, the annual concentration cycle of airborne microorganisms would have a lower amplitude.

The fungal distribution described above are those that will grow on culture media. The great majority of the fungal spores that are in the atmosphere are either dead or cannot be cultured (Fig. 4.7).

As can be seen and must be emphasized from Figs. 4.6 and 4.7, the wide scatter in the data shows that many observations must be made before any resonable conclusions may be made about the atmospheric loadings of microorganisms over time!

4.2.5. Particle Size Distribution

The particle size of extramural airborne microorganisms can vary from the size of a single virus or bacterium to large particles or debris "rafting" many kinds and sizes of microorganisms. Many of the rafted bacteria, both urban and rural,

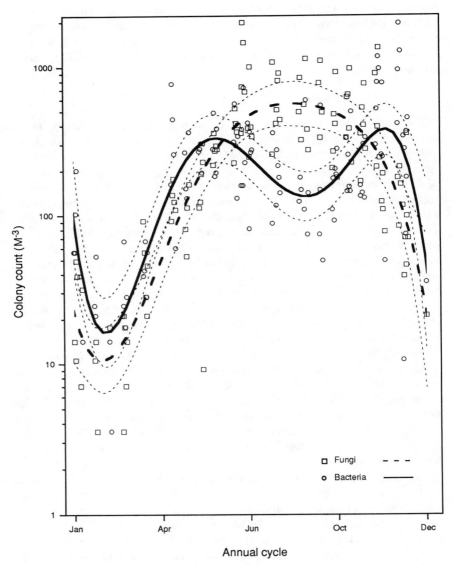

Figure 4.6. Annual concentration cycle of airborne bacteria and fungi on a 400-ft-high building in Montreal, Canada in 1951. Dotted line is the 99% confidence limit. (Adapted from Kelly and Pady, 1964.)

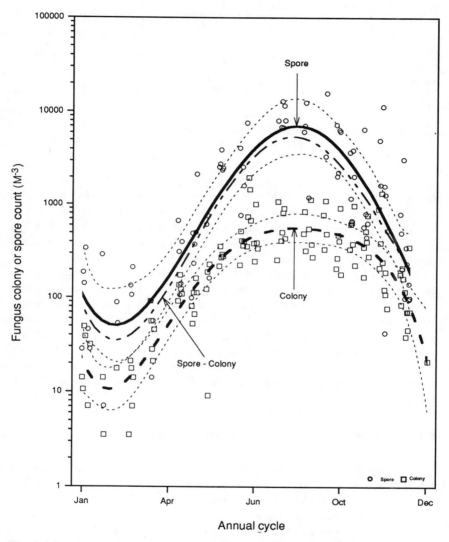

Figure 4.7. Annual concentration cycle of airborne viable fungi total fungal spore counts on a 400-ft-high building in Montreal, Canada in 1951. Dotted line is the 99% confidence limit. (Adapted from Kelly and Pady, 1954.)

Table 4.4. Particle size distribution (percentage) of bacterial aerosols at four diverse locations in Sweden.

Location	Particle Size[a] (μm)			
	>8	5.5–8	3.3–5.5	<3.3
Rural	37.8	22.7	19.8	19.6
Seacoast	48.9	18.0	16.1	16.9
City park	48.2	25.5	17.5	8.9
City street	53.2	24.5	14.9	7.7

[a]Aerodynamic particle size categories of Andersen sampler.

Source: Bovallius et al. (1978).

could be blown off plants as has been shown in the laboratory where bacteria were rafted on plant and fungal debris and released from plants by small puffs of wind (Lighthart et al., 1992a, 1992b). An analysis of the particle size distribution from samplers taken 2 m above the ground at four locations in Sweden, two rural and two urban, showed that 40% in the rural and 50% in the urban sites had particles containing viable bacteria ≥8 μm (Bovallius et al., 1978; Table 4.4; a bacerium is usually about 1 μm in diameter). It is reasoned that the large particles are of local origin because an 8 μm diameter particle has a sedimentation velocity in still air of about 0.20 cm s^{-1}, and a 3.3⁻μm particle has a sedimentation velocity in still air of 0.002 cm s^{-1}; thus, an 8-μm particle would take about 15 min to settle 2 m, whereas a 3.3-μm particle would take over a day. The data in Table 4.4 also indicates that there are more locally produced large particles in the urban areas compared to rural locations, as evidenced by the relatively smaller proportion of small particels in the urban locations.

4.3. Intramural (Indoor) Bioaerosol Distribution

A variety of adverse human health effects have been attributed to indoor exposure to airborne microorganisms (Table 4.5), including allergic reactions (Bernstein and Safferman, 1973; Croft et al., 1986; Gravesen, 1979), respiratory distress (Castellan et al., 1984; Croft et al., 1986; Jacobs, 1989; Sattar and Ijaz, 1987; Zeterberg, 1973), hypersensitivity reactions (Flaherty et al., 1984; Samson, 1985; Woodward et al., 1988), and infectious diseases (Sattar and Ijaz, 1987). Although algae (Bernstein and Safferman, 1973), bacteria (Bollin et al., 1985; Castellan et al., 1984, 1987; Croft et al., 1986), fungi (Gravesen, 1979; Woodward et al., 1988; Zeterberg, 1973), and viruses (Hierholzer, 1990; Sattar and Ijaz, 1987) have been associated with these effects, isolation and identification of an etiologic agent is often not readily accomplished in indoor surveys (Morey and Feeley, 1990).

Table 4.5. Adverse human health effects caused by indoor exposure to various airborne microorganisms.

Type of Microorganism	Identification	Adverse Health Effects	Indoor Sources and Amplification Sites
Algae	*Chlorella* *Chlorococcum* *Schizothrix* *Anabaena*	Allergic reactions	House dust, aquariums, humidifiers
Bacteria	*Bacillus* sp.	Hypersensitivity pneumonitis	Dry surfaces
	Legionella pneumophila	Pneumonia, Pontiac fever	Hot water
	Micropolyspora faeni	Hypersensitivity pneumonitis	Decomposing organic matter
	Mycobacterium tuberculosis	Tuberculosis	Humans
	Pseudomonas spp.	Pneumonia	Humans, water
	Staphylococcus spp.	Rashes, fever	Humans, household pets
	Streptococcus spp.	Fever, colds	Humans
Endotoxin	Gram negative bacteria	Fever, malaise, respiratory distress	Humidifiers, cotton dust, house dust
Fungi	*Aureobasidium pullulans*	Hypersensitivity, pneumonitis	Waterspray, painted surfaces
	Alternaria sp.	Allergic reactions	Carpet
	Aspergillus fumigatus	"Farmer's lung," allergic reactions	Hay, household pets
	Fusarium sp.	Allergic reactions	Standing water
	Cladosporium sp.	Allergic reactions	Building materials, furnishings
	Penicillium spp.	Allergic reactions, toxicosis	Building materials, furnishings, air handling systems, plants
	Stachybotrys atra	Toxicosis	Wetted cellulose, jute-backed carpet
Parasites	*Acanthamoeba* sp.	Eye infections	House dust
	Naegleria fowleri	Meningitis	House dust
Viruses	Rhinoviruses	Common cold	Humans
	Varicella zoster	Chicken pox	
	Paramyxoviruses	Measles Mumps	
	Poliovirus	Polio	

85

4.3.1. Historical Perspective

The outbreak of respiratory illness at the 1976 American Legion convention in Philadelphia focused attention on a common source exposure to a bacterium later classified as *Legionella pneumophila* (Winn, 1988). Retrospective investigations revealed that this bacterium was likely responsible for other earlier respiratory disease outbreaks (Osterholm et al., 1983; Thacker et al., 1978). Legionnaire's disease is characterized by malaise, headache, and muscle ache followed by a rapid rise in fever and chills. Although morbidity rates are generally less than 5% for exposed individuals, mortality may approach 10–15% (Dennis, 1990; Winn, 1988). In the American Legion outbreak, acute respiratory illness was documented in 182 conventioneers and 29 deaths were reported (Winn, 1988). A 1965 outbreak at St. Elizabeth's Hospital in Washington, D.C. resulted in a case–fatality ratio of 17.3% (Thacker et al., 1978) and an outbreak among immunosuppressed patients in an Iowa hospital reported a case–fatality ratio of 46% (Helms et al., 1983).

Pontiac fever, a self-limiting nonpneumonia influenza, is another manifestation of inhalation of *Legionella* spp. originally discovered after severe illness struck health department employees in Pontiac, Michigan. Because the attack rate was so high (95%), an airborne route was suggested as the mode of transmission for the disease-causing bacterium (Glick et al., 1978). A defective air conditioning system was implicated as the source, and investigators were eventually able to cause illness in laboratory animals exposed to the air conditioning condenser water (Winn, 1988).

Severe illnesses have also been caused by exposure to airborne fungi. Although fungal infections resulting from inhalation of spores or other elements generally results in localized lung infections, systemic infections have also been reported. *Aspergillus flavus* was discovered as the cause of invasive pulmonary aspergillosis in immunosuppressed patients exposed to the fungus via a contaminated ventilation system (Tobin et al., 1987) and indoor *Penicillium* spp., *Aspergillus* spp., and *Aureobasidium pullulans* have been implicated in changes of the occupants' peak pulmonary flow rate (Tobin et al., 1987).

Fragments of microbial cells including cell wall segments, flagella, and genetic material and by-products of metabolism are transported as bioaerosols and have caused adverse health problems (Sorenson et al., 1987). Inhalation of toxin-containing spores may lead to "pulmonary mycotoxicosis" due to the absorption of toxins through mucous membranes of the respiratory tract (Tobin et al., 1987). Mycotoxins are products of fungal metabolism and approximately 30–70% of fungi belonging to a toxin-producing genus are toxin producers when isolated from environmental sources (Tobin et al., 1987). Mycotoxins are generally of low molecular weight and highly soluble. Dermal exposure to toxin-producing fungi, therefore, may result in skin irritation and respiratory distress or lead to severe systemic toxicosis (Tobin et al., 1987). Some mycotoxins may also act as immunosuppressives and increase the likelihood of opportunistic and secondary bacterial infections (Croft et al., 1986). Bacterial endotoxins, produced by several genera of Gram-negative bacteria, have also been implicated in respiratory distress and other adverse health effects (Tobin et al., 1987). Byssinosis, a lung disorder caused by the inhalation of cotton dust, has been correlated to endotoxin (Castellan et al., 1984; 1987).

4.3.2. Distribution Within Indoor Environments

Microorganisms are ubiquitous in the indoor environment. The amplification (growth) and dispersal of bacteria, viruses, parasites, algae, and fungi with potential for adverse human health effects in indoor environments is often insidious and not revealed until occupants experience illness. Cases of illness resulting from indoor exposure to airborne microorganisms has been documented in hospitals, nursing homes, residences, schools, offices, and industrial/agricultural settings.

4.3.2.1. Hospital/Medical Facilities

Health care facilities are especially conscious of the exposure of patients to airborne pathogenic and opportunistic microbial agents. Hospitals use monitoring to identify nosocomial infections, trace dissemination of airborne microorganisms, detect personnel, instruments, and activities that generate airborne contamination, evaluate the efficiency of air cleaning equipment, and assess hazards of ventilation systems (Sayer et al., 1972). Green et al. (1962) reported levels of airborne microbial contaminants ranging from 160 to > 2500 m^{-3} in hospitals depending on the functions performed in the area, human activity, surface contamination, and ventilation. No attempts were made by the researchers to correlate hospital-acquired infections with the airborne counts or exposure of patients while confined in that study. Cases of illness, however, have been reported resulting from indoor exposure in hospitals, and researchers have demonstrated aerosolization of pathogenic *L. pneumophila* from contaminated hospital humidifiers (Zuravleff et al., 1983), sink outlets (Helms et al., 1983), and shower spray heads (Bollin et al., 1985). In experimental trials with contaminated humidifier water, Zuravleff et al. (1983) demonstrated retrieval of airborne *L. pneumophila* with aerobiological sampling and seroconversion of exposed guinea pigs. The aerosolization of 1–5-μm-diameter droplets containing legionella bacteria from hospital shower heads and hot water faucets was shown by Bollin et al. (1985). Monitoring for *Legionella* sp., however, is generally limited to potentially contaminated water and is not routinely included in air surveys (ACGIH, 1989).

4.3.2.2. Residential Indoor Air

People without medical confinement may spend up to 22 h a day inside (Spengler and Sexton, 1983), resulting in exposure to indoor airborne contaminants.

Adverse health effects resulting from indoor exposure to microorganisms are often difficult to correlate with specific etiologic agents due to the seemingly random and vague symptomology and the numerous uncontrolled variables within residential environments. Cases of flulike symptoms, headaches, diarrhea, dermatitis, and general malaise were experienced by residents of a suburban Chicago house over a 5-year period, yet no clinical diagnosis could explain the cause. It was finally determined that the illnesses were caused by indoor exposure to a mycotoxin produced by *Stachybotrys atra* growing in wood fiber board of the house (Croft et al., 1986).

Cladosporium, the most common fungus isolated in indoor air environments (Gravesen, 1979; Hirsch and Sosman, 1976), until recently was not considered an indoor contaminant. This organism, however, has been implicated in allergic reactions, raising questions as to its role in adverse health effects in indoor environments. Gravesen (1978) reported the presence of *Cladosporium* sp. in 100% of the airborne samples taken in 100 houses, many occupied by allergy sufferers. Although no standards currently exist for indoor concentrations of airborne microorganisms, counts of 3000 *Cladosporium* spores m^{-3} of air has been presented as a threshold level for allergic reactions (Gravesen, 1979). The presence of 100 *Alternaria* spores m^{-3} has also been cited as a level resulting in reactions in sensitive individuals (Gravesen, 1979). Dose–response data for other microorganisms is not presently available (ACGIH, 1989).

Scott et al. (1982) reported on the bacterial flora of over 200 English households with numerous genera isolated from surfaces. Wet areas were associated with the isolation of enteric bacteria, whereas Gram-positive bacilli, micrococci, and pseudomonads were more frequently isolated from dry surface sites with some potential for aerosol dispersal. Similar results were obtained by Finch et al. (1978) in a survey of 21 houses where Gram-positive cocci and *Bacillus* spp. were commonly isolated. No investigation, however, was done to ascertain the significance or sources of these organisms.

Fungi were targeted for isolation and identification in over 150 houses during winter months by Su et al. (1992). These researchers correlated the isolation of airborne *Fusarium* sp., a fungus that requires high humidity for growth, with standing water and water collection in houses. Airborne *Penicillium* spp. and *Aspergillus* spp., common soil fungi, were correlated with houses having crawl spaces rather than occupied basements. Increased reporting of respiratory complaints were correlated with houses reporting airborne *Cladosporium* spp., *Epicoccum* spp., *Aureobasidium* spp., and yeasts.

4.3.2.2. Nonresidential—Schools, Office Environments, Hotels

Symptoms of malaise, fever, and cough eventually leading to a diagnosis of hypersensitivity pneumonitis in employees resulted in a diagnosis of indoor exposure at a fungus-contaminated multistory office complex (Woodward et al.,

1988). *Aureobasidium pullulans* in open waterspray chambers and corrugated cardboard was determined to be the cause of those illnesses and the presence of open water chambers in hydroculturing of ornamental plants in a bank office were cited as source of airborne *Aspergillus* spp. (Samson, 1985).

The presence of fungal spores, animal hair, pollen, and particulates were also linked to the reporting of respiratory distress in employees of a multistory biological research facility where maintenance of the air handling system was cited as the reason for dispersal of numerous airborne contaminants (Samimi, 1990).

4.3.2.3. Industrial and Agricultural

Complaints by employees at a textile-producing facility serviced by a humidification system revealed bacterial endotoxin as the cause of hypersensitivity pneumonitis (Flaherty et al., 1984). *Micropolyspora faeni,* a thermophilic actinomycete, has been implicated in cases of illness resulting from exposure to decomposing organic matter such as compost or municipal garbage. Exposure to this organism may result in hypersensitivity pneumonitis and other allergic reactions.

Aspergillus, Penicillium, and *Monilia* are other fungal genera that have been implicated in adverse reactions of agricultural workers. *Aspergillus fumigatus* is a fungus with known, pathogenic effects including "farmer's lung," a debilitating respiratory disease often contracted in barnyard settings where contaminated hay is handled (Gravesen, 1979).

4.3.3. Sources and Sinks in Indoor Air

Anywhere there is moisture is a potential source of indoor bacteria, algae, fungi, and fungal by-products. This includes the air handling system, building materials and furnishings, and the occupants, pets, and plants (Fig. 4.8). Investigators have suggested that relative humidity less than 70% will inhibit growth of fungi on construction materials (Coppock and Cookson, 1951), and Grant et al. (1989) reported that although indoor humidity fluctuates, relative humidity > 80% may only occur indoors during a brief period in the winter. Airborne and surface-associated fungi, therefore, would be subjected to periods of low water activity throughout the year. White (1990) maintains, however, that fungal growth is sustained with ambient humidity as low as 20% indoors.

4.3.3.1. Air Handling Systems

Humidifiers, ventilators, and cool-mist vaporizers have been directly linked to illness indoors (Table 4.5). Bacteria, fungi, algae, and free-living amoebas have been isolated from water reservoirs of these systems and are often found in high concentrations. When these devices are used, water droplets and particulates in the reservoirs are aerosolized (Highsmith and Rodes, 1988). Often,

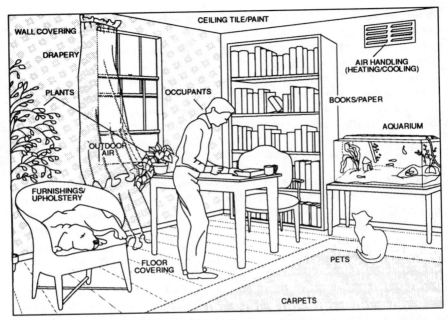

Figure 4.8. Illustration of sources and amplification sites of indoor airborne microorganisms including outdoor air through an open window, the air handling system, draperies, carpeting, upholstered furniture, wallpaper, plants, occupants, and household pets. (Illustration courtesy of Shirley Burns.)

attempts to decontaminate these portable units by the homeowners are unsuccessful. The potential for growth and dispersal of microorganisms via heating and cooling systems in homes and large buildings is of concern because equipment is less accessible and maintenance is expensive (Burge et al., 1987). In a study of more than 4300 office workers, increased symptoms were reported in buildings with humidified or chilled air (Burge et al., 1987), and Finnegan et al. (1984) showed similar results with significantly higher numbers of upper respiratory and mucous membrane complaints from workers in air conditioned offices. Burge et al. (1987) reported the highest complaint rates from workers in office serviced by local induction units followed by those supplied by central induction/fan coil units. Buildings with natural or mechanical ventilation have recorded the fewest reports of employee health effects (Finnegan et al., 1984; Burge et al., 1987).

Fungal and bacterial cells can grow on air handling filters resulting in increased airborne counts. Martikainen et al. (1990) report that microorganisms could actively grow on air handling filters when relative humidity was greater than 75%, necessitating changing of filters depending on climate and site. Poorly maintained air conditioning filters also can result in gaps for passage of unfiltered air (Samimi, 1990).

Cooling system waters have also been implicated as potential sources of aerosolized *Naegleria fowleri* and *Acanthamoeba* (ACGIH, 1989). Nasal mucosa is the initial site for *N. fowleri* infections and aerosolized *Naegleria* antigens may also lead to allergic reactions.

4.3.3.2. Building Materials and Furnishings

Building materials and furnishings may provide a niche for bacteria, fungi, algae, and free-living protozoa (White, 1990). Carpeting collects dust, pet dander and hair, other debris, and harbors moisture, providing an ideal environment for the growth of bacteria and fungi (Table 4.5). Human activity (e.g., walking and vacuuming) has been shown to reentrain fungal spores from contaminated carpet, thereby increasing potential for respiratory exposure (Buttner and Stetzenbach, 1993). *Alternaria* spp. are the most common fungi isolated from carpet dust, and monitoring has shown that carpeting enhances the survival of this organisms in the winter (Hirsch and Sosman, 1976). The composition of the carpet fibers has also been linked with fungal spore survival and growth with a greater occurrence of *Alternaria* and *Monilia* in woolen carpeting compared to synthetics. Carpet pile depths of > 0.5 in. were also shown to be sites for growth of *Cladosporium, Alternaria,* and *Monilia.* Airborne *Pseudomonas aeruginosa* were isolated from a house where the carpeting had recently been shampooed and was still wetted (Finch et al., 1978). Sources of toxigenic *Penicillium* and mycotoxin-producing *Stachybotrys atra* are found to include moisture cellulose (e.g., ceiling tiles and paper products) and has been isolated from jute-backed carpeting (Gravesen, 1979). Unfortunately, this fungus does not compete well with other fungi on isolation medium commonly used for indoor air surveys and may be overlooked during monitoring. Studies have demonstrated that inhalation exposure doses can produce toxicosis, and low numbers of toxigenic fungal spores can result in severe health problems (Sorenson et al., 1987).

The presence of soiled, painted, or papered surfaces supports the growth of fungi when lower moisture was present. *Aureobasidium* and *Phoma* have been associated with deterioration of painted surfaces (Gravesen, 1979).

Skin testing for allergic reactions to algae and air sampling for airborne algal cells have resulted in an association of both green algae (*Chlorella* and *Chlorococcum*) and blue-green algae (*Schizothrix* and *Anabena*) with adverse human health effects (Bernstein and Safferman, 1973). Although the extent of allergic reactions due to algal exposure has not been fully investigated, house dust and aeration of aquariums have been proposed as possible sources. House dust has also been implicated as a source of *Acanthamoeba,* a free-living amoeba isolated from eye infections. Whereas this amoeba is usually linked with infections following injury to the eye, some cornea infections have been traced to contact lens solutions contaminated with dust (ACGIH, 1989). House dust is also a source of fungi and mites that may cause allergic reactions (ACGIH, 1989).

4.3.3.3. Occupants

Building occupants and their activity are also considered sources of microorganisms (Table 4.5). Human source microorganisms are dispersed from nasal and oral surfaces, skin, clothing, and hair of building occupants. Bacteria, viruses, and parasites are primarily the human source microorganisms of concern in indoor environments. Numerous viruses are transmitted through the air particularly via droplet nuclei produced in the nose and throat of infected individuals (Zeterberg, 1973). The viruses are expelled during coughing and sneezing and transmitted person-to-person directly to a new host or as a result of contamination of surfaces (Sattar and Ijaz, 1987; Zeterberg, 1973). Mumps, measles, chicken pox, polio, and the common cold are some examples of viral diseases that can be transmitted through the air (Atlas, 1984). Comparisons of human associated microbial flora in indoor air samples versus outdoor air samples have been used as an indication of high occupancy rate and poor ventilation of commercial buildings and schools. The presence of household pets has also been linked with increased incidence of airborne *Aspergillus fumigatus* (Hirsch and Sosman, 1976).

References

Adams, A. P., M. Garbett, H. B. Rees, and B. G. Lewis. 1980. Bacterial aerosols produced from a cooling tower using wastewater effluent as makeup water. J. Water Pollut. Control Assoc. 52(3):498–501.

Adams, A. P., and J. C. Spendlove. 1970. Coliform aerosols emitted by sewage treatment plants. Science 169:1218–1220.

Adams, A. P., J. C. Spendlove, E. C. Rengers, M. Garbett, and B. G. Lewis. 1979. Emission of microbial aerosols from vents of cooling towers: I. Particle size, source strength and downwind travel. Dev. Indust. Microbiol. 20:769–779.

American Conference of Governmental Industrial Hygienists (ACGIH). 1989. Guidelines for the assessment of bioaerosols in the indoor environment. ACGIH, Cincinnati, OH.

Atlas, R. M. 1984. Microbiology: Fundamentals and applications. Macmillan Publishing Co., New York.

Aylor, D. E., and F. J. Ferrandino. 1985. Escape of urediniospores of *Uromyces phasioli* from a bean field canopy. Ecol. Epidemiol. 75:1232–1235.

Aylor, D. E., and G. S. Taylor. 1983. Escape of *Peronospore tabacina* spores from a field of diseased tobacco plants. Ecol. Epidemiol. 74:525–529.

Bernstein, M. E., H. M. Howard, and G. C. Carroll. 1973. Fluorescence microscopy of Douglas fir foliage epiflora. Can. J. Microbiol. 19:1129–1130.

Bernstein, I. L., and R. Safferman. 1973. Clinical sensitivity to green algae demonstrated by nasal challenge and *in vitro* tests of immediate hypersensitivity. J. Allergy Clin. Immunol. 51:22–28.

Blanchard, D. C., and L. D. Syzdek. 1972. Concentration of bacteria in jet drops from bursting bubbles. J. Geophys. Res. 77(27):5087–5099.

Bollin, G. E., J. F. Plouffe, M. F. Para, and B. Hackman. 1985. Aerosols containing *Legionella pneumophila* generated by shower heads and hot-water faucets. Appl. Environ. Microbiol. 50:1128–1131.

Bovallius, A., B. Bucht, R. Roffey, and P. Ånäs. 1978. Three-year investigation of the natural airborne bacterial flora at four localities in Sweden. Appl. Environ. Microbiol. 35(5):847–852.

Bovallius, A., B. Bucht, R. Roffey, and P. Ånäs. 1978. Long-range air transmission of bacteria. Appl. Environ. Microbiol. 35(6):1231–1232.

Burge, S., A. Hedge, S. Wilson, J. H. Bass, and A. Robertson. 1987. Sick building syndrome: a study of 4373 office workers. Ann. Occup. Hyg. 31:493–504.

Butterworth, J., and H. A. McCartney. 1991. The dispersal of bacteria from leaf surfaces by water splash. J. Appl. Bacteriol. 71:484–496.

Buttner, M. P., and L. D. Stetzenbach. 1993. Monitoring airborne fungal spores in an experimental indoor environment to evaluate sampling methods and the effects of human activity on air sampling. Appl. Environ. Microbiol. 59:219–226.

Camann, D. E., B. E. Moore, H. J. Harding, and C. A. Sorber. 1988. Microorganism levels in air near spray irrigation of municipal wastewater: the Lubbock infection surveillance study. J. Water Pollut. Control Fed. 60:1960–1970.

Castellan, R. M., S. A. Olenchock, J. L. Hankinson, P. D. Millner, J. B. Cocke, C. K. Bragg, H. H. Perkins, and R. R. Jacobs. 1984. Acute bronchoconstriction induced by cotton dust: dose related responses to endotoxin and other dust factors. Ann. Intern. Med. 101:157–163.

Castellan, R. M., S. A. Olenchock, K. B. Kinsley, and J. L. Hankinson. 1987. Inhaled endotoxin and decreased spirometric values. 1987. N. Engl. J. Med. 317:605–610.

Certes, A. 1884. Sur la culture à l'abri des germes atmosphérique des eaux et des sédiments rapportés par les expéditions du Travailleur it du Talsiman. Compt. Rend. 98:690–793.

Clark, R. P., and S. G. Shirley. 1973. Identification of skin in airborne particulate matter. Nature (Lond.) 246:39–40.

Constantine, D. G. 1969. Airborne microorganisms: Their relevance to veterinary medicine. In R. L. Dimmick and A. B. Akers (eds.), An introduction to experimental aerobiology. Wiley–Interscience, New York.

Coppock, J. B., and E. D. Cookson. 1951. The effect of humidity on mould growth on construction material. J. Sci. Food Agric. 2:534–537.

Croft, W. A., B. B. Jarvis, and C. S. Yatawara. 1986. Airborne outbreak of trichothecene toxicosis. Atmos. Environ. 20:549–552.

Dennis, P.J.L. 1990. An unnecessary risk: Legionnaires' disease. In P. R. Morey, J. C. Feeley, Sr., and J. A. Otten (eds.), Biological contaminants in indoor environments. American Society for Testing and Materials, Philadelphia, PA.

Dimmick, R. L., H. Wolochow, and M. A. Chatigny. 1979a. Evidence that bacteria can form new cells in airborne particles. Appl. Environ. Microbiol. 37:924–927.

Dimmick, R. L., H. Wolochow, and M. A. Chatigny. 1979b. Evidence for more than one division of bacteria within airborne particles. Appl. Environ. Microbiol. 38:642–643.

Elliott, L. F., T. M. McCalla, and J. A. Deshazer. 1976. Bacteria in the air of housed swine units. Appl. Environ. Microbiol. 32:270–273.

Fannin, K. F., J. C. Spendlove, K. W. Cochran, and J. J. Gannon. 1976. Airborne coliphages from wastewater treatment facilities. Appl. Environ. Microbiol. 31:705–710.

Fannin, K. F., A. C. Vana, and W. Jukubowski. 1985. Effect of activated sludge wastewater treatment plant on ambient air densities of aerosols containing bacteria and viruses. Appl. Environ. Microbiol. 49:1191–1196.

Finch, J. E., J. Prince, and M. Hawksworth. 1978. A bacteriological survey of the domestic environment. J. Appl. Bacteriol. 45:357–364.

Finnegan, M. J., C. A. C. Pickering, and P. S. Burge. 1984. The sick building syndrome: prevalence studies. Brit. Med. J. 289:1573–1575.

Fischer, B. 1886. Bakteriologische untersuchungen auf einer reise nach Westindien. Z. Hyg. 1:421–464.

Flaherty, D. K., F. H. Deck, J. Cooper, K. Bishop, P. A. Winzenburger, L. R. Smith, L. Bynum, and W. B. Witmer. 1984. Bacterial endotoxin isolated from a water spray air humidification system as a putative agent of occupation-related lung disease. Infect. Immunol. 43:206–212.

Fulton, J. D. 1966a. Microorganisms of the upper atmosphere. III. Relationship between altitude and micropopulation. Appl. Microbiol. 14(2):245–250.

Fulton, J. D. 1966b. Microorganisms of the upper atmosphere. IV. Microorganisms of a land air mass as it traverses an ocean. Appl. Microbiol. 14(2):241–244.

Fulton, J. D. 1966c. Microorganisms of the upper atmosphere. V. Relationship between frontal activity and the micropopulation at altitude. Appl. Microbiol. 14(2):245–250.

Fulton, J. D., and R. B. Mitchell 1966. Microorganisms of the upper atmosphere. II. Microorganisms in two types of air masses at 690 meters over a city. Appl. Microbiol. 14(2):232–2236.

Glick, T. H., M. B. Gregg, B. Berman, G. Mallison, W. W. Rhodes, Jr., and I. Kassanoff. 1978. Pontiac fever. An epidemic of unknown etiology in a health department. I. Clinical aspects and epidemiologic aspects. Am. J. Epidemiol. 107:149–160.

Gordon, J. E. (ed.). 1965. Control of communicable diseases in man. American Public Health Association, Washington, DC.

Grant, C., C. A. Hunter, B. Flannigan, and A. F. Bravery. 1989. The moisture requirements of moulds isolated from domestic dwellings. Internat. Biodet. 25:259–284.

Gravesen, S. 1978. Identification and prevalence of culturable mesophilic microfungi in house dust from 100 Danish homes, Allergy 33:168–272.

Gravesen, S. 1979. Fungi as a cause of allergic disease. Allergy 34:135–154.

Greene, V. W., D. Vesley, R. G. Bond, G. S. Michaelsen. 1962. Microbiological contamination of hospital air. I. Quantitative studies. Appl. Microbiol. 10:561–566.

Gregory, P. H. 1973. The microbiology of the atmosphere, 2nd ed. John Wiley & Sons. New York.

Helms, C. M., M. Massanari, R. M. Zeitler, S. Streed, M. J. R. Gilchrist, N. Hall, W. J. Hausler, J. Hirano, S. S., E. V. Nordheim, D. C. Arny, and C. D. Upper. 1983. Lognormal distribution of epiphytic bacterial populations on leaf surfaces. Appl. Environ. Microbiol. 44(32):695–700.

Hierholzer, J. C. 1990. Viruses, mycoplasmas as pathogenic contaminants in indoor environments. In P. R. Morey, J. C. Feeley, Sr., and J. A. Otten (eds.), Biological contaminants in indoor environments. American Society for Testing and Materials, Philadelphia, PA.

Highsmith, V. R., and C. E. Rodes. 1988. Indoor particle concentrations associated with the use of tap water in portable humidifiers. Environ. Sci. Technol. 22:1109–1112.

Hirano, S. S., E. V. Nordheim, D. C. Arny, and C. D. Upper. 1982. Lognormal distribution of epiphytic bacterial populations on leaf surfaces. Appl. Environ. Microbiol. 44:695–700.

Hirsch, S. R., and J. A. Sosman. 1976. A one-year survey of mold growth inside twelve homes. Ann. Allergy 36:30–38.

Imshenetsky, A. A., S. V. Lysenko, and G. A. Kazakov. 1978. Upper boundary of the biosphere. Appl. Environ. Microbiol. 35(1):1–5.

Jacobs, R. R. 1989. Airborne endotoxins: an association with occupational lung disease. Appl. Ind. Hyg. 4:50–56.

James, R. R., J. C. Miller, and B. Lighthart. 1993. The effect of *Bacillus thuringiensis* var. *kurstaki* on a beneficial insect, the Cinnabar moth (Lepidoptera:Arctiidae). J. Econ. Entomol. 86:334–339.

Jones, B. L., and J. T. Cookson. 1983. Natural atmospheric microbial conditions in a typical suburban area. Appl. Environ. Microbiol. 45:919–934.

Kelly, C. D., and S. M. Pady. 1953. Microbiological studies of air over some nonarctic regions of Canada. Can. J. Bot. 31:90–106.

Kelly, C. D., and S. M. Pady. 1954. Microbiological studies of air masses over Montreal during 1950 and 1951. Can. J. Bot. 32:591–600.

Lee, R. E., Jr. 1973. Relationship between viable bacteria and air pollutants in an urban atmosphere. Amer. Ind. Hyg. Assoc. J. 34(4):164–170.

Lighthart, B. 1967. A bacteriological monitoring of sewage on passage through an activated sludge sewage treatment plant. Doctoral dissertation, University of Washington, Seattle, WA.

Lighthart, B. 1984. Microbial aerosols: Estimated contribution of combine harvesting to an airshed. Appl. Environ. Microbiol. 47:430–432.

Lighthart, B., and A. S. Frisch, 1976. Estimation of viable airborne microbes downwind from a point source. Appl. Microbiol. 47:430–432.

Lighthart, B., B. T. Shaffer, and B. Marthi. In press 1994. Artificial wind puff liberation of microbial bioaerosols deposited on plants. Aerobiologia.

Lighthart, B., J. C. Spendlove, and T. G. Akers. 1979. Factors in the production, release and viability of biological particles. Pp. 11–22. In R. L. Edmonds (ed.), Aerobiology: the ecological systems approach. US/IBP Synthesis Series 10. Dowden, Hutchinson and Ross, Inc., Stroudsburg, PA.

Lindemann, J., H. A. Constantinidou, W. R. Barchet, and C. D. Upper. 1982. Plants as sources of airborne bacteria, including ice nucleation active bacteria. Appl. Environ. Microbiol. 44(5):1059–1063.

Lloyd, A. B. 1969. Dispersal of Streptomycetes in air. J. Gen. Microbiol. 57:35–40.

Martikainen, P. J., A. Asikainen, A. Nevalainen, M. Janunen, P. Pasanen, and P. Kalliokoski. 1990. Microbial growth on ventilation filter materials. Proceedings of the 5th International Conference on Indoor Air and Climate, Toronto, Canada. 3:203–206.

Meier, F. C. 1935a. Microorganisms in the atmosphere of arctic regions. Phytopatol. (Abstract) 25:27.

Meier, F. C. 1935b. Collecting microorganisms in the arctic atmosphere: with field notes and material by C. A. Lindbergh. Arctic Monthly 40:5–20.

Meier, F. C. 1936. Collecting microorganisms from wind above the Caribbean Sea. Phytopathol. (Abstract) 26:102.

Miller, J. C. 1990. Field assessment of the effects of a microbial pest control agent on nontarget Lepidoptera. Amer. Entomol. Summer. pp. 135–139.

Millner, P. D., D. A. Bassett, and P. B. Marsh. 1980. Dispersal of *Aspergillus fumigatus* from sewage sludge compost piles subjected to mechanical agitation in open air. Appl. Environ. Microbiol 39:1000–1009.

Morey, P. R., and J. C. Feeley. 1990. The landlord, tenant, and investigator: their needs, concerns, and viewpoints. In P. R. Morey, J. C. Feeley, Sr., and J. A. Otten (eds.), Biological contaminants in indoor environments. American Society for Testing and Materials, Philadelphia, PA.

Osterholm, M. T., T. D. Chin, D. O. Osborne, H. B. Dull, A. G. Dean, D. W. Fraser, P. S. Hayes, and W. N. Hall. 1983. A 1957 outbreak of Legionnaires' disease associated with a meat packing plant. Amer. J. Epidemiol. 117:60–67.

Pady, S. M., and C. D. Kelly. 1953a. Studies on microorganisms in arctic air during 1949 and 1950. Can. J. Bot. 31:107–122.

Pady, S. M., and C. D. Kelly. 1953b. Numbers of fungi and bacteria in transatlantic air. Science 117:607–609.

Pady, S. M., and C. D. Kelly. 1954. Aerobiological studies of fungi and bacteria over the Atlantic ocean. Can. J: Bot. 32:202–212.

Pady, S. M., C. D. Kelly, and N. Polunin. 1948. Arctic aerobiology, II. Nature (Lond.) 162(4114):379–381.

Pady, S. M., and C. L. Kramer. 1967. Diurnal periodicity of airborne bacteria. Mycologia 59:714–716.

Preece, T. P., and C. H. Dickinson. 1971. Ecology of leaf-surface microorganisms. Academic Press, New York.

Randall, C. W., and J. O. Ledbetter. 1966. Bacterial air pollution from activated sludge units. Amer. Ind. Hyg. Assoc. J. 27:506–519.

Riemensnider, D. K. 1967. Reduction of microbial shedding from humans. Contamin. Control Aug:19–20.

Roberts, J. W. 1973. The measurement, cost and control of air pollution from unpaved roads and parking lots in Seattle's Duwamish Valley. Doctoral dissertation, University of Washington, Seattle, WA.

Samimi, B. S. 1990. Contaminated air in a multi-story research building equipped with 100% fresh air supply ventilation systems. Preceding of the 5th International Conference on Indoor Air Quality and Climate, Toronto, Canada. 4:571–576.

Samson, R. A. 1985. Occurrence of moulds in modern living and working environments. Eur. J. Epidemiol. 1:54–57.

Sattar, S. A., and Ijaz, M. K. 1987. Spread of viral infections by aerosols. CRC Crit. Rev. Environ. Control. 17:89–131.

Sayer, W. J., N. M. MacKnight, and H. W. Wilson. 1972. Hospital airborne bacteria as estimated by the Andersen sampler *versus* the gravity settling culture plate. Amer. J. Clin. Pathol. 58:558–562.

Scott, E., S. F. Bloomfield, and C. G. Barlow. 1982. An investigation of microbial contamination in the home. J. Hyg. Camb. 89:279–293.

Sellers, R. F., and J. Parker. 1969. Airborne excretion of foot-and-mouth disease virus. J. Hyg. Camb. 67:671–677.

Sorber, C. A., H. T. Bausum, S. A. Shaub, and M. J. Small. 1976. A study of bacterial aerosols at a wastewater irrigation site. J. Water Pollut. Control Fed. 48:2367–2379.

Sorenson, W. G., D. G. Frazer, B. B. Jarvis, J. Simpson, and V. A. Robinson. 1987. Trichothecene mycotoxins in aerosolized conidia of *Stachybotrys atra*. Appl. Environ. Microbiol. 53: 1370–1375.

Spengler, J. D. and K. Sexton. 1983. Indoor air pollution: a public health perspective. Science 221:9–17.

Spendlove, J. C. 1957. Production of bacterial aerosols in a rendering plant process. Public Health Rept. 72:176–180.

Spendlove, J. C. 1974. Industrial, agricultural, and municipal microbial aerosol problems. Dev. Ind. Microbiol. 15:20–27.

Su, H. J., A. Roynitzky, H. A. Burge, and J. D. Spengler. 1992. Examination of fungi in domestic interiors by using factor analysis: correlations and associations with home factors. Appl. Environ. Microbiol. 58:181–186.

Sywassink, W. Johnson, L. Wintermeyer, and W. J. Hierholzer. 1983. Legionnaires' disease associated with a hospital water system: a cluster of 24 nosocomial cases. Ann. Intern. Med. 99:172–178.

Teltsch, B., and E. Katzenelson. 1978. Airborne enteric bacteria and viruses from spray irrigation with wastewater. Appl. Environ. Microbiol. 35(2):290–296.

Thacker, S. B., J. V. Bennett, T. F. Tsai, D. W. Fraser, J. E. McDade, C. C. Shepard, K. H. Williams, Jr., W. H. Stuart, H. B. Dull, and T. C. Eickhoff. 1978. An outbreak in 1965 of severe respiratory illness caused by the Legionnaires' disease bacterium. J. infect. Dis. 138:512–519.

Timmons, D. E., J. D. Fulton, and R. B. Mitchell. 1966. Microorganisms of the upper atmosphere. I. Instrumentation for isokinetic air sampling at altitude. Appl. Microbiol. 14(2):229–231.

Tobin, R. S., E. Baranowski, A. P. Gilman, T. Kuiper-Goodman, J. D. Miller, and M. Giddings. 1987. Significance of fungi in indoor air: report of a working group. Can. J. Public Health 78:S1–14.

Wells, W. F. 1934. On air-borne infection. Study II. Droplets and droplet nuclei. J. Hyg. Camb. 20:611–627.

Wells, W. F., and M. F. Wells. 1942. Air-borne infection as a basis for a theory of contagion. Pp. 99–101. In Aerobiology. No. 17:289.

White, W. C. 1990. An overview of the role of microorganisms in "building-related illnesses." Develop. Ind. Microbiol. 31:227–229.

Winn, W. C., Jr. 1988. Legionnaires' disease: historical perspective. Clin. Microbiol. Rev. 1:60–81.

Wright, T. J., V. W. Greene, and H. J. Paulus. 1969. Viable microorganisms in an urban atmosphere. J. Air Pollut. Control. Assoc. 19(5):337–341.

Woodward, E. D., B. Friedlander, R. J. Lesher, W. Font, R. Kinsey, and F. T. Hearne. 1988. Outbreak of hypersensitivity pneumonitis in an industrial setting. J. Amer. Med. Assoc. 259:1965–1969.

Zeterberg, J. M. 1973. A review of respiratory virology and the spread of virulent and possible antigenic viruses via air conditioning systems. Part 1. Ann. Allergy 31:228–234.

Zobell, C. E. 1949. Marine microbiology. Chronia Botanica Inc., Waltham, MA.

Zuravleff, J. J., V. L. Yu, J. W. W. Shonnard, J. D. Rihs, and M. Best. 1983. *Legionella pneumophila* contamination of a hospital humidifier. Amer. Rev. Respir. Dis. 128:657–661.

5

Deposition, Adhesion, and Release of Bioaerosols

H. Hollis Wickman

Symbols Used

Symbol	*Definition*
A = Hamaker constant	p^0 = equilibrium liquid vapor pressure
$\mathbf{A}(t)$ = random force	
a, b = ellipse axes	Q = electric charge
C_s = Cunningham slip correction factor	q_e = electron charge
	R = gas constant
D = particle–surface distance	Re = Reynolds number
D_E = Einstein diffusion coefficient	r = particle radius
d = particle diameter	r_m = mean radius of curvature
E_g = gravitational energy	St = Stokes number
E_i = collector efficiency	T = temperature
E_T = thermal energy	t = time
Fr = Froude number	U = contact potential
$F_{\mathrm{p.o.}}$ = pull-off force	U_0 = free-stream velocity
f_g = gravitational force	V = volume
f_v = friction force	v = particle velocity
f_s = LvdW force	v_c = critical capture velocity
g = gravitational constant	v_g = deposition velocity
h = characteristic distance	v_s = terminal velocity
k_B = Boltzmann constant	x_s = stopping distance
LvdW = London–van der Waals	$\langle x^2 \rangle$ = mean square particle displacement
m = particle mass	
N_g = gravity number	Y = compressibility modulus
Pe = Peclet number	y_c = critical parameter for impaction on spherical collector
p = pressure	

α = dynamical shape constant
γ = surface tension, or surface energy
ϵ_0 = permittivity of free space
η = air viscosity

ν = kinematic viscosity
ρ = particle mass density
σ = Schmidt number
τ = relaxation time

5.1. Introduction

Bioaerosols typically consist of a range of particle sizes and shapes. Further, each particle, even if it is a single, "simple" bacterium, may have a complex surface morphology. The degree of hydration as well as the size of the particle or cluster often changes with time after the aerosol is generated. This can lead to significant changes in its mechanical, elastic, or surface properties. For such reasons, it is not possible to provide a detailed picture of all types of bioaerosol–surface interactions. To approximate the situation, it is necessary to introduce several simplifying assumptions, which are guided by knowledge of less complex aerosol systems.

To a good approximation, bioaerosols may be viewed as thermodynamically unstable suspensions of complex colloidal particles entrained in a gas. The particles can vary in size from a virus, 0.1 μm diameter or less, to a single bacterium, about 1 μm in diameter, to agglomerations of material of density of 1 g cm^{-3} and about 50 μm in diameter. The thermodynamic instability arises from a number of factors which limit the airborne lifetimes of individual particles. For example, the spontaneous adsorption of molecules and moisture on the particle surface stabilizes the particle in a thermodynamic sense (reducing its surface energy), but also makes the aerosol heavier so it sediments more rapidly. Agglomeration, involving adhesive collisions of smaller aerosols, likewise produces larger particles. Alternatively, conditions may exist where evaporation of water from an original aerosol droplet leads to very small and buoyant particles. These will be moved by the air motion and thermal diffusion, and eventually come in contact with a surface, which could be liquid or solid.

The motion of bioaerosols to and through the near surface region is commonly termed *impingement*. When near the surface, the particle experiences London–van der Waals (LvdW) and electrostatic forces, which are normally attractive, in which case the particle remains fixed to the surface, so that *deposition* and *adhesion* occur. While the particle is on the surface, additional adhesion forces may become important. At a later time, wind and mechanical or other forces may be strong enough to overcome adhesion and result in removal of the particle from the surface. This is termed *release*.

The general theory of particle adhesion has been reviewed by earlier authors such as Corn (1966) and Krupp (1967) and is also dealt with in recent texts devoted to aerosol physics (Hinds, 1982; Reist, 1984). The subject of aerosol deposition, particularly spores and pollen, has been treated in a useful fashion by Chamberlain (1967). In recent years, as lithographic features on integrated circuits have encompassed the colloid size range, aerosol adhesion and removal has become a key issue in the microelectronics industry and has emerged as a major determinant of chip yield. Several of the topics discussed in a review series in that context (Mittal, 1988, 1981) are also relevant here.

The literature of deposition and adhesion of bioaerosols *per se* is rather limited. Indoor, health-related investigations focus on filtration techniques with less attention to details of microscopic adhesion. Outdoor plant studies or wind-tunnel investigations often employ pollen or spores as model aerosols (Paw U, 1983) impinging on grass or plant surfaces. Recently, genetically modified bacteria (such as ice-minus bacteria) have been aerosolized and deposited on plant surfaces (Giddings, 1988; Burris, 1989). Such developments have begun to focus more interest on bioaerosol–surface interactions.

Because impingement can be rather complex, explicit calculations for even a single particle under ambient conditions are rarely attempted. The situation is even more complicated in the case of bioaerosols, such as individual viruses, bacteria, agglomerated clusters of aerosolized living bacteria, fungal spores, pollens, and so on which may impinge on "nonideal" structures such as leaf surfaces (Chamberlain, 1967; Schrödter, 1960; Ingold, 1960). Further, the living cells may adhere to the surface via cell secretions and chemical bonds leading to bioadhesion (Marshall, 1986; Bongrand, 1988).

In view of these complications, we have chosen here to emphasize the microscopic and molecular processes that are generally applicable, though their unambiguous enumeration in specific circumstances remains somewhat of a challenge. We only briefly deal with continuum fluid mechanics or aerosol dynamics transport models used to simulate macroscopic motion of particle plumes. A thorough description of aerosol motion using continuum mechanics is available elsewhere in a number of excellent standard treatises (Levich, 1962; Fuchs, 1964; Hidy and Brock, 1970; Happel and Brenner, 1986). In these treatments, it is often suitable to include particle–surface interaction in a phenomenological manner in the form of nonreflecting walls or particle sinks to which particles move or diffuse at some rate and upon contact are irreversibly lost (as in filtration) from the air mass (Nazaroff and Cass, 1989). The discussion given below connects with such descriptions in its consideration of particle–surface interactions occurring in the near-surface, or boundary layer region of such models.

The current description will draw attention to the fact that adhesive forces are commonly dominated by molecular structure and organization near and within the contact region of the two surfaces and depend to a lesser extent on the bulk makeup of the materials. The surface region is characterized by a thickness D, roughly the equilibrium particle–surface separation. This distance D varies in width from a few to tens of angstroms. This length scale focuses our attention on the microscopic or molecular makeup of the surface layers of interacting particles and bulk surfaces. It also motivates development of a model system, whose surface is defined at the molecular level, which will improve our knowledge of bioaerosol–surface interactions and which may be appropriate to direct adhesion force measurement. To this end, candidate membrane surfaces are identified and initial experiments with such surfaces summarized (Leckband et al., 1992a).

Forces between particles and surfaces have been measured by a number of techniques. Examples include spinning a particle off a leaf surface using centrifugal forces or blowing a particle from a surface with air forces (Corn, 1961a, 1961b). Some of these macroscopic techniques are briefly described below. However, we emphasize newer techniques which can provide direct measurements of surface–surface interactions under equilibrium conditions. One of these is the surface force measurement apparatus (SFA) developed by Israelachvili and co-workers (Israelachvili and Adams, 1978). Applications of the SFA technique to complex biological surfaces are just beginning to appear (Leckband et al., 1992a). Very recently, forces between a supported spherical silica colloid (3.5 μm in diameter) and a silica surface in saline solution have been measured using an atomic force microscope (AFM) (Ducker et al., 1991). Such developments suggest that direct measurements involving aerosol particles and surfaces may be anticipated in the very near future.

5.2. Deposition

Transport of aerosols to surfaces involves (i) kinetic and hydrodynamic factors which bring the particle to the surface region, (ii) proximity effects arising from the presence of the wall, and (iii) specific surface–particle interactions (Levich, 1962; Hidy and Brock, 1970). Figure 5.1 provides a schematic picture of an aerosol particle undergoing Brownian diffusion en route to deposition on a surface, with subsequent contact deformation, and condensation of moisture leading to a meniscus and capillary forces which enhance the overall adhesion. Initially, it is convenient to consider the deposition processes separately from the other issues.

The transport of particles to surfaces has been widely investigated both theoretically and experimentally and is described in a significant literature. In much of this work, details of particle–surface interactions are not normally an issue. In part, this is due to an assumption that adhesion forces are generally large and, hence, a particle reaching a surface simply adheres, or sticks. In part, it is due also to the fact that direct information about aerosol–surface interactions has been unavailable or difficult to obtain. More recent treatments of deposition often describe transport processes and also include specific consideration of particle–surface interactions. Examples include Ounis et al. (1990), Shapiro et al. (1990), Ying and Peters (1991), and Gupta and Peters (1985). These descriptions are mathematically detailed and involve approximations appropriate to different parti-

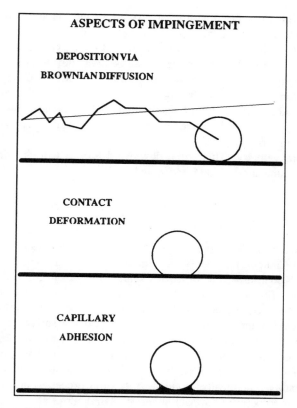

Figure 5.1. Deposition and particle–surface interactions: Brownian deposition, contact deformation (schematic), and capillary adhesion.

cle size ranges, flow conditions, and so forth. Before considering such treatments, it is useful to look at simpler models and limiting conditions. These provide physical insight and a rationale for dealing with the various transport processes in a quasi-independent fashion.

Size and mass considerations. Particles larger than about 50 μm in diameter and density at least 1 gm cm^{-3} have sufficient mass to cause them to sediment rapidly owing to gravity. As the particles become smaller, they sediment more slowly and can be carried to a surface roughly following air streamlines. As a result of their momentum, some particles will follow trajectories that deviate from the airstream and cause them to impact a surface directly. At relatively low particle surface velocities, the particles may stick, whereas at higher velocities, a number are observed to bounce. This particle bounce increases with the size of the particle up to the large particle region. Small particles, below about 1 μm in diameter, do not sediment but instead remain suspended and diffuse randomly in the air, which can be moving. The overall air motion and particle diffusion can then bring the aerosol to surface contact and deposition. Evidently, in this small-particle region, which includes the size of many bacteria and bioaerosols, the influence of particle–surface forces can be a significant factor in addition to the diffusive motion. In the case of turbulent flow, particles of various sizes can be ejected from the main airflow through the near-surface region and impact the surface and adhere (Wells and Chamberlain, 1967; Davies, 1966; Friedlander and Johnstone, 1957). It is convenient to restate this qualitative behaviour using some common terminology.

Deposition velocity. This is the downward flux of particles divided by the atmospheric concentration and is denoted by v_g:

$$v_g = \frac{\# \text{ deposited cm}^{-2}/\text{s}^{-1}}{\# \text{ cm}^{-3} \text{ in air}} . \tag{5.1}$$

A recent review of dry deposition by Nicholson (1988a, 1988b) and the early summary of Chamberlain (1967) provide useful experimental data.

Inertial deposition. Aerosol particles larger than about 3 μm in a gas stream may possess sufficient inertia, that is, momentum, so that their trajectories intercept a collector surface placed in the flow. In laminar conditions, the gas streamlines would not intercept the collector surface. If the surface of the collector retains the particles, they are separated from the stream and the process is called inertial deposition. At one extreme, the particle is retained simply by adhesive forces. At the extreme, the particle may strike the surface with such force that it is irreversibly (chemically) bonded to it. Both processes are termed inertial impaction, or sometimes "impact adhesion" (Brenner et al., 1981; Wall et al., 1988).

Gravitational settling. Particles of sufficient mass will sediment and come to rest on lower surfaces. The duration of the settling period may depend on air motion, but the force responsible for the settling is gravity, which is independent of hydrodynamic or surface–particle interactions.

Convection. Temperature gradients produce air motion that may carry suspended aerosols to surfaces.

Diffusion. All particles possess a thermal energy that is expressed as diffusive, or Brownian motion. An average thermal velocity relative to the suspending medium can be assigned, thus providing a mechanism for motion to a surface.

Eddy diffusion. Momentum and density fluctuations of small subregions (eddies) of a fluid in turbulent flow bring particles to the sublayer region so that deposition can occur. In an extreme limit, thermal diffusion can be negligible compared to eddy diffusion as a deposition mechanism (Friedlander and Johnstone, 1957). In general, both thermal diffusion of particles and eddy diffusion operate to bring particles to surfaces adjacent to turbulent fluid flow. The description of deposition involves chance events and requires a stochastic analysis.

The near-surface region. This is taken to be the region in which the particle is subject to direct interaction with the surface as a result of van der Waals forces. These interactions are felt over distances of a few to a hundred or more particle diameters and set a length scale for the region. The surface region so defined corresponds approximately to the hydrodynamic boundary layer, characteristic of the fluid in the absence of any suspended colloid particles (Schlicting, 1960). Within this boundary layer, the fluid viscosity increases sufficiently so that the region acquires properties distinct from the bulk fluid or atmosphere. Clearly, this region arises because of interaction and collision of the fluid molecules with the wall. Boundary layer effects are most pronounced at large Reynolds number (defined below) flow and in dense media (Corino and Brodkey, 1969). The depth of this fluid viscous sublayer region varies with airflow conditions and geometry. In the limit of laminar flow at low Reynolds numbers, it may be somewhat less than particle-related near-surface region. However, because of the lack of precision in either definition, we will often refer to the two regions interchangeably.

In addition to kinetic factors, the electrodynamic interactions between a particle and surface produce attractive forces on particles whose result can be motion of the particle relative to the surrounding air mass, leading to deposition. The most important particle–surface interactions are as follows:

London–van der Waals forces. These arise from interactions between permanent and induced electric dipole moments, as well as fluctuating electric dipole moments existing even in neutral, spherically symmetric atoms. The forces arising from fluctuating moments are also known as dispersion forces. Dispersion forces are often, but not always, the most important contribution to the net particle–surface electrodynamic interaction for uncharged surfaces. In hydrated bioaerosol particles, the interactions arising from dipole–dipole interactions may compete with or exceed dispersion forces.

Electrostatic forces. Unbalanced electrical charge on the surface of the particle can result in significant forces, leading to adhesion or repulsion. A number of mechanisms, including ionization and collisions, exist to charge particles (Hinds, 1982; Reist, 1984). When an electric field is present in the flow region, an additional mechanism for deposition is present. Charged or polarizable particles experience a force which is a function of the electric field strength and which depends on the geometry and charge distributions. A complete treatment of deposition in the presence of flow and electric fields is beyond the scope of the current discussion. Elementary electrostatic interactions that underlie this deposition mechanism are outlined in later sections dealing with static adhesion phenomena.

In summary, transport of an aerosol to the surface qualitatively consists of two events, occurring on greatly different length scales: Macroscopic, or bulk, transport occurs as movement on centimeter to atmospheric length scales, as a result of processes such as convection and laminar or turbulent flow. Microscopic transport, of more concern here, refers to transport of particles from the near-surface region, at most 1 mm or so, onto the surface. To describe the particle motions within the near-surface region in the presence of interactions, it is necessary to set up an equation of motion. The interaction with the surface will appear as a force in this equation. Such equations also provide a framework for more specific discussion of topics like impaction, diffusion, and rebound. Sedimentation due to the force of gravity is the simplest example of aerosol dynamics and is discussed in the following section.

5.2.1. Sedimentation

Equations from elementary mechanics describe the motion of a particle in a viscous medium such as air. A useful model consists of a monodisperse aerosol that is uncharged and at sufficiently low number density so that collisions between the particles, possibly leading to coagulation, are unlikely. Coagulation, also known as agglomeration, is a subject of considerable importance in its own right and involves a number of the topics in the present discussion.

The low number density also means that surface buildup of particles, or their dynamic release, is negligible. The particle densities are assumed near 1 g cm^{-3}, typical for bioaerosols. The particles are taken to be rigid spheres. This is a questionable assumption for a bacterium. However, while airborne, it is often a reasonable approximation. Nonspherical shapes (such as prolate or oblate ellipsoids) can be accounted for by correction factors (Schrödter, 1960; Fuchs, 1964) applied to expressions given below. A more serious reservation about the spherical shape arises when the particle attaches to a surface. Here, elastic or plastic deformation, capillary condensation at the particle–surface region, and other effects act to deform the particle. This issue is taken up in the discussion of adhesive or contact deformation.

The simplest hydrodynamic conditions are those of relatively slow movement of the air mass or the particle, so that turbulence is not present. A well-known criterion for such laminar flow is that the Reynolds number for the particle suspension is of the order of unity or less. The Reynolds number is

$$\mathrm{Re} = \frac{\rho v d}{\eta}. \tag{5.2}$$

Here, ρ is the air density (1.29 kg m^{-3}, STP), v is the particle speed, d is the particle diameter, and η is the air viscosity (1.8 × 10^{-3} N s m^{-2}). For particle diameters less than 100 μm, direct calculation shows Re < 1.0 under ordinary conditions where the air speed is less than 13 m s^{-1} (~30 mph).

The low Reynolds number regime also means that so-called inertial effects of the particle are not important (Happel and Brenner, 1965; Fuchs, 1964). Light particles are carried along by the air and have negligible motion relative to it.

Heavier particles, though they may move with respect to the surrounding air mass, do so at constant speed, with the airflow around the particle being laminar, and the air resistance given by Stokes Law. The latter (derived under the current assumptions) states that the friction force on a particle is linearly proportional to the particle velocity,

$$f_v = 6\pi\eta r v, \tag{5.3}$$

with r the particle radius.

In these circumstances, the simplest particle equation of motion which also includes possible fluctuating molecular, aerodynamic or other forces is the Langevin equation:

$$m\frac{dv}{dt} = f_g + f_v + f_s + \mathbf{A}(t), \tag{5.4}$$

where v is the aerosol velocity, m is its mass, f_g is the gravitational force, f_v is the frictional or viscous force experienced by the particle as it moves through the fluid, f_s is the LvdW force of interaction with the wall, and $\mathbf{A}(t)$ represents random forces experienced by the particle as a result of collisions with surrounding air molecules. This term is negligible for large particles but is dominant for small particles and is responsible for Brownian motion. Because it is a random force, the methods for solving the stochastic Langevin equation (Gupta and Peters, 1985) differ from those used to handle the deterministic ordinary differential equation in the absence of $\mathbf{A}(t)$.

Solutions to Eq. (5.4) describe the motion of the particle and are a function of the size and mass of the aerosol. However, some useful criteria for determining whether sedimentation or diffusion dominates the motion of a particle in a stationary air mass may be established without complete solution of the equation. One approach is to compare the displacement energy, due to the forces f_g and f_v acting over a characteristic distance, with the thermal energy of the particle, $E_T = (3/2)k_B T$, k_B is the Boltzmann constant, 1.387×10^{-23} J$-$K^{-1}. This is a form of the Equipartition of Energy Principle from classical statistical mechanics, which states that a free particle at temperature T has associated with every translational degree of freedom an energy $k_B T/2$. For a particle able to move in three dimensions, the thermal energy is $(3/2)k_B T$. This energy appears as kinetic energy of motion, $(1/2)mv^2$, so the mean square speed in the absence of friction or gravity is $3k_B T/m$. For an aerosol of diameter 1 μm and density 1 g cm^{-3}, the mean thermal speed is 0.48 cm s^{-1}. Collisions with air molecules greatly reduce this speed.

An energy to be compared with E_T is obtained by computation of the action of the forces f_g or f_v taken over a characteristic particle movement distance h, say 1 cm. The energy gained by the particle in falling a distance h is $mgh =$

E_g. The condition for sedimentation as the primary motion is, therefore, that the gravitational energy change E_g gained while moving the distance $h = 1$ cm downward be greater than the center-of-mass thermal energy E_T, which is available to keep the particle in random motion. With the particle properties given earlier, an explicit form for the energy is easily obtained. Neglecting the air density and denoting the particle density by ρ, the gravitational energy that a particle would lose in falling a distance h is $E_g = (4/3)\pi r^3 \rho g h$. The condition for sedimentation, $E_g > E_T$, is, therefore,

$$\frac{4}{3}\pi r^3 \rho g h > \frac{3}{2}k_B T. \tag{5.5}$$

In terms of particle radius, relation (5.5) is

$$r > \left(\frac{9k_B T}{8\pi\rho g h}\right)^{1/3} \tag{5.6}$$

This sedimentation criterion is incomplete in that it does not provide information on the rate of settling. To obtain this, one must directly solve Eq. (5.4) in which the time variable appears explicitly. This yields a terminal velocity v_s, with which a particle sediments, as well as the time τ and distance x_x required to reach the terminal velocity. For a spherical particle in the colloid size range $r \leq 100$ μm, moving at speeds below 100 m s^{-1}, the retarding frictional force is given by Stokes Law, Eq. (5.3). A steady velocity obtains when the frictional force is equal to the gravitational force mg leading to

$$v_s = \frac{2r^2 g\rho}{9\eta} \equiv \tau g. \tag{5.7}$$

In Eq. (5.7), τ is known as the *relaxation time* of the particle. This is the time taken to "relax" to a speed that is $1/e$ of the original speed. The *stopping distance* x_s is the distance a particle would travel if injected with an initial speed v_0 into the atmosphere of viscosity given above. This distance is obtained by solving Eq. (5.4) with only f_v present. The result is

$$x_s = \frac{mv_0}{b}(1 - e^{bt/m}), \tag{5.8}$$

with $b = 6\pi\eta r$. When $t = \infty$,

$$x_s = \frac{mv_0}{b} = v_0\tau. \tag{5.9}$$

Table 5.1 provides a conventional tabulation of stopping distances, terminal velocities, and stopping distances for a range of spherical aerosols of density 1 g cm^{-3}. Also given in Table 5.1 is the ratio E_T/E_g. We see that thermal effects dominate the gravitation effects for particles below approximately 0.02 μm. This result agrees with a similar table in the standard text of Fuchs (1964) where arguments using Brownian diffusion constants lead to equivalent conclusions.

Table 5.1 shows that bioaerosols of diameter larger than approximately 10 μm sediment rapidly. Details of their interaction with surfaces will not be critical in the deposition process but remain important in the context of adhesion, determining whether a particle deposited by gravity on a leaf remains fixed or can be dislodged easily and continue further motion to the ground. Particles less than roughly 10 μm will move slowly and will typically reach the surface region by overall air motion. Because the near-surface region has a small linear dimension, even the relatively slow diffusive motion can be an adequate means of transport in the boundary layer.

Removal of the assumption of spherical shape modifies the foregoing results somewhat. In general, terminal velocities, stopping distances, and so on are reduced, and the definitions given above must be modified. For example, Schrödter (1960) in a discussion of ellipsoidal fungal spores gives the expression

$$v_s = \frac{2x_s g}{9\eta} b^3 \sqrt{a^3} \sqrt{b^2}, \tag{5.10}$$

where a and b are the axes of the ellipse. An alternative and common approach when the distortion from spherical is not great is to modify Stoke's law to the form

$$f = 6\pi\eta\alpha rv, \tag{5.11}$$

where r is the radius of a sphere of equal volume and α is the dynamical shape constant. Tabulations of α have been quoted by Chamberlain (1967) and are

Table 5.1. Aerosol kinematic parameters.

Diameter d (μm)	Stopping Distance x_s (cm)	Terminal velocity v_s. (cm)	Relaxation Time τ (s)	E_T/E_g ($h = 1$ cm)
0.05	5.64×10^{-14}	7.44×10^{-6}	7.259×10^{-9}	0.968301
0.10	9.03×10^{-13}	2.98×10^{-5}	3.04×10^{-8}	0.121038
0.50	5.64×10^{-10}	0.000744	7.59×10^{-7}	0.000968
1.00	9.03×10^{-9}	0.002975	3.04×10^{-6}	0.000121
5.00	5.64×10^{-6}	0.074378	7.59×10^{-5}	9.68×10^{-7}
10.00	9.03×10^{-5}	0.297511	0.000304	1.21×10^{-7}
50.00	0.056449	7.437766	0.007590	9.68×10^{-10}
100.00	0.90319	29.75106	0.030358	1.21×10^{-10}

reproduced in Table 5.2, owing to their possible relevance to agglomerated dimers, trimers, and so on of micron-sized bacterial bioaerosols. Additional formulas for shape constants are given in some detail by Fuchs (1964).

5.2.2. Diffusion

The thermal energy of aerosol particles leads to a random motion which is assumed (as a first approximation) to be superimposed on the hydrodynamic flow and particle motion due to external forces such as gravity, electric fields, and so on. This Brownian motion is described by the Einstein equation, which relates the particle mean square displacement $\langle x^2 \rangle$ during time t to the diffusion constant D_E,

$$\langle x^2 \rangle = 2D_E t. \tag{5.12}$$

The diffusion constant is given in simplest approximation (Stokes' law viscous drag, neglect of slip correction) by

$$D_E = \frac{k_B T}{6\pi\eta r}. \tag{5.13}$$

Equation (5.13) can be derived using arguments based on thermal motion, and the phenomenological Fick's first law of diffusion. Fuchs (1964) provides an improved estimate of D_E, but, for our purposes, Eq. (5.13) suffices. Direct calculation of average root-mean-square displacement using Eqs. (5.12) and (5.13) illustrates the very rapid increase of the random motion with decreasing particle size. Table 5.3 from Fuchs (1964), provides the diffusion coefficient, the mean speed, and the average distance traveled by a particle in 1 s for a range of particle sizes relevant to bioaerosols.

From Table 5.3, it is clearly seen that small particles will traverse dead layers near surfaces much more rapidly than larger particles. We shall see in the

Table 5.2. Dynamical shape factors.

Shape	Axial Ratio	α
Ellipsoid	4	1.28
Cylinder	1	1.06
Cylinder	2	1.14
Cylinder	3	1.24
Cylinder	4	1.32
Two spheres touching	2	1.14
Three spheres touching as a triangle	—	1.20
Three spheres in line	3	1.37
Four spheres in line	4	1.57

Table 5.3. *Brownian diffusion parameters.*

$r(\mu m)$	$D(cm^2\ s^{-1})$	$\langle v \rangle$ (cm s^{-1})	$\langle x \rangle$ (cm)
1×10^{-2}	1.35×10^{-4}	157	1.31×10^{-2}
2×10^{-2}	3.59×10^{-5}	55.5	6.75×10^{-3}
5×10^{-2}	6.82×10^{-6}	14.0	2.95×10^{-3}
1×10^{-1}	2.21×10^{-6}	4.96	1.68×10^{-3}
2×10^{-1}	8.32×10^{-7}	1.76	1.03×10^{-3}
5×10^{-1}	2.74×10^{-7}	0.444	5.90×10^{-4}
1	1.27×10^{-7}	0.157	4.02×10^{-4}
2	6.10×10^{-8}	0.0555	2.78×10^{-4}

following section that the importance of adhesive forces relative to other forces, such as gravity, increases as particle size decreases below approximately 1 μm. The result is that a particle reaching a surface by diffusion will remain attached to it in the absence of mechanical or other action to remove it. Because of this, calculations of diffusive deposition involve a specification of the original particle concentration and the surface geometry. As a function of time, the particles diffuse to the surface, which acts as a sink for the particles. In these circumstances, standard differential equations, such as Fick's second law, may be set up and solved for appropriate boundary conditions. For an initial uniform concentration, the near-surface concentration decreases to near zero; new particles diffuse to the surface and are trapped. A concentration gradient is established whose width depends on time. Eventually, all particles are adsorbed on the surface (assuming that surface forces do not change as result of adsorption—a sometimes risky assumption, as will be seen). The time taken for removal of particles by diffusion to the surface is a strong function of the container geometry. The foregoing analysis assumes that the surface acts as a perfect sink, that is, that the adhesive energies are very much greater than the thermal energies of the particles. If the adhesion energies are of the order of the thermal energies, there will be an increase in particle concentration near the walls (positive adsorption). The concentration profile will be given by a Boltzmann distribution. This more complex situation, common in the case of fluid particle suspensions, is infrequently encountered in aerosol adsorption, owing to the strong LdvW interactions.

5.2.3. Laminar Deposition

A previous section dealt with the case of a particle injected into quiescent air. Depending on its mass, the particle either slowed to a constant velocity owing to gravity (larger particles) or rapidly assumed diffusive motion (small particles). These events are now considered in combination with laminar motion of the air mass (flow Re < 1) over an object (collector), taken as a sphere of radius R, with R of the order 1 cm. Airflow speeds are assumed modest, below 100 cm s^{-1}.

An alternative, common practical collector geometry is a cylinder, appropriate to filter collectors. Deposition in the context of filtration is an important technical area which has been much studied; see Hinds (1982) for a useful summary. However, the physical events occurring in deposition are equally illustrated by the simple spherical geometry.

Very small particles ($r < 0.1$ μm), have negligible inertia and are carried with the airstream. They would not often reach collector surfaces in the absence of their diffusional motion. Thus, the mechanism of deposition of such particles in laminar flow is *diffusional deposition*. This was schematically indicated in Figure 5.1. Somewhat larger particles ($r > 0.2$ μm) also follow the streamlines but because of their larger size, intercept the surface. This deposition is known as *interception* and is an additional deposition mechanism for these particles. Medium size particles ($r > 1$ μm) have sufficient inertia so that they cannot follow exactly the airstream and reach the collector by the *inertial deposition* or *inertial impaction* mechanism. This is shown qualitatively in Fig. 5.2, which shows the flow of a monodisperse stream of particles. A particle is identified by its position y normal to the center of stream. Particles with values of y up to a critical value y_c will impact the sphere. Particles with y greater than y_c will miss the sphere. We note in passing that conditions can be adjusted so that as the size of the particle is allowed to become smaller, particles below a critical size will flow around the sphere and fail to impact; that is, diffusion and interception are minimized. This is the basis of particle separation using impactors. In practical measurements, an impactor or a series of impactors consisting of an inlet tube and one or more plates are employed to measure impaction as well as aerosol particle size distributions (Hinds, 1982).

The sedimentation of large particles ($r > 50$ μm) through the air provides an additional mechanism known as *gravity* deposition. Finally, the collector may

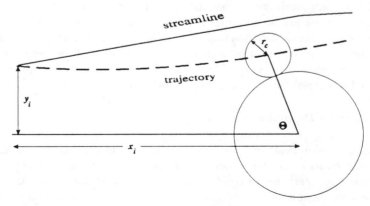

Figure 5.2. Inertial deposition of a particle on a spherical collector. A limit trajectory is shown for a monodisperse particle stream.

be charged or and electric field may be present; hence, *electrostatic deposition* is an additional mechanism. In practice, particles can be deposited as a result of more than one mechanism, and the importance of different mechanisms will vary with the airspeed. In most cases, it is a very good approximation to consider the total deposition to be a sum of the individual mechanisms. In a later section, we mention simulations which address this issue.

The various types of flow and deposition have been much studied, both theoretically and experimentally. A complete analysis is rather complex owing to the fact that the flows and deposition depend on the particle and collector size and geometry, the viscosity of medium, the fluid density, and velocity field (Fuchs, 1964; Yeh and Liu, 1974). The objective of theoretical descriptions is to establish dimensionless equations of motion of the widest range of validity, as shown by experimental confirmation. A rigorous review of calculations and approximations for a number of geometries is available in Hidy and Brock (1970). Useful descriptions of inertial deposition are given by May and Clifford (1967), Langmuir and Blodgett (1946), Tardos et al. (1978), and Fernandez de la Mora and Rosner (1982).

An important objective in simulating aerosol flow around a collector is to determine how effective the collector is in removing particles from the airstream. The aerosol *collection efficiency*, E_i, of an object in the path of an aerosol stream may be defined as the number of particles of a given size retained on the object relative to the amount that would pass through its projected area in the absence of the collector. The subscript indicates the collection efficiency for different deposition mechanisms, as introduced earlier. Aerosol collection efficiencies depend on the geometry of the collector and vary strongly with particle size. In the case of simple interception, the collection efficiency on a spherical collector in the simplest geometrical approximation is

$$E_{int} = \frac{y_c^2}{R^2},$$
(5.14)

where y_c is the critical distance for interception and R is the radius of the sphere. This geometrical expression can be improved to include the particle size dependence, and an expression given by Friedlander (1957) is

$$E_{int} = (1 + N_I)^2 - \frac{3}{2}(1 + N_I) + \frac{1}{2}\left(\frac{1}{1 + N_I}\right),$$
(5.15)

where N_I is d/R with d the particle diameter.

Inertial deposition efficiencies are commonly plotted as a function of the Stokes number, defined as the ratio of the stopping distance, χ_s, to a length, l, characteristic of the collector,

$$St = \frac{x_s}{l}.$$ (5.16)

If the collector is a sphere or cylinder, l is the diameter. Both experiment and theory yield results which are schematically shown in Figure 5.3 for a spherical collector. Qualitatively, it is seen that the collection efficiency decreases with decreasing particle size. Small particles tend to stay in suspension and avoid the collector. We shall later see that the efficiency for larger particles again decreases owing to bounce.

An empirical expression for collector efficiency for inertial deposition is (Fuchs, 1964):

$$E = \begin{matrix} (1 - 1.214/St)^2, St \geq 1.214; \\ E = 0, St < 1.214. \end{matrix}$$ (5.17)

Gravity deposition is approximated by the expression

$$E_g = \frac{Fr}{(1 + Fr)}.$$ (5.18)

where is Fr is a dimensionless gravity, or Froude number, $U_0^2/2gl$, U_0 is the airflow speed far in front of the collector. It may be worth mentioning that the dimensionless parameters Re, St, and Fr have the utility that when they are equal

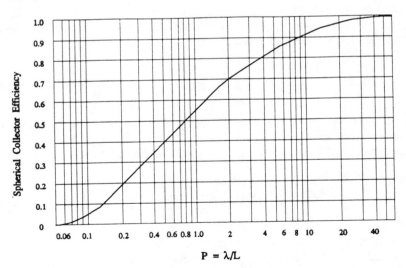

Figure 5.3. Approximate variation of spherical collector efficiency as a function of the Stokes number of impinging particles.

for different aerosol systems, the aerosols behave the same in terms of their aerodynamic motion. Finally, the influence of Brownian diffusion is given by

$$E_{BD} = 3.988 \, Pe^{-2/3}, \qquad (5.19)$$

where Pe is the dimensionless Peclet number, $2lU_0/D_E$, and D_E is Brownian diffusion constant (Levich, 1962). An expression for electrostatic deposition will be given in a following section.

The listing of semiempirical expressions for the various contributions to deposition suggest they act independently of each other. Because a particle is only deposited once, and more than one mechanism may apply to the situation, it is clear that a simple sum of contributions will overestimate the total deposition. Thus, it is of considerable interest to consider the combined effects of diffusion (and other mechanisms) and flow, whether laminar or turbulent. This is a more difficult problem for which solutions under steady-state conditions have been given by a number of authors (Levich, 1962; Davies, 1966). More recently, this problem has been reviewed by Gupta and Peters (1985), denoted GP. These authors identify limitations in earlier treatments, such as the superposition assumption or the use of equilibrium conditions to allow solutions of convective–diffusive equations. To circumvent these problems, GP use a stochastic Langevin equation similar to Eq. (5.4). They employ a Brownian dynamics method to integrate the Langevin equation for several conditions of flow around a spherical collector. On the one hand, they find agreement with analytical results for steady-state conditions valid for different ranges of Stokes' numbers. In the case of combined mechanisms of deposition, the GP method yields collector efficiencies that are, as expected, less than the sum of the efficiencies of individual deposition mechanisms. The machine calculations are consistent with relation

$$E_T = 1 - \Pi_i \, (1 - E_i), \qquad (5.20)$$

where the product involves the various deposition mechanisms (Spurny and Pich, 1965). The improvement in accuracy over simple superposition found by GP was approximately 5% for the example studied. The power of the GP technique lies in being able to handle both the steady-state and time-dependent collection under the combined influence of diffusive, inertial, and electrostatic deposition. Results for these conditions are also presented and show the influence of charge deposition on accretion rate. The original paper and later work dealing with the angular dependence of aerosol diffusional deposition (Gupta and Peters, 1986) should be consulted for details of the molecular dynamics calculations.

Deposition of bio-related aerosols of plants, moss, grass, trees, and so on has been studied experimentally for many years. These experiments have shown that deposition under laminar conditions depends on surface roughness, humidity, particle concentration gradients, and other factors, sometimes in a complex manner. Generally, only the deposition velocity is measured. Clough (1975) reported wind-tunnel measurements of deposition of various particles to the moss *Hypnum cupressiforme*. Radioactive tagging was used to show that in the range of wind speeds normally encountered, larger particles are much more efficiently collected than the smaller fraction. Typical aerosols involved in this and similar investigations include *Lycopodium* spores (diameter; 35 μm), polystyrene spheres (3 μm), oleic acid droplets (0.5 μm), and Aitken nuclei (0.08 μm). Garland and Cox (1982) noted that previous laboratory and field measurements had produced conflicting information regarding the deposition velocity of particles in the range 0.05–1.0 μm. These authors used a concentration gradient method to measure deposition of small particles on grass and found that deposition velocities did not exceed 0.1 cm s^{-1} on average. Some measurements suggested a small upward flux of particles. In cases such as these, involving small particles where diffusion can be an important mechanism for deposition, specific adhesion forces between particles and surfaces can be an important factor. Such forces are not determined via deposition velocity measurements.

Little (1977) presents results of studies of deposition of 2.75-, 5.0-, and 8.5-μm polystyrene aerosols to a range of smooth and hairy leaf surfaces. At high speeds (500 cm s^{-1}) and larger particle diameters, evidence for bounce off is found. Particle collection is most efficient at the tips and edges of leaves where flow is turbulent. Bounce off is much more pronounced for stems < 2 mm in diameter than for other surfaces.

5.2.4. Turbulent Deposition

At higher wind speeds and larger Reynolds numbers, airflow becomes turbulent. This flow is commonly divided into three regions: the turbulent core, the transition or buffer region, and the sublayer region discussed earlier. Because the sublayer is next to the surface, particles must cross this interface in the process of deposition. Depending on the degree of turbulence and hydrodynamic conditions, particles may diffuse across the boundary by Brownian diffusion, or they may be ejected across the boundary layer by the random mass transport mechanism of eddy diffusion. Turbulent flow occurs at high Reynolds numbers, Re $\geq$ 5000, so this behavior would not be applicable to typical conditions of deposition of sprayed bioaerosols. However, turbulent flow of bioaerosols may be encountered during spraying as a result of high-pressure nozzle flow. Further, for the atmosphere as a whole, the characteristic length scale is very large, even though the velocity of aerosol movement is small. These conditions suffice to define large Reynolds numbers, so that laminar-type currents of air do not generally occur in the atmosphere and are uncommon in the layers that are significant for the dissemination of bioaerosols over large distances (Schrödter, 1960).

The subject of turbulence remains an active research area of great complexity and of special recent interest in the context of chaos (Ruelle, 1990). Here, we mention a few studies that are representative of the literature and which introduce several of the basic features of turbulent aerosol deposition.

A useful critique of descriptions of deposition from turbulent flow next to surfaces has been given by Cleaver and Yates (1975), denoted CY. This analysis leads to a model in which fluid fluctuations lead to the downsweep of particles directly to the wall. This downsweep view (Klein et al., 1967) removed a complication in earlier models which involved a two-step process consisting of particle movement to the sublayer and then projection across the sublayer, a distance similar to stopping distances discussed earlier. Classic high-speed photographs of colloid deposition from flowing turbulent liquid give support to this downsweep deposition mechanism (Corino and Brodkey, 1969).

Analyses of turbulent deposition, including as that of CY, are rather complex, involve a number of approximations, and necessarily employ fluid mechanics terminology that we have not emphasized in the present discussion. We are again content to summarize the physical ideas as an introduction to such descriptions.

In CY, the deposition of particles by inertial downsweep is described by a dimensionless flux

$$N_i^+ = \frac{N}{cu_*},$$
(5.21)

where N is the actual flux, c is the concentration of particles, and u_* is the wall friction velocity defined by τ/ρ, with τ the shear stress at the wall (dyn cm^{-2}) and ρ the air density. N_i^+ is found to be a function of t_p, the particle relaxation time (Sec. 5.2.1). Data may be plotted as a function of $t_p^+ = t_p u_*^2/\nu$, with ν the kinematic viscosity. Recall that t_p is proportional to the particle mass. A schematic representation of deposition data from a number of sources in terms of N_i^+ is shown in Fig. 5.4. This figure shows that the deposition flux decreases with decreasing particle relaxation, that is, particle size, as expected on qualitative grounds.

In the CY approximation,

$$N_i \sim t_p^+ \, (t_p^+ \le 0.1).$$
(5.22)

However, experimentally it is observed that turbulent deposition increases for values of t_p^+ below approximately 0.1. This is because diffusion becomes increasingly important as particle size decreases. CY, therefore, add the following linear diffusion flux term:

$$N_d^+ = \frac{0.084kT}{3\pi\rho\nu^2 d},$$
(5.23)

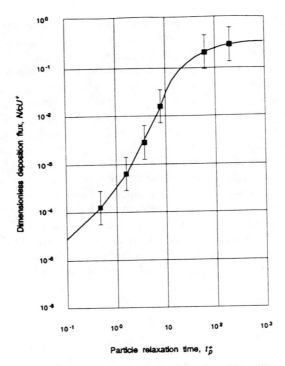

Figure 5.4. Deposition flux versus particle relaxation time [from data given in Cleaver and Yates (1975)].

where d is the particle diameter. Assuming the two source terms are additive, the total flux is

$$N^+ = N_i^+ + N_d^+ \simeq \frac{8.5d^2u_*^2}{400v^2} + \frac{0.084kT}{3\pi\rho v^2 d}. \qquad (5.24)$$

The minimum in the above equation roughly represents the crossover from turbulent inertial deposition to turbulent diffusive deposition. This can be written in dimensionless form as

$$\mathrm{Re}\,\sigma^{1/3} \simeq 1.11, \qquad (5.25)$$

where Re is the particle Reynolds number and σ is the Schmidt number defined by $\sigma = v/D_E$. Empirically, it is found that diffusional deposition rates are correlated with Schmidt numbers consistent with Eq. (5.25) (Wells and Chamberlain, 1967).

In conclusion, we note that as in the case of laminar deposition, random or

Brownian diffusion remains the primary deposition mechanism for small particles, below roughly 0.1 μm. The discussion assumed that the diffusive and inertial deposition processes were independent and, hence, additive. This superposition assumption is an approximation that is not generally applicable, also noted in the discussion of laminar deposition. A similar situation also exists for turbulent deposition (Gutfinger and Tardos, 1979).

5.3. Adhesion

Specific discussion has not yet been provided concerning the forces which cause particles to adhere to surfaces, filters, collectors, and so on. An inquiry into the origin of these forces is necessary for a better understanding of the deposition process, particularly because techniques are now available which make direct measurement of these adhesion forces possible (Israelachvili, 1985). These techniques hold promise for unraveling LvdW forces, electrostatic forces, and even molecular lock-and-key mechanisms of adhesion that are immediately relevant to bioaerosols and to bioadhesion in general (Marshall, 1986; Bongrand, 1988).

The electrodynamic forces by which an aerosol and a surface interact are of two types: LvdW interactions that are always present, and electrostatic interactions which require an imbalance in charge on the aerosol or the surface, or the presence of an external field. These at-a-distance interactions persist when the particle is in contact with a surface and can then be supplemented by additional adhesive interactions which depend on the nature of the interface between the particle and the surface. For example, capillary condensation in the interfacial region can lead to meniscus formation and additional, significant adhesion. In this case, when a liquid is present, ions in solution may adsorb to the surfaces leading to double-layer forces which can be adhesive or repulsive depending on surface charges. The theory of adhesion forces and electrical double-layer forces in colloidal fluids is known as the Derjaguin–Landau–Verwey–Overbeek (DLVO) model. The DLVO theory is discussed in standard colloid texts (Hiemenz, 1986) and in discussions of intermolecular forces in liquids (Israelachvili, 1985). Finally, as the particle rests on the surface and is subject to the various forces, elastic or plastic deformation can occur, increasing the particle–surface contact area, and strengthening the adhesive interaction.

This brief summary illustrates the complex character of adhesion. In nearly all cases, the net interaction is the sum of several terms whose relative strength may vary with environmental conditions. As an initial approximation, we again assume that the various contributions are superposed. In this way, the interactions can be discussed separately and compared for specific conditions. The discussion involves a number of formulas, some of which are for convenience provided in Tables 5.4 and 5.5. The equations in the tables, and most others quoted below, are in SI units and are consistent with the notation of Israelachvili (1985), with

Table 5.4. *Formulas arising in particle–surface LvdW adhesion model system: spherical particle of radius* r *at distance* D *from a plane surface.*

1. Interaction energy and force in terms of Hamaker constant A.	$\text{Energy} = W(D) = \dfrac{Ar}{6D}, \text{Force} = F(D) = \dfrac{Ar}{6D^2}$
2. Hamaker constant in terms of Lifshitz theory	$A = \dfrac{3}{4}k_B T \left(\dfrac{\epsilon_1 - \epsilon_3}{\epsilon_1 + \epsilon_3}\right)^2 + \dfrac{3\hbar\omega}{16\sqrt{2}} \dfrac{(n_1^2 - n_3^2)^2}{(n_1^2 + n_3^2)^{3/2}}$
3. Hamaker constant in terms of London–van der Waals parameters.	$A = \pi^2 C \rho_1 \rho_2,$ $C = (u_1^2 \alpha_{02} + u_2^2 (\alpha_{01}) + \dfrac{u_1^2 u_2^2}{3 k_B T} + \dfrac{3}{2} \dfrac{\alpha_{01}\alpha_{02} h \nu_1 \nu_2}{(\nu_1 + \nu_2)}$

r the particle radius instead of R. However, much of the literature uses cgs units. As a result, either or both may be used in specific calculations given below. In many cases, the LvdW interactions dominate, so these are discussed first.

5.3.1. London–Van der Waals Forces

The physical basis of van der Waals intermolecular interactions may be developed along two lines, as discussed in standard texts (Mahanty and Ninham, 1976; Israelachvili, 1985; Hiemenz, 1986). The early Bradley–Hamaker–London view is based on microscopic, molecular properties (Bradley, 1932; Hamaker, 1937; London, 1937). Interactions between individual, isolated molecules are assumed to be pairwise additive. This leads to a determination of the short-range cohesion/adhesion forces between macroscopic objects. Unfortunately, the pairwise additivity assumption is not generally valid because the interaction of a pair of molecules can depend on the presence of a third molecule in the immediate

Table 5.5. *Formulas arising in particle–surface electrostatic adhesion model system: spherical particle of radius* R *at distance* D *from a plane surface.*

4. Electrostatic force due to contact potential U (Krupp, 1967)	$F = \pi\epsilon_0 r \left(\dfrac{U^2}{D}\right)$
5. Electrostatic force due to charge Q on metal particle next to metal surface (Krupp, 1967). $D \ll R$	$F = \dfrac{Q^2}{16\pi\epsilon_0 [\gamma + (1/2)\ln(2r/D)]^2 rD},$ $\gamma = \text{Euler's constant} \approx 0.5772$
6. Electrostatic force due to charge Q distributed in a shell of depth δ on the particle (Krupp, 1967). $D, \delta \ll R$.	$F = \dfrac{Q^2}{16\pi\epsilon_0 Dr}$ $\dfrac{\ln(1 + \delta/D)}{[\gamma + (1/2)\ln(2r/D)][\gamma + (1/2)\ln(2r/(D + \delta))]}$
7. Electrostatic image force due to charged particle in contact with conducting surface. $D > R$	$F = \dfrac{Q^2}{4\pi\epsilon_0(2r + 2D)^2}$

neighborhood. This difficulty is overcome with the later Lifshitz theory, which focuses on continuum bulk materials properties, particularly the frequency-dependent dielectric constant (Dzyaloshinskii, et al., 1961). An approximate bridge between the two approaches is the molecular polarizability (depending on atomic properties) to which the bulk dielectric constant is related.

The important concept in either the London or Lifschitz case is that atomic or molecular charge distributions produce electric fields which, in turn, polarize adjacent molecules and lead to a potential energy of interaction. Differentiation of this potential as a function of separation between objects leads to the force between them. For macroscopic objects, calculations reduce to an expression involving a geometrical factor and a materials-dependent factor, the Hamaker constant. It is common to refer to the forces predicted by either of these theories as van der Waals forces. The strength of the interactions depends strongly on the dielectric medium in which the objects are immersed, and in some cases LvdW forces are repulsive. LvdW adhesion forces are closely related to surface energies or surface tension. Thus, at the conclusion of this section, we provide expressions relating surface tension to the Hamaker constants used to estimate forces of attraction between macroscopic bodies.

London–van der Waals picture. The simple LvdW approach focuses on molecular origins of adhesion and remains useful as an introduction to the subject. It also provides a way to compute interactions due to materials containing permanent molecular dipoles. This can be important for hydrated materials. The technically more sophisticated Lifschitz approach leads to an integral expression for the adhesion force for uncharged molecules without large electric dipole moments. This situation approximates many common inorganic materials, and current treatments of particle adhesion often simply quote a useful approximation [cf. Eq. (5.35)] based on the Lifschitz approach (Bowling, 1985; Ranade, 1987). This is appropriate for dry powder adhesion but is suspect in the context of bioaerosols or other highly hydrated materials.

The microscopic LvdW picture is based on the familiar Lennard-Jones or 6–12 potential, used to approximate the electrostatic interaction between two separated atoms:

$$V(r) = \frac{B}{r^{12}} - \frac{C}{r^6} = V_R + V_A .$$

(5.26)

In this equation, r refers to the interatomic distance, and V_A and V_R represent attractive and repulsive energy terms. The repulsive term develops when the outer electron clouds of adjacent molecules begin to overlap. Although often taken as a phenomenological parameter, it is possible to determine this term, as done, for example, in simulations of tribology and adhesion at the molecular level (Landman et al., 1990). In the LvdW theory, the constant representing an attractive potential in Eq. (5.26) is denoted C_{12} for dissimilar atoms, and is a sum of three terms (SI units):

$$C_{12} = C_{\text{DEBYE}} + C_{\text{KEESOM}} + C_{\text{LONDON}} = C_D + C_K + C_L \tag{5.27}$$
$$= \frac{1}{(4\pi\epsilon_0)^2}\left((u_1^2\alpha_{01} + u_2^2\alpha_{02}) + \frac{u_1^2 u_2^2}{3}k_B T + \frac{3\alpha_{01}\alpha_{02}h\nu_1\nu_2}{2(\nu_1 + \nu_2)}\right)$$

with α_{0i} the molecular electronic polarizability of atom i, u_i the permanent electric dipole moment of molecule i, and ν_i a frequency characteristic of the electronic energy of outer electrons of molecule i.

The Debye and Keesom terms are semiclassical interactions involving parameters such as electric dipole moments which can be calculated as averages of operators over molecular wavefunctions. The Keesom term involves temperature. The London term depends on the instantaneous charge distributions and frequencies of oscillation of electron motion within an atom or molecule. These frequencies are associated with electronic energy levels by the relation $E = h\nu$. Hence, a frequency parameter appears in all expressions for the London dispersion interaction. In the original London theory, each atom was characterized by a single frequency. Later theories generalized the London approach to involve an integration over the characteristic frequencies of an atom and also to include solvent effects (McLachlan, 1963a, 1963b).

Hamaker constants. To obtain the interaction between two macroscopic bodies of volumes V_1 and V_2, of the same material, the additivity of interactions between atoms in the two bodies allows an integration over the two macroscopic volumes. With the volume density of the atoms denoted ρ_1 and ρ_2, the interaction energy is

$$E(V_1, V_2, D) = -\int\frac{\rho_1 C_{11}\rho_1}{r_{12}^6}\,dV_1\,dV_2$$
$$= (\text{const}) \times f(V_1, V_2) \tag{5.28}$$
$$= A_{12} \times F_{\text{geometry}}(D).$$

In Eq. (5.28), the parameter D is introduced to denote the distance between the macroscopic bodies, and the materials-specific constant A_{12} is the Hamaker constant. It is conventionally defined as

$$A_{12} = \pi^2 C_{12}\rho_1\rho_2. \tag{5.29}$$

The geometrical factor F is provided below for important cases relevant to aerosols. When the London term dominates, the Hamaker constant becomes

$$A_{11} = \frac{3\alpha^2\rho^2 h\nu}{64\epsilon^2} = (\text{const}) \times \text{frequency}. \tag{5.30}$$

This equation emphasizes the dependence of the Hamaker constant on a characteristic electronic frequency for the atoms in the material.

For all but combinations of highly polarizable materials, the Hamaker constant for two dissimilar materials is given to good approximation by the root mean square of the two Hamaker constants for the individual materials:

$$A_{12} = \sqrt{A_{11}A_{22}}. \tag{5.31}$$

The final important property of Hamaker constants that should be noted is that they vary depending on the medium separating the two materials. A useful relation combining Hamaker constants for material 1 and material 2 interacting across a medium 3 is (Israelachvili, 1972)

$$A_{132} \simeq (\sqrt{A_{11}} - \sqrt{A_{33}})(\sqrt{A_{22}} - \sqrt{A_{33}}). \tag{5.32}$$

Calculations of Hamaker constants can be carried out readily using the simpler versions of the original macroscopic Lifshitz theory. For reviews, see Mahanty and Ninham (1976), Parsegian and Weiss (1981), and Israelachvili (1985). One approach relates the molecular polarizability α to the complex electric permitivity $\epsilon(\omega)$ of the medium, which, in turn, is expressed as a function of the index of refraction n. Expressions similar to those appearing in the generalized LvdW picture can be converted to sums or integrals over the measurable frequency-dependent permitivity to obtain explicit expressions for the Hamaker constants. An excellent experimentalist's introduction to calculations using the Lifshitz real permitivity $\epsilon(i\zeta)$ is given by Hough and White (1980). Using the more complete theory, Israelachvili (1985) gives expressions for similar and different media interacting across a third medium. For similar materials acting across medium 3, the result is

$$A = \frac{3}{4}kT\left(\frac{\epsilon_1 - \epsilon_3}{\epsilon_1 + \epsilon_3}\right)^2 + \frac{3h\nu\,(n_1^2 - n_3^2)^2}{16\sqrt{2}\,(n_1^2 + n_3^2)^{3/2}}. \tag{5.33}$$

The first term corresponds to the Keesom and Debye terms of Eq. (5.27).

It is natural to ask whether the Hamaker constants obtained by the microscopic and macroscopic methods can be related. In general, this is not directly possible. However, when the dispersion forces [second term in Eq. (5.33)] dominate, and when the indices of refraction are not large, the ordinary Hamaker constant can be related, using the Lifschitz approach, to an average electronic frequency. The result is (Krupp, 1967)

$$A = \frac{3h\nu}{4\pi}. \tag{5.34}$$

In current discussions of LvdW interactions of particles on surfaces, Eq. (5.34) is commonly used in combination with the appropriate geometrical factor from Table 5.8 to yield

$$F_{LvdW} = \frac{h v R}{8 \pi D^2}.$$ (5.35)

In applications of this equation, the frequency variable is quoted in electron volts. Typical values range from 0.6 to 0.9 eV for hydrocarbon or polymeric materials to 9.0 eV for a metal such as silver. A tabulation of LvdW constants in these units may be found in Visser (1972).

The previous discussion has dealt with so-called nonretarded Hamaker constants. That is, the particles are sufficiently close so that interactions between them are instantaneous. At larger separations, the finite time required for the electric field fluctuations to propagate from one material to another leads to a breakdown in assumptions in the theory. The main result of retardation is that the forces fall off more rapidly and vary as r^{-7}, rather than r^{-6}. For most circumstances involving aerosols, and for measurements involving particles separated by less than 50 nm, retardation effects are minimal.

The preceding theories, and experimental measurements discussed in a later section, provide information about the Hamaker constants for a wide range of materials. A summary of values for typical materials is given in Table 5.6. The values given have been calculated by Israelachvili using Eq. (5.33), and are in good agreement with measured values. Table 5.6 also illustrates the influence of separating medium on the force of interaction between materials. The upper entries correspond to interaction across vacuum or, roughly, air, and the lower

Table 5.6. Nonretarded Hamaker constants for common materials[a] (medium 1 interacting with medium 2 across medium 3).

Medium 1	Medium 3	Medium 2	$A_{132}/(10^{-20}\text{J})$
Water	Vacuum	Water	3.7
N-Dodecane	Vacuum	n-Dodecane	3.0
Hydrocarbon	Vacuum	Hydrocarbon	7.1
PTFE	Vacuum	PTFE	3.8
Polystyrene	Vacuum	Polystyrene	6.5
Metals	Vacuum	Metals	25–40
Mica	Vacuum	Mica	10
Mica	Water	Mica	2.0
Polystyrene	Water	Polystyrene	1.4
Water	Hydrocarbon	Water	0.3–0.5
N-Dodecane	Water	n-Dodecane	0.44
Water	n-Dodecane	Water	0.44

Source: Israelachvili (1985).

entries are for similar situations in which water or a hydrocarbon is present. The qualitative effect of an intervening medium for these materials is to greatly reduce the Hamaker constant and, hence, adhesion force.

Geometrical factors. The geometrical factor F in Eq. (5.28) may be evaluated in closed form for several cases relevant to aerosols. Table 5.7 gives the factors for the cases of two spheres, a sphere and a flat surface, and two flat surfaces. The distance between the objects is denoted D. Table 5.7 and previous expressions can be combined to obtain the force between such objects. For example, the force between a spherical particle of radius R and a surface a distance D away is given by

$$f = \frac{AR}{6D^2}. \tag{5.36}$$

Aerosol-surface adhesion force. The energetics of interaction between an aerosol particle and a surface are summarized in Fig. 5.6. This shows schematically how the interaction energy varies with distance between the particle and a surface. The repulsive and adhesive contributions to the total energy are indicated separately. The force between the particle and surface is obtained by differentiation of the potential, $F = -dV/dr$ is shown schematically in Fig. 5.6. The force depends on the separation distance D as indicated in Table 5.8, under the assumption that the particle–surface separation is very small compared to the particle diameter. With the data of Tables 5.7 and 5.8, it is possible to estimate the LvdW force between a particle and a surface separated by air. We assume a particle radius of 1 μm and a particle–surface separation at contact of 3 Å. An effective Hamaker constant of 1×10^{-19} J is estimated from Table 5.7 and roughly represents an organic–metal interaction. With these parameters, the force is found to be 18 mdyn. With the same Hamaker constant and D, adhesion forces for spherical particles range from a few to a thousand millidynes as the particle size ranges from 0.1 μm to 50 μm. These numbers provide something of a benchmark to which other forces, capillary, electrostatic, gravity, and so on, can be compared.

Importance of surface structure. The preceding discussion has provided an overview of the methods by which van der Waals forces may be calculated. Before leaving this topic, we draw attention to a result, implicit in these theories,

Table 5.7. Geometrical factors multiplying Hamaker constants for simple geometries. D is the distance between objects, assumed small compared to R.

Geometry	Factor
Sphere (R_1)–sphere (R_2)	$R_1R_2/6D^2(R_1 + R_2)$
Sphere (R)-wall	$R/6D^2$
Wall–wall	$1/6\pi D^3$ per unit area

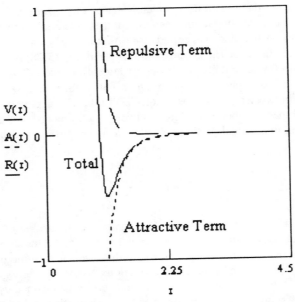

Figure 5.5. Schematic variation of particle–surface interaction energy as function of separation (arbitrary units).

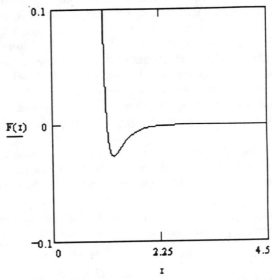

Figure 5.6. Particle–surface interaction force as a function of separation (arbitrary units).

that provides some insight into factors affecting adhesion between small particles. The point of interest is the relation between the equilibrium separation between adhering objects and the material volume of the objects that is effective in determining the adhesion force. As shown in previous discussions, the attractive force between a particle and a surface varies inversely with separation between them. It is found empirically that the force required to remove particles from surfaces corresponds to an equilibrium separation of some 2–3Å between the objects. That is, substitution of $D = 2.5$ Å and use of Hamaker constants tabulated earlier provides a useful estimate of the force of adhesion of particles on surfaces. At the small distances over which adhesion forces become important, it turns out that the surface region which determines the adhesive interactions has a depth which is equivalent to the separation distance between the macroscopic objects. Thus, when a particle and surface are separated by some 3 Å, the surface structures within layers 3–10 Å deep in the particle and in the flat surface predominantly determines the adhesion force. A practical example of this is the use of hydrophobic polymer coatings only a few molecules thick on micron-size hydrophilic particles to reduce or eliminate agglomeration of the particles. This somewhat nonintuitive result depends on the fact that the van der Waals forces fall off rather rapidly with distance and is also embodied in a useful approximation in adhesion theory called the Derjaguin approximation. The Derjaguin approximation allows simplifying geometrical approximations to be used in the analysis of surface forces when the radii of curvature of the surfaces are large compared to their separation (Israelachvili, 1985). An example is noted in Sec. 4.2. Thus, when considering adhesion of materials, we are primarily concerned with the surface structure of the objects. In the case of bioaerosols, we are led to an examination of the surface properties of these objects. In the context of model systems, one looks for materials with the simplest type of surface. We return to this point in Sec. 5.2 where direct force measurements on crystalline S-layers from bacteria are discussed.

Adhesion forces and surface tension. The preceding expressions for the force of attraction between two objects, including two flat surfaces, have been expressed either in terms of Hamaker constants or optical constants arising in the Lifschitz theory. Knowing the forces involved, it is possible to compute the work required to separate two bodies initially in contact. In this case, it is conventional to speak of the work of cohesion if the two materials are the same and work of adhesion if the materials are different. In a similar manner, recall that surface tension is defined as the energy required to expose unit areas of new surface via the following procedure. A body of original dimensions 1 cm × 1 cm × 2 cm is separated into two equal volume cubes of 1 cm on a side. In so doing, two new surfaces of unit area have been created. The work involved in the process is defined as the work of adhesion/cohesion W, and by definition, $W = 2\gamma$. By some standard, but nontrivial, geometrical arguments, it is possible to relate the adhesion force between two objects at "zero" separation to the work of cohesion/ adhesion. The result is

$$F = 2\pi \left(\frac{r_1 r_2}{r_1 + r_2} \right) W. \tag{5.37}$$

With the relation between surface tension and work of adhesion/cohesion, the following expressions give the adhesion forces, in terms of conventional surface tension, between macroscopic bodies (Israelachvili, 1985):

$$F = 2\pi r \gamma_{SL} \text{ (two identical spheres in liquid)}, \tag{5.38}$$
$$F = 2\pi r \gamma_S \text{ (two identical spheres in vacuum)}, \tag{5.39}$$
$$F = 4\pi r \gamma_S \text{ (sphere on surface in vacuum)}, \tag{5.40}$$
$$F = 4\pi r \gamma_{SV} \text{ (sphere on surface in vapor)}. \tag{5.41}$$

In the above equations, r refers to the radius of the sphere involved, and conventional notation for surface tension has been used.

5.3.2. Electrostatic Forces

Charging of aerosols and enhanced electrostatic deposition is readily illustrated by room filters or flue stack collectors (White, 1963). There are a variety of mechanisms by which aerosols may become charged, a number of which are discussed in clear fashion by Hinds (1982), Reist (1984), and Loeb (1958). The three broad classes of charging mechanisms most generally recognized are *static electrification, diffusion charging,* and *field charging.* Static electrification occurs when particles are separated from bulk (solid or liquid) and other surfaces, or when different materials are brought into contact. Examples include electrolyte charging, spray electrification (sometimes observed as charged mist surrounding waterfalls), and contact charging. In the latter important case, when dissimilar materials having different energies for their outermost labile electrons are brought into contact, charge is transferred, with electrons flowing from higher electrochemical potential to lower. Diffusion charging refers to the diffusive flow of naturally occurring ions to aerosols, and subsequent charge transfer. Field charging requires an electric field to produce large concentrations of electrons and ions which diffuse to aerosols moving through a plasma region.

The specific discussion of charging mechanisms which follows is restricted to contact charging because our primary interest here is in particles on or near a surface. References cited earlier may be consulted for additional information. Some comment on the amount of charge that a particle may acquire is appropriate. Depending on circumstances, the total free-electron charge on a 1-μm fluidlike bioaerosol particle in the atmosphere can range from zero electron charge units to an upper limit of approximately 10^4 [the limit for positive charge is somewhat higher owing to the greater energy required for ion emission from surfaces (Hinds, 1982)]. The reason for the upper limit is that for liquid aerosols, as the charge builds up, a so-called Rayleigh limit is reached where the repulsive forces cause the particle to fragment into smaller particles of lower charge. A continuous distribution of charges is present because diffusional collisions with ambient ions tends to distribute charge among particles according to a Boltzmann distribution. The mean charge on a particle will depend on the concentration of ions present, whether a source of ions or electrons is present, whether an electric field is present, and so forth. Under normal ambient conditions, a 1-μm particle is likely to have at most two or three elementary charges.

Given that aerosol charging mechanisms exist and that collector surfaces can be charged, it remains to indicate how this influences the deposition process. It is necessary to calculate the forces between charged particles and surfaces and to evaluate this force as a mechanism for electrostatic deposition. The cases of (i) a charged particle experiencing simple coulomb interactions, (ii) charging due to contact potential differences, and (iii) the effect of electric fields on the deposition process are considered.

Coulomb forces. In principle, the calculation is an exercise in static electricity. For simplicity, we assume the particle is metallic as is the collector. This assumption is motivated by the fact that bioaerosols are highly hydrated, water has a high dielectric constant, and charge seems to rapidly equilibrate on such aerosols (Fuchs, 1964). When the particle is far from the surface, the force can be estimated by considering the particle to be a point charge interacting with its image charge induced in the conductor. The resulting force is given by entry 7 of Table 5.5:

$$F = \frac{Q^2}{4\pi\epsilon_0(2r + 2D)^2},$$

(5.42)

where Q is the charge on the particle, r is its radius, and D is the distance between the particle and the surface. Unfortunately, complications arise when the particle approaches the surface, in which case D and r become comparable. In this range, the electric field becomes more complex, particularly when the particle is polarizable and contains molecules with permanent dipole moments or high-Z atoms. The field lines become distorted and a much more complex solution is required (Krupp, 1967; Davis, 1969; Robinson and Fowlkes, 1989). Krupp (and references cited therein) discusses the situation and provides an expression valid for a metallic particle when $D << r$. This is the contact region of interest here. Typically, $D = 3$Å and $r = 1$ μm. Krupp's expression is given in entry 6 of Table 5.5. Krupp also provides a somewhat more complicated expression, not reproduced here, which is valid in the intermediate region, $D \sim r$, and which at larger separations, $D > r$, reduces to Eq. (5.42). Robinson and

Fowlkes (1989) describe a multipole expansion technique which should lead to similar results.

Discussions of adhesion commonly use Eq. (5.42) to approximate the electrostatic contribution to adhesion (Bowling, 1985; Ranade, 1987) when the particle is contacting a surface. The error introduced in using the simple image force expression for particles contacting a surface may be illustrated by the case just mentioned and a particle charge of $2 \times 10^3 \ q_e$. The image force expression yields a force of 0.2 mdyn, whereas Krupp's expression yields the value 3.1 mdyn. In this case, one method would indicate a electrostatic force that is small compared to ordinary LvdW adhesion forces, whereas the more exact expression predicts that the electrostatic interaction will be comparable with the dispersion force. In general, for particles in the 1–5 μm range and significant charges in the range $2000q_e$–$5000q_e$, the electrostatic forces will dominate or be comparable to LvdW forces. This level of charge is not uncommon and, under conditions of electrostatic filtration, charging to much higher levels occurs. At the other extreme, ambient monovalent or divalent ions in the atmosphere produce much lower charges on aerosols, only a few electron charges, and electrostatic interactions are negligible compared with dispersion forces.

Contact potentials. A major electrostatic interaction arises as a result of the contact potential developed between any two dissimilar materials. A difference in the electrochemical potentials of labile electrons in the two materials results in charge flow between them, leading to a double layer and an attractive force. Entry 4 of 5.5 provides the appropriate expression. Indeed, entries 4 and 5 are essentially equivalent. One emphasizes the view of previously charged material interacting. The other emphasizes an imbalance in charge arising from charge flow upon contact arising from the contact potential, U, entry 4. The potential U is normally less than 0.5 V. Using entry 4, a particle radius of 1 μm, and separation of 4Å, we find the contact potential adhesion force is 2 mdyn. This is appreciable, and again shows that contact electrostatic adhesion may dominate other sources. In practice, no measurement of contact potentials for bioaerosols have been made. Also given in Table 5.5 is a further approximate expression due to Krupp involving a dielectric sphere in which charge is distributed in a surface region of depth $\delta \ll D, r$. This approximates conditions that may be encountered in the case of bioaerosols. Its validity does not seem to have been tested experimentally.

Influence of electric fields. When an electric field is present, the situation becomes more complex. If the particles are charged, they experience a force due to the field. The field may be directed to the surface and hence enhance deposition. A rather broad range of conditions and experimental geometries can be of interest for bioaerosols. In some conditions, such as wet scrubbers, the collector can be a large spherical water droplet. In other cases, the collector can be a flat surface of variable dielectric constant (metal to insulator), or commonly a cylindrical geometry consisting of a tube with an anode (or cathode) wire in the center. In these cases, with an inlet flow established, a transport equation can be set up and appropriate boundary conditions imposed to determine the enhanced deposition rate in the collector. Just how this enhancement occurs depends on the polarizability of the particle, the surface, diffusion rates, turbulent or laminar conditions, collector geometry, and so on. It is not possible to deal further with such calculations, but Table 5.8 lists some examples of particle–collector interactions that must be reckoned with in practical cases. Formulas for mechanisms 1–5, in dimensionless form, appropriate to flow and deposition calculations are given by Kraemer and Johnston (1955) and Nielsen and Hill (1976).

In conclusion, we note that the problem of electrostatic interactions, electric field effects, and dielectric colloids also arises in the context of electrographic printing technology. This literature also provides useful expressions and techniques for calculations of particle–surface interaction forces (Hartmann et al., 1976; Robinson and Fowlkes, 1989).

5.3.3. Capillary Adhesion

Aerosols have a hydrophobic/hydrophilic surface character that varies with relative humidity and degree of hydration. Many investigations have empirically demonstrated that measured adhesion forces can be a function of relative humidity (Hinds, 1982). If the surface is hydrophilic, moisture can condense in the contact region between the particle and the surface. This accumulation of moisture will eventually (minutes to hours, depending on conditions) form an annular meniscus as shown in Fig. 5.7.

The relation between the vapor pressure of the liquid in the meniscus and the curvature of the meniscus air–liquid interface is given by the Kelvin equation

$$p = p^0 \exp\left(\frac{\gamma \overline{V}}{RT r_m}\right) \tag{5.43}$$

in which p is the curved surface vapor pressure, p^0 is the equilibrium vapor pressure, $\overline{V}$ is the molar volume of the liquid, and r_m is the mean curvature, $r_1 r_2/(r_1 + r_2)$, which is negative in the geometry shown, and $r_2 \approx a$. For finite radii,

Table 5.8. Electrostatic interactions.

1. Coulomb force	Both surface and aerosol are charged.
2. Charged particle image force	Aerosol charged, image force induced in neutral surface.
3. Charged collector image force	Surface charged, image force induced in neutral aerosol.
4. External electric field force	Aerosol charged. Electric field induces charge in surface.
5. Induced polarization dipole force	Uncharged aerosol and surface. Electric field induces charge in both.
6. Contact potential	Particle and surface have different electrochemical potentials, leading to charging upon contact.

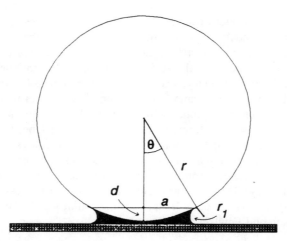

Figure 5.7. Capillary adhesion. The symbols are defined in the text.

p is less than p^0, water will condense, and the curvature will adjust until the pressure p is in equilibrium with the partial pressure of water corresponding to the relative humidity. Because of the negative radius of curvature, the pressure inside the meniscus is decreased below atmospheric pressure an amount

$$\Delta p = \gamma\left(\frac{1}{r_1} + \frac{1}{r_2}\right) \simeq \frac{\gamma}{r_1} \tag{5.44}$$

because r_2 is much larger than r_1. In Fig. 5.7, r_1 is shown, d is the height of the meniscus, r is the radius of the sphere, and a is defined as shown. The negative pressure acts over a circular area $A = \pi a^2 \approx 2\pi rd$ to produce the capillary adhesion force

$$F \sim (\gamma/r_1)2\pi rd \simeq 4\pi r\gamma \cos\theta. \tag{5.45}$$

In Eq. (5.45), the substitution $d \approx 2r_1 \cos\theta$, valid at small θ, has been used. Note that the final expression for the adhesive force does not include a parameter characteristic of the meniscus radius of curvature. In the case of complete wetting, $\cos\theta = 1$, the expression is

$$F \simeq 4\pi r\gamma. \tag{5.46}$$

It is useful to estimate the capillary adhesion force for a particle of radius 1 μm and for water with $\gamma = 72$ dyn cm^{-1}. In this case, Eq. (5.46) predicts an adhesive force of 90.5 mdyn, an appreciable force that would dominate the other contributions to adhesion that we have considered thus far.

Equation (5.56) has been verified experimentally in a number of studies. These include the work of Gillespie and Settineri (1967), using glass spheres and water or mineral oil, and McFarlane and Tabor (1950), also using glass spheres and various liquids: water, glycerol, decane, octane, alcohol, benzene, and aniline. O'Brien and Hermann (1973) derived an expression for the case of differing materials:

$$F = 2\pi r\gamma \left(\cos \theta_1 + \cos \theta_2\right). \tag{5.47}$$

They used macroscopic plano-curved surfaces ($r = 24.2$ cm) and found excellent agreement between the measurements and the forces predicted by Eq. (5.46). At the other extreme of size, Fisher and Israelachvili (1980, 1981a, 1981b) have investigated the validity of Eq. 5.45 down to very small separations and meniscus radii, that is, to the region where the finite size of individual molecules may make the macroscopic equation suspect. Using molecularly smooth cylindrical mica surfaces ($r = 1$ cm), they were able to confirm Eq. (5.45) for hydrocarbons with meniscus radii down to 0.5–1.0 nm. The corresponding limit for liquid water was much higher, about 5 nm. The reasons for this difference were not completely resolved but are thought to arise in part from the long-range, hydrogen-bonded water structure.

Although the applicability of the macroscopic equation for smooth surfaces under clean conditions seems clear, there are a number of measurements of capillary adhesion in less pristine circumstances where hysteresis effects and sharp changes in adhesion as a function of relative humidity have been observed (McFarlane and Tabor, 1950; Corn, 1961a, 1961b; Zimon, 1982; Hinds, 1982). These effects can be complicated functions of the asperity (smoothness) of the surface, particle shape, deformation, and so forth. As a result, it is not possible to treat them in a systematic fashion.

5.3.4. Contact Deformation

The adhesion forces on real particles are sufficient to deform the contact region and lead to an increase in contact area with time. This continues until a quasi-equilibrium state is reached. Owing to the deformability of cells and bioaerosols, contact deformation can be an important consideration for these materials. The contact area is circular for an isotropic, elastically deformed sphere. It is assumed for simplicity that the surface deformations are negligible compared to deformations of the particle. The deformations represent stored elastic strain energy in the particle. With time, plastic or viscoelastic flow may also occur so that energy is lost as heat or consumed in the form of new molecular conformations, and the particle is permanently distorted. For the present, only elastic deformations are considered. Plastic deformation can be important in the context of impact adhesion and rebound, considered in the following section.

The analysis of elastic contact deformation began with Hertz, and this topic remains a subject of great technical and engineering interest (Johnson, 1985). Later theories of adhesion and deformation were developed to account more accurately for the contact deformation and to derive expressions for the "pull-off" force that is a function of the aerosol particle materials properties. Primary theories are those of Johnson, Kendall, and Roberts (1971), denoted JKR, Muller, Yuschenko, and Derjaguin (1980), Derjaguin, Muller, and Toporov (1975), denoted DMT, Muller, Yuschenko, and Derjaguin (1982), denoted MYD, and Hughes and White (1979, 1980), denoted HW. These theories have been reviewed by Horn, Israelachvili, and Pribac (1987) in the light of microscopic measurements of contact deformation by the surface force measurement technique (Sec. 4.2). Tsai, Pui, and Liu (1991), denoted TPL, have recently revisited this problem in the context of aerosol contact adhesion and have also reviewed the earlier literature.

The starting point for discussions of contact deformation is a description of the deformations and stresses in the contact region for two spheres of equal radii (allowing one sphere radius to become indefinitely large recovers the sphere–plane geometry). Figure 5.8 from Horn et al. (1987), provides a very useful summary of the salient features of current primary theories. In this figure, R is the particle radius. Column 1 summarizes the results obtained by Hertz (1881); column 2, those of JKR; column 3 those of DMT; and the last column, the results of MYD. Figure 5.8 depicts the stresses present in the bodies, the shape under compressive load, the shape under zero load, and the predicted adhesion or pull-off force. The pull-off force depends on the surface energy γ and the radius r of the spheres, but (surprisingly) not on the contact radius a (defined below). A point of interest is the difference in pull-off force that is predicted by the several theories. The difference in prediction is thought to arise because of different materials properties. A number of authors have concluded that for hard solids of small radius and low surface energy, the DMT theory is applicable, whereas for softer materials with large surface energy and radius, the JKR theory is appropriate. The latter predicts that soft materials will always have a smaller pull-off force than hard materials experiencing similar adhesion forces. This is understood on physical arguments as arising from the availability of stored elastic energy which is released when the particle is separated from the surface. The MYD or HW theories derive expressions that depend on materials properties and predict pull-off forces that range between the JKR and DMT extremes. The difference in predicted forces for the JKR and DMT theories amounts to 25%, see Fig. 5.8, and the following discussion.

More recently, TPL have reviewed the subject of contact deformation and have arrived at a somewhat different picture. They define an adhesion parameter similar to MYD and find that, in general, the contact area is not reduced at the pull-off point, as predicted, for example, by JKR. The contact area remains essentially constant. Further, they find that the pull-off force for soft materials

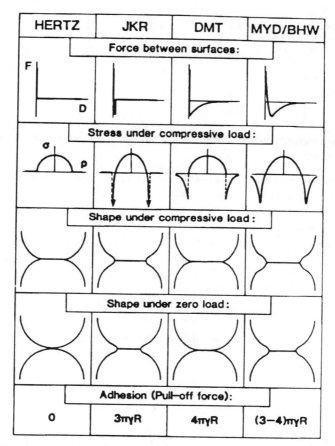

Figure 5.8. A comparison of the main features of various theories of the deformation and adhesion of two spheres.

is proportional the adhesion parameter, which includes ranges such that the pull-off force for a soft material can exceed pull-off forces for hard materials.

These points are now considered in more quantitative terms. The geometry of the particle and surface in contact is given in Fig. 5.9. A completely rigid body contacts the surface in a single point, Fig. 5.9A. In this case, the force of attraction is the same as the so-called pull-off force, $F_{p.o.}$, which is defined as the force needed to remove a particle. This force for a rigid particle is given by

$$F_{p.o.} = 4\pi\gamma r. \tag{5.48}$$

This equation is reminiscent of the capillary force equation, but, here, γ is the specific surface energy or surface tension of the surface and the particle, assumed

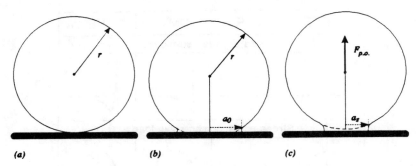

Figure 5.9. (A) Rigid sphere on rigid surface. (B) Deformed sphere on rigid surface. (C) Deformation (schematic) occurring upon pull-off from surface.

equal. Recall an earlier equation for the adhesion force between a sphere of radius r and surface, in terms of the Hamaker constant A and distance between sphere and surface, D:

$$F = \frac{Ar}{6D^2}.$$ (5.49)

From Eqs. (5.48) and (5.49), the relation between surface energy and Hamaker constant is

$$\gamma = \frac{A}{24\pi D^2}.$$ (5.50)

Owing to the adhesion force, real particles undergo some distortion and the initial point of contact increases to a circular area of radius a. The particle shape is distorted (in vertical cross section through the center of the particle), producing a "contact profile." This profile and further details of the analyses are functions of materials properties of the particle, primarily the Young's modulus E (dyn cm^{-2}), the Poisson ratio ν, and a composite modulus K equal to (2/3) [(1 $- \nu^2$)/E]$^{-1}$. The more rigid the material, the larger the Young's modulus and the smaller the equilibrium contact area, a_0. The object of the theoretical analyses is to compute the equilibrium contact radius a_0 and the pull-off force. For small particles of hard materials, where the DMT picture is valid, the pull-off force is the same as the simple adhesion force, Eq. (5.48). However, for softer particles, the elastic energy stored in the particle owing to adhesive compression reduces the force needed to remove the particle. In this case, the JKR theory is applicable and the pull-off force is given by

$$F_{p.o.} = 3\pi\gamma r.$$ (5.51)

The force to remove the particle is reduced from the rigid particle case and is independent of the materials properties of the particle. The materials properties appear in the expression derived by JKR for the contact area just before pull-off. With the equilibrium contact radius area denoted by a_0, the value a at pull-off is given by $a_s = a_0/4^{1/3}$. The parameter a_0 is given by $(24\pi\gamma R^2/K)^{1/3}$. Thus, the contact area decreases to a_0, at which point pull-off occurs abruptly.

TPL discuss the earlier theories and present an analysis that is reminiscent of the work of MYD. The latter authors parameterized the problem in terms of a continuous parameter μ_3 defined by

$$\mu_3 = \frac{8}{\pi}\left[\frac{4\gamma^2 r}{9\pi K^2 D_0^3}\right]^{1/3} = \Pi'. \tag{5.52}$$

The parameter Π' is introduced by TPL. In the work of MYD, small values of μ_3, $\mu_3 \ll 1$, correspond to soft materials and lead to the DMT results. Large values of μ_3, $\mu_3 \gg 1$, correspond to soft materials and reproduce the JKR results. TLP arrive at a somewhat different result. The argument is that as soft particles deform and increase their surface area, the pull-off force or adhesion force should increase and become proportional to the contact area. The force of adhesion between two surfaces of area πa^2 is, in terms of the Hamaker constant,

$$F = \frac{\pi a^2 A}{12D\pi^3} \simeq \frac{Ar^2}{12D^3}. \tag{5.53}$$

TLP argue that a correct parameterization and evaluation of adhesion forces should tend to reproduce Eq. (5.53) for soft particles. Exact agreement is not expected because Eq. (5.53) involves no materials parameters and, hence, cannot accurately describe elastic aerosols. Note that in some discussions of contact adhesion, materials elastic properties are neglected, the particle is assumed to deform to produce a circular contact area, and Eq. (5.53) is used as the measure of contact adhesion pull-off forces (Bowling, 1985).

TLP have used a numerical method which the generalizes the analytical approach of Muller et al. They introduce a parameter Π defined by

$$\Pi = \left[\frac{25A^2 r}{144D_0^7 K^2}\right]^{1/3} \tag{5.54}$$

The same limiting conditions of large or small Π approximately reproduce the JKR and DMT results, respectively. The basic result of TLP is

$$
\begin{aligned}
F_{p.o.} &= 0.95\,(2\pi\gamma r) \text{ when } \Pi \ll 1 \\
&= 0.4\,(2\pi\gamma r)\Pi \text{ when } \Pi \gg 1
\end{aligned}
\tag{5.55}
$$

Equation (5.55) reproduces earlier results for hard materials and provides a materials-dependent value for the pull-off force for soft materials.Direct calculation of Π for a soft materials like plastic yields a value of approximately 10. Thus, a larger lift-off force is predicted for a soft material than a hard material, in contrast to the earlier theories.

Recent experiments have been performed to test the JKR and DMT theory. Perhaps most direct are those based on the surface force apparatus (Horn et al., 1987; Chen et al., 1991, and references cited therein). Another experiment used contacting elastomeric lenses of polydimethylsiloxane (PDMS) (Chaudhury and Whitesides, 1991). The surface force apparatus provides a measurement of the deformation and the pull-off forces. Materials used were surfaces coated with hydrocarbon monolayers in fluid, crystalline, and amorphous states, as well as clean mica surfaces. The work of Horn et al., (1987) and Chen et al. (1991) concluded that the JKR theory provides a better description of the deformed shapes of elastic theories than the DMR theory. The work of Chaudhury and Whitesides using PDMS also was consistent with JKR theory.

In summary, the balance of experimental data are consistent with the JKR predictions for pull-off forces and surface energies. However, the JKR theory predicts infinitely high stresses at the contact boundary; recall Fig. 5.8. This difficulty cannot be resolved within the continuum approximation of either theory and requires a more microscopic approach, such as that of Hughes and White (1979), who conclude that the problem is complex and model dependent and that the correct expression for the pull-off force probably lies between the JKR and DMR limits. The recent TLP theory, however, predicts forces beyond these limits. It is clear that this interesting area, of considerable importance for bioaerosols, is one of continuing development. Additional experiments to resolve extant issues are desirable.

5.3.5. Impact Adhesion and Rebound

A particle impinging on a surface may either "stick" or rebound. The conditions for these events depend on a number of factors discussed in earlier sections. For mechanically stiff particles and surfaces, and slow approach speeds, or larger, softer particles, the particle will stick to the surface when normal adhesion forces are present. At higher approach speeds, it is observed that the particles "bounce" (Paw U, 1983). At still larger approach speeds, the particles have sufficient energy to impact, plastically deform, and become chemically bonded to the surface owing to the large pressures developed at the interface. This is impact adhesion. The description of these events depends on specific materials properties and, like most areas of adhesion, can become rather complex. In the following, we consider some general aspects of the impact, with emphasis on establishing conditions for crossover from adhesion to rebound.

The following collision mechanics discussion follows a useful analysis given by Dahneke (1971, 1973). The model assumes an incoming spherical particle which moves normal to, interacts with, and either sticks to or rebounds normally from a surface (always having much greater mass than the particle). If there were no energy loss between the particle and the surface, the particle would simply rebound in a mechanical elastic collision. In practice, energy is lost during the impaction by a variety of mechanisms which underscore the difficulties in achieving accurate theoretical descriptions of inelastic impact. Among the loss mechanisms are elastic and plastic deformation, internal friction, heat, radiation of acoustic waves, and work done in deforming small asperities on both particle and surface. The object of the derivation is to obtain an expression for the critical capture velocity, v_c, defined as the maximum velocity a particle may have and still be captured by the surface. Particles with velocities greater than v_c will bounce.

To describe the situation, it is necessary to specify the initial and rebounding velocities beyond the range of the surface interactions. These are v_i and v_r, respectively. Also needed are the well depths, E_i and E_r, as shown in Fig. 5.5, for both incoming and rebounding particles. The depth of the wells will be different owing to mechanisms mentioned earlier, as well as contact charging.

To account for nonadhesive energy loss upon impact, Dahneke defines a coefficient of restitution as the ratio of v_i to the rebound velocity at the moment of rebound:

$$e = \frac{v_i}{(v_r)_{\text{contact}}} .$$

(5.56)

This phenomenological coefficient of restitution satisfies $e = v_r/v_i < 1$. Rotation of the particle could be included as an additional energy term and a rotational coefficient of restitution defined. Upon rebound, the particle must exchange kinetic energy for potential energy as it climbs out of the potential well, so its final velocity away from the wall is reduced. Solving for v_r in terms of parameters previously defined leads to

$$\frac{v_r}{v_i} = \left(e^2 + \frac{E_r - e^2 E_i}{mv_i^2/2} \right)^{1/2} .$$

(5.57)

The velocity at which the rebound kinetic energy is equal to zero is used to define the critical velocity for capture, v_c. From Eq. (5.56), this is

$$v_c = \left(\frac{2}{me^2} (E_r - e^2 E_i) \right)^{1/2} .$$

(5.58)

Capture will occur when v_i is less than v_c, otherwise rebound will occur. When E_i and E_r are equal, the capture velocity is

$$v_c = \left(\frac{2E}{m}\frac{1-e^2}{e^2}\right)^{1/2}.$$

(5.59)

Equation (5.59) tends to isolate loss of velocity to nonadhesive factors and illustrates the fact that the Dahneke definition of coefficient of restitution is different from the classical definition (Wall et al., 1988). If E_r is much larger than E_i, Eq. (5.59) reduces to

$$v_c = \left(\frac{2E_r}{me^2}\right)^{1/2}$$

(5.60)

In this case, the adhesion effects, such as contact deformation, can be important even if the coefficient of restitution is near unity.

The foregoing phenomenological description shows that the critical capture velocity depends on E_i, E_r, and e. If the well depths and coefficient of restitution are independent of particle size, that is, mass, it is seen that larger particles rebound more than smaller particles. This relates to an earlier discussion of laminar deposition collection efficiency which was seen to increase with particle size, up to about 10 μm. At larger sizes, rebound is often important, and collection efficiencies again decrease for particle impinging with velocities greater than the critical capture velocity.

Further discussion of particle rebound requires consideration of materials properties of the surface and particle, and assumptions about the mechanisms of energy loss. Thus, one may invoke arguments based on contact adhesion or contact charging described in earlier sections to estimate E_i and E_r to obtain v_c. A critique of such calculations may be found in Wall et al. (1988) or Tsai et al. (1990). Useful earlier papers dealing with impact adhesion, deformation, and rebound include Brenner et al. (1981) and Rogers and Reed (1984). In no case has attention been given to highly deformable particles typical of bioaerosols. In these cases, only the qualitative behavior illustrated by Dahneke-type models is available for guidance.

5.3.6. General Aerosol Deposition

The preceding sections have identified the major interactions that arise in discussions of aerosol deposition. The interactions in the near-surface region have been emphasized because it is here that adhesion forces are most strongly felt. A general theory of aerosol deposition would account for these direct microscopic interactions, as well as the indirect, macroscopic, hydrodynamic interactions between a particle and a surface. Many approaches to this problem have been taken over the years (Fuchs, 1964; Hidy and Brock, 1970; Happel and Brenner, 1965). However, a number of complications, both mathematical and physical, arise in attempts to construct a theory valid over wide ranges of gas or aerosol densities. In many cases, the issue of specific particle–surface adhesion is not developed because it is sufficient to consider the surface as a sink: When the colloid reaches the surface by whatever mechanism, it is removed permanently. We comment briefly on some of the physical issues arising in recent work, selected somewhat arbitrarily, dealing in a general fashion with particle–surface interaction. This includes the discussion of transport and deposition given by Shapiro, Brenner, and Guell (1990), denoted SBG, and a review of the influence of gas dynamics on particle interactions by Ying and Peters (1991).

As noted by SBG, a significant challenge in dealing with transport of particles to surfaces is to account for both the microscopic particle–surface interactions as well as the hydrodynamic transport properties of the system. For example, treating the particle–surface interaction as a simple sink overlooks the fact that Brownian particles have thermal energy of motion, not all will remain on the surface, and, depending on the relative size of kT_B and the potential well depth, a Boltzmann distribution for particle concentration near the surface will be established. Conditions for validity of the particle–sink picture were established as a part of the SBG investigation.

SBG offer a description of diffusive and convective transport of Brownian particles in fluid flowing next to a solid surface. They stress the importance of different length scales corresponding to the near-surface microscopic region (λ), and the overall hydrodynamic system size length scale ($L \gg \lambda$). A particle–surface interaction similar to that of Fig. 5.5 is assumed, and the microscopic length scale l is taken as the range of this interaction. A perturbation theoretic description is then developed in terms of the small parameter $\delta = \lambda L \ll 1$. By matching (diffusions/transport) equations valid in the near-surface region with transport equations valid at macroscopic distances from the surface, solutions are obtained which depend on λ, δ, and the depth of the LvdW interaction potential (relative to kT_B). Conditions were established for "shallow," "deep," and "very deep" LvdW potential wells, and matching solutions obtained for the latter two cases.

In the case of very deep wells, SBG recover a solution that effectively corresponds to the perfect sink assumptions of macroscopic descriptions of transport and deposition [e.g., Nazaroff and Cass (1989)]. This can be viewed as irreversible adsorption of particles. For a deep well, adsorption is observed, but now it is reversible and conditions equivalent to a Boltzmann concentration distribution near the surface are found. The concentration increase near the surface is defined in terms of a surface excess of particles. SBG carry out an explicit estimate of deposition for a LvdW potential corresponding to Brownian deposition conditions and particles in the 0.1-μm range. With Hamaker constants corresponding to values quoted elsewhere in this chapter, they find that the particle concentration effectively vanishes near the surface, and conditions corresponding to the perfect-sink solutions are reproduced. This result is consistent with conventional use of the perfect sink assumption. It should be remembered, however, that as particles deposit onto collectors, the effective particle–surface potential changes from the initial conditions. It should not be concluded that the perfect-sink model is applicable to arbitrary aerosol deposition conditions.

In summary, the SBG approach provides an improved mathematical description of Brownian deposition under a range of conditions, some of which are typical of smaller bioaerosols. Such continuum descriptions, as well as Monte Carlo approaches (Gupta and Peters, 1985), provide important insight into microscopic aspects of aerosol deposition.

The SBG approach focuses on conditions of relatively high gas density and continuum hydrodynamics. It is also of interest to consider the more general situation of widely varying gas density, including low-density, gas dynamics region. Although specific LvdW or electrostatic interactions are independent of gas density, the indirect gas dynamics interactions are not. The problem of interparticle and particle–surface gas dynamics interactions has been recently reviewed by Ying and Peters, denoted YP, in terms of resistance (Brenner and O'Neill, 1972) and mobility (Batchelor, 1976) matrices that describe the relationships between the forces and torques acting on spherical particles and their influence on the particle translational and angular velocities. The major focus of YP is on a description of how gas dynamics modulates particle–surface interactions as a function of gas density, that is, the Knudsen number. One motivation of their work is the idea that complex bioaerosols can be modeled as a collection of simple objects such as spheres, ellipsoids, and so on. The interactions of the complex bodies are modulated by the gas molecules around them, which produce the forces which vary widely from the continuum (hydrody-namic) regime (low Kn) to the free-molecule regime (large Kn). By careful consideration of the gas dynamics for simple geometrical objects, the interactions between larger complex bodies may be worked out [see also Chan and Dahneke (1981)]. YP show that for continuum or near-continuum (small Kn) flows, the force matrices are well known. For intermediate and large Kn (free-molecule) flow, only asymptotic forms to describe particle–surface interactions at infinite separation are known. The authors also conclude that Monte Carlo and Brownian dynamics simulations will be important tools in simulating gas dynamics interac-tions over wide ranges of gas densities. It is likely that as these techniques evolve, interactions involving complex bioaerosol particles will be investigated and additional information about deposition mechanisms obtained.

5.4. Aerosol Adhesion and Release Measurements

Normally it is not possible to measure adhesion forces *in situ* while deposition is occurring. (An approximation to this is encompassed in the force measurement technique described below.) Thus, adhesion has traditionally been deduced by determining how much force is required to remove a particle from a surface. Because this is equivalent to the release of the particle, both adhesion force measurements and release are conveniently considered together.

From common experience, it is sometimes possible to clean a surface simply by blowing off offending particles with an airstream. In sophisticated form, this is the aerodynamic method for adhesion measurement. Another common technique involves centrifugation. These two methods are examples of what can be termed macroscopic measurements. The forces measured by macroscopic methods are in the millidyne range or larger. These methods focus on the minimum force required to separate a particle from a surface and do not provide a continuous measurement of particle–surface forces as a function of separation. To accomplish the latter requires more sophisticated instrumentation. Such methods are referred to as microscopic techniques because the forces measured can be as low as microdynes and the distance resolution is an angstrom or less. Examples of microscopic force measurement apparatus discussed below are the atomic force microscope (AFM) and the surface force apparatus (SFA).

5.4.1. Macroscopic Adhesion Measurements

Most approaches share the common feature that attached particles are subjected to an increasing force whose value is noted at the time the particles detach from a surface (Boehme, et al., 1962). Measurement methods include weighing, pendulum, aerodynamic, inclined plane, centrifuge, vibration, and electric fields. Of these, the weighing, aerodynamic, centrifuge, and, possibly, vibration methods are most relevant to uncharged bioaerosol adhesion and are discussed here. The aerodynamic and centrifuge techniques have been most widely employed in the context of bioaerosols or related particles. The centrifuge technique has provided the most consistent experimental data [e.g., Kordecki and Orr (1960)]. Some of the methods observe individual particles, by optics or photography. More commonly, collections of particles are observed and the fraction remaining after a certain force is tabulated. As noted in previous sections, the actual forces required to remove a particle have a significant range, $0.001 \leq f < 10$ dyn.

Weighing. Two surfaces in contact are separated while monitoring their apparent weight. One of the first measurements of adhesion employed a sensitive quartz balance to measure the force to separate spherical quartz spheres (Bradley, 1932). The measured force was a few dynes for spheres of diameters in the millimeter range. With time, this general method has increased in sophistication and has evolved into the microscopic surface force apparatus described below.

Aerodynamic. An airstream passes over a particle-covered surface. When the aerodynamic drag experienced by the particles exceeds the adhesion, the particles will be removed from the surface. Figure 5.10 shows a typical velocity profile buildup (Schlichting, 1960). The drag force is related to the easily measured flow rate in the bulk airstream. Difficulties arise in knowing precisely the flow

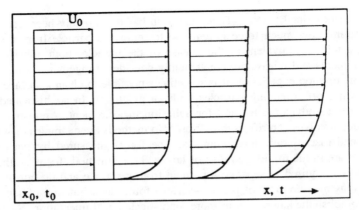

Figure 5.10. Approximate airflow over surface. The effective aerodynamic force decreases as the surface is approached (laminar flow).

rates in the near-surface or boundary layer region where the particles reside. Differences in geometry and other experimental details also complicate intercomparison of data. An approximate expression for the drag force exerted on a particle has been given by Zimon (1982),

$$F = \frac{CAv^2 (\rho_p - \rho_f)}{2}.$$ (5.61)

where C is the drag coefficient, A is the cross-sectional area presented by the spherical particle, v is the fluid velocity, and ρ_p and ρ_f are the densities of the particle and fluid, respectively. In the Stokes flow regime, C is given by 24/Re. For a linear velocity distribution in the boundary layer of thickness δ, normally larger than the particle radius, and with an average fluid velocity v_0, fluid viscosity μ, and particle diameter d, the force is given by

$$F = \frac{3\pi\eta v_0 d^2}{2\delta}.$$ (5.62)

The expression confirms intuition and relates the force to the cross-section area of the particle. In comparisons of aerodynamic adhesion measurements with centrifuge techniques, it is found that agreement is greatest for particles with diameters larger than 100 μm (Corn, 1966).

Centrifuge. A rotor surface to which particles adhere is spun in a centrifuge. The force experienced by the particles is

$$F = m\omega^2 R,$$ (5.63)

where m is the particle mass, ω is the angular speed of the rotor, and R is the radial distance to the particle. In practice, a submonolayer of particles is spread on a surface, the rotor spun in the centrifuge, and a measurement of the mass remaining, or a count of the number particles remaining (the residue) is made. Repetition of the process provides a plot of residue versus speed (or force). Differentiation of the plot yields the frequency distribution of the particle–surface force or adhesion (Boehme et al., 1962; Kordecki and Orr, 1960).

Aylor (1975) used both the centrifugal and the aerodynamic methods to study the detachment of conida from *Helminthosporium maydis* cultures on corn plant leaves. The detachment force in this case was approximately 10 mdyn. In these fungi, the ease of detachment argues for physical adhesion mechanisms of attachment between spore and stalk.

Vibration. Oscillatory motion leads to large variations in the acceleration and, hence, force on particles attached to a surface. One variation is to use a loud-speaker coil as a displacement generator. In any case, for a particle of mass m, the normal force is easily found. The displacement is given by $x(t) = A \sin \omega t$, the maximum acceleration is $a = \omega^2 A$, leading to a maximum force $f = \omega^2 A m$. This expression shows that for a given frequency, the force decreases with particle size. However, at frequencies in the megahertz range, detachment forces can, in principle, exceed those available with centrifuges. At these frequencies, side problems such as elastic deformation of either particle or surface and cavitation (if liquid is present) make the technique less attractive. Reported work deals with particles larger in diameter than 20 μm (Derjaguin and Zimon, 1961).

5.4.2. Surface Force Measurements

Two primary techniques are available for the study of forces between particles and surfaces. Atomic force microscopy (AFM) (Binnig et al., 1986; Sarid, 1991) is derived from earlier surface microscopy methods, primarily scanning tunneling microscopy (STM) (Binnig and Rohrer, 1986). The surface force apparatus (SFA) derives from experiments devoted to direct determination of LvdW forces between macroscopic objects (Bradley, 1932; Derjaguin et al., 1956; Israelachvili and Tabor, 1973; Israelachvili and Adams, 1978). In the contact form of AFM, a probe tip of atomic dimensions is suspended on a sensitive cantilever leaf spring and then positioned over and near to a surface so that it experiences LvdW attraction or repulsion forces. As the tip is moved over a surface and encounters variations in the surface structure, the force changes and the cantilever deflects. By design, the cantilever spring constant is sufficiently small (below about 1 N m^{-1}) so that the tip does not displace atoms from the surface under study. Optical techniques are generally used to monitor the cantilever deflection which reflects the topography and yields a surface microscopy with lateral and vertical resolution in the subnanometer range. As a direct measurement of force, the lateral position of the tip can be held constant, and the force measured as a function of vertical distance to the sample. The leaf spring can be calibrated with a force sensitivity in the microdyne range. Motion is achieved with piezoelectric ceramic supports. This technique has recently been used to measure the force between a silica particle and a surface in aqueous media (Ducker et al., 1991). It has also been used to image and manipulate biological macromolecules (Hansma et al., 1992).

The SFA developed by Israelachvili and co-workers predates AFM but involves somewhat similar technology. It samples larger surface areas. The SFA technique employs two crossed cylindrical prisms, of radius about 1 cm and length 1 cm. The geometry of two crossed cylinders, each of radius R, is locally equivalent to a sphere of radius R near a flat surface. The cylinders are covered with atomically smooth mica on which layers of interest have previously been deposited. The shapes of the surfaces and the separation D between them is measured to within 0.1 nm by employing an optical interference technique using "fringes of equal chromatic order" (FECO). These fringes can be viewed through the eyepiece of a spectrometer or recorded on a video camera and allow surface separations and shape changes to be continually monitored in real time. In addition, the FECO technique makes it possible to observe directly the capillary condensation of water around two contacting surfaces. Because this occurs only when the surfaces are hydrophilic, but not when they are hydrophobic, this phenomenon can be used to monitor any changes in the hydrophilic–hydrophobic balance of surfaces exposed to different atmospheric conditions, that is, varying humidity.

Figure 5.11 shows a schematic of a typical surface force apparatus. The two surfaces are made to approach or separate from each other using mechanical and piezoelectric distance control stages. The upper and lower surfaces are movable and the lower surface is suspended from the end of a cantilever spring (of stiffness $K = 100$–1000 N m^{-1}, 1–10 mdyn Å^{-1}). At sufficiently short distances, the surfaces experience LvdW interactions which are registered by deflection of the leaf spring and detected optically. In this way, a determination of the forces of interaction as a function of distance can be achieved. The resolution in distance that can be achieved is less than 1 Å, and the force sensitivity is less than 1 mdyn. In many cases, the interaction energy between two surfaces varies with distance in a manner similar to that shown in Fig. 5.5. The force between the two surfaces is given by the derivative of this potential: $F = -(dU/dD)$, where U represents the potential and D is the separation. The force curve is similar as shown in Fig. 5.12. In fact, for the geometry of the SFA, it is possible to relate directly the measured force F to the interaction energy U. The relation is $F/R = 2\pi U$ and arises from the geometry of large surfaces separated by small distances, which leads to the Derjaguin approximation (Israelachvili, 1985). Figure 5.12 shows a schematic plot of force versus separation, with measured F/R as the left ordinate and the interaction potential as the right ordinate. The equilibrium separation is denoted by D_0.

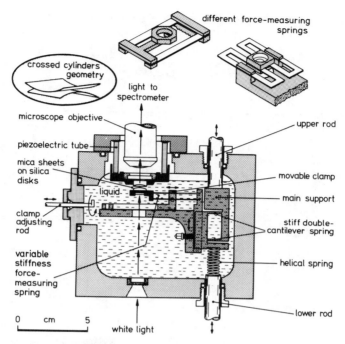

Figure 5.11. Force measurement apparatus. The force between two crossed cylindrical prisms is determined as a function of their separation.

Recent SFA experiments have been carried out on many different systems, starting with mica surfaces in air and in liquids, as a test of DLVO theory (Israelachvili and Adams, 1978). The apparatus has been applied to tribology problems and has revealed local order in liquids near interfaces (Israelachvili, 1987). Experiments relevant to bioaerosols include work with surfaces related to outer membranes of microorganisms. For example, a series of experiments has been carried out using lipid bilayers deposited on mica surfaces by Langmuir–Blodgett techniques (Marra, 1986). Very recently, measurements on materials derived directly from bacterial surfaces have been initiated, with an objective of gaining information under conditions relevant to bioaerosols. These measurements are discussed in a subsequent section.

5.4.3. Release

The mechanisms whereby adsorbed particles may be released as aerosols are numerous. In plants, for example, pollens are released by mechanical or aerodynamical means (Ingold, 1960). In general, an external force must be exerted on the particle to overcome the local adhesion forces, including LvdW, electrostatic,

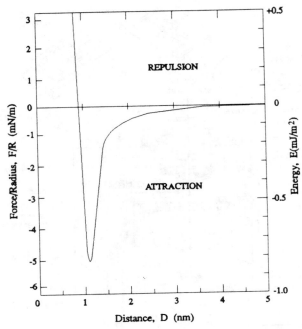

Figure 5.12. Force and interaction energy as a function of particle–surface separation (schematic).

capillary, and so on. The forces can be exerted by mechanical means, as in simple dusting, motions of leaves, and so on. Air or other gas motions are most important, and in such cases forces depend on aerodynamic drag. Generally, the movement of a particle begins with rolling and sliding (Bagnold, 1941). Natural release corresponding to the experimental geometries of inclined planes, pendulums, and centrifugation is rare. Oscillatory motion may be important in combination with mechanical motion as in shaking of a rug. In principle, these cases can be dealt with by methods already described. One method not considered thusly is the use of an electric field to remove particles. A consideration of this technique parallels somewhat the earlier discussion dealing with electrostatic forces.

Interest in soil erosion, radioactive contaminants, traffic aerosols, and other environmental issues has focused a great deal of interest on aerosol release or resuspension. In addition, fungal spore release is an interesting topic from several perspectives. Only a few references are indicated here that may serve as an introduction to such work. Ingold (1960), Schrödter (1960), and Aylor (1978) discuss suspension and dispersal of fungal spores. Corn and Stein (1965) discuss resuspension from glass and metal surfaces. A review of models of particle resuspension is given by Smith et al., 1982. Additional reviews of particle resuspension are provided by Sehmel (1980) and Nicholson (1988a, 1988b). The focus of nearly all of this work is on natural sources of resuspension. Aylor (1978) provides a useful picture of wind gusting as a means of removal. An initial gust of wind produces a velocity profile such that a large wind velocity exists near the surface. In a short time, however, a steady state is reached where the velocity is very small near the surface in the viscous boundary layer. An approximate view of the variation in airspeed or aerodynamic force with distance from the surface was shown in Fig. 5.10. An explicit calculation, with dimensions for a laboratory experiment, may be found in Corn (1966). This picture is also valid in the context of natural aerosol release and dispersion.

Interest in an electric field as a possible release mechanism arises from a consideration of the forces that are exerted on a charged particle resting on a conducting surface and exposed to an electric field. A brief summary of the salient points involved in electrostatic release of particles is as follows. First, small conducting particles become charged and experience a force that is proportional to the square of both the diameter and the electric field. The forces are basically the inverse of forces that arise in the context of enhanced deposition in the presence of electric fields (Shimada et al., 1989, and discussion in previous sections). Fuchs (1964) mentions evidence that small aerosols, even if nominally insulators, respond to electric fields as if they were conductors. This might occur because of rapid or hopping motion of charge carriers over the small dimensions of the colloid. For particles smaller than roughly 10 μm and substantial electric fields, greater than 10^5 V cm^{-1}, calculations suggest and experiments roughly corroborate that electric field forces exceed normal adhesion forces (Cooper and Wolfe, 1990) for particles in the micron-size range including insulators, such as SiO_2, as well as metals. The field strength required to remove particles increases with decreasing particle size and becomes a limiting factor in applications of this technique, which is not highly developed. In air at STP and a gap of 1 cm, the breakdown field is approximately 30 kV cm^{-1}. Applications to bioaerosols in the micron-size range would require alternate gases of higher dielectric constant than air, or possibly vacuum conditions.

Finally, we note that electrostatic release is a key factor in the xerographic process, which involves the removal of charged insulating particles, typically 10 μm in diameter, by electric fields (Hays, 1988). Here the model is that of a uniformly charged, spherical insulator resting on a conducting surface. In these circumstances, an additional polarization force proportional to the square of both the electric field and the particle diameter is present, leading to the expression (Hartmann et al., 1976; Davis, 1969)

$$F_E = -\beta QE - 4\gamma\pi\epsilon_0 R^2 E^2, \tag{5.64}$$

where β and γ are constants that depend on the dielectric constant of the particle. For a typical dielectric constant of 4, β, and γ are equal to 1.6 and 0.25, respectively. The polarization force is always directed toward the collector, so the electric field that produces the maximum detachment force is given by

$$E = \frac{\beta Q}{4\gamma\pi\epsilon_0 R^2}. \tag{5.65}$$

We assume that the aerosol is charged, is nominally insulating, and that the collector geometry is a planar conducting surface, which is necessary to support an electric field. The particle should be insulating to minimize charge leakage to the collector. Application of an electric field produces a force on the particle which is the sum of two terms. One is an electrostatic force arising because of the interaction of the charged particle with the electric field. This is proportional to the particle charge and to the electric field. The other is a polarization force which is proportional the square of both the charge and the electric field.

Removal of particles by shaking, that is, the use of acceleration, is effective only for larger particles, as can be seen by a simple calculation. For the case of a spherical particle on a surface and acceleration normal to the surface, the removal force is $f = ma = (4/3)\pi R^3 \rho$. This must exceed the LvdW adhesion force of $AR/6D^2$. With the Hamaker constant for polystyrene, a radius of 1 μm, and a separation of 5 Å, the acceleration must exceed 820,000g; whereas for a radius of 10μm, the acceleration is 8200g. Hence, mechanical shaking is generally ineffective for removal of bioaerosols.

5.5. Bioaerosol Surfaces

The outer cell envelope (often referred to as the plasma membrane) of either prokaryotes or eukaryotes has a complex molecular architecture which is species-specific. Spores and pollen also present molecularly complex exterior surfaces, but here the dominant material is a mechanically stable polysaccharide matrix which provides rigidity and protection. The various cell envelopes usually share the following features: (1) a di-acyl phospholipid bilayer, in which are embedded various proteins, many of which span the lipid bilayer, (2) an "exterior" characterized by lipid bilayer loci occupied by peptidoglycan polymers (protein molecules to which polysaccharide molecules are attached) that extend into the surrounding medium and are implicated in cell–cell recognition and adhesion activity, and (3) a supporting or interwoven subsurface of cytoskeletal proteins (also peptidoglycans) that may extend into the cytoplasm a distance greater than the plasma membrane thickness. In certain bacteria of note, *Archaebacteria, Mycoplasma*, feature (3) is absent. In some cases, the exterior of the cells are covered by two-dimensional ordered protein layers called S-layers (Sleytr and Messner, 1988). These much simpler structures represent an interesting model surface for adhesion studies and are discussed in more detail below.

The outer membrane surfaces contain ionizable groups and most bacteria, or single cells such as erythrocytes, carry a net negative charge at neutral pH. In aqueous suspension, the soluble polysaccharide molecules on the surface are extended and provide a steric interaction with surfaces with an effective range of tens to thousands of angstroms. At these distances, LvdW interactions are negligible, but DLVO interactions can be important. The case of a suspended aerosol is different. Here, there will be less water to hydrate the surface. As a result, polymeric material on the outer surface will tend to be collapsed onto the particle surface. Little direct information is available concerning the roughness of the collapsed surface polymers, but it is expected to be of the order of 10–20 Å. Under these conditions, LvdW and electrostatic forces are dominant and the theory of previous sections is applicable.

5.5.1. Bacterial S-Layers

Regular surface protein arrays, or S-layers, are paracrystalline two-dimensional arrays of protein monomers that, when present, constitute the outermost layer of the cell envelope. They occur in a wide variety of prokaryotes, particularly *Archaebacteria* and *Eubacteria*. A schematic rendition of major features of (a) *Archaebacteria,* (b) Gram-positive, and (c) Gram-negative membranes possessing S-layers is given in Fig. 5.13. Also shown are typical membrane molecular units, such as lipids, peptidoglycans, and so on. CM denotes the cell membrane; CW denotes the rigid cell wall layer in Gram-positive organisms, which is composed of peptidoglycan and/or other polymers; OM denotes the outer membrane; PG denotes peptidoglycan layer; and S denotes the crystalline S-layer.

Freeze-fracture and related ultrastructural techniques have been used to obtain electron micrographs of replicas of intact crystalline S-layers, Fig. 5.14. Figure 5.14 shows an electron micrograph of a metal carbon replica of a freeze-etched bacterium, *Desulfotomaculum nigrificans* (Sleytr et al., 1986). Different strains of this and other bacteria have been found to display considerable diversity in the molecular weights of the S-layer subunits and in the two-dimensional geometry of the subunits. Figure 5.14 shows a square lattice S-layer from *D. nigrificans* (Sleytr et al., 1986). Crystalline arrays with hexagonal (p6), square (p4) and oblique (p2) lattices have been observed for *D. nigrificans*. The individual S-layer subunits are single, homogeneous protein or glycoprotein molecules, whose molecular weights (among known bacteria) range from about 40,000 to 200,000 (Sleytr et al., 1988). Their appearance or disappearance depends on environmental and growth conditions, as in laboratory conditions, S-layer-deficient mutants may supplant the wild species. S-layers have been shown to function as protective coats, or outer permeability barriers, promoters for cell adhesion and surface recognition, and as frameworks which determine and maintain cell shape or envelope rigidity. Their function in adhesion is of particular interest in the context of bioaerosols. Because the S-layers form the outer envelope of the bacterium,

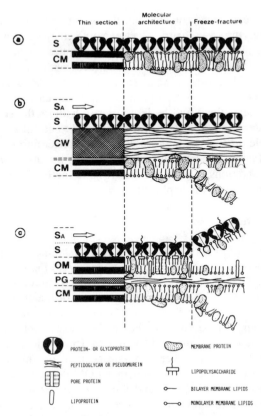

Figure 5.13. Schematic representation of (A) *Archaebacteria*, (B) Gram-positive, and (C) Gram-negative cell structures. [From Sleytr and Glauert (1982).]

it is this surface that initially contacts substrates. In addition, owing to the range of LvdW forces, it is the composition of the S-layer that will largely determine the nature of the particle–surface interactions.

Of particular interest here is the fact that S-layers can be isolated from the bacterial membrane and reassembled in vitro to form roughly square sheets of dimension 1 to 6 μm on a side, Fig. 5.15 (Sleytr et al., 1986). These dimensions are appropriate for use with the surface force measurement apparatus described in a preceding section.

5.5.2. Bacterial S-Layer Adhesion

Bacterial S-layers provide a relatively simple and well-defined biological surface material that can be used to investigate directly bacteria–surface interactions. These interactions should be representative of microscopic interactions between

Figure 5.14. Electron micrograph replica of S-layer surface of *D. nigrificans*. Bar. 150 nm. [From Sleytr et al. (1986).]

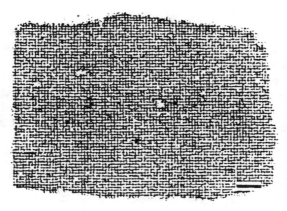

Figure 5.15. Negatively stained self-assembly sheet product from *D. nigrificans*. Bar, 1 um. [From Sleytr et al. (1986).]

bioaerosols and surfaces. In addition, they should provide information about changes in interactions that occur while a bioaerosol is resting on a surface. Initial measurements performed using the S-layer from *Bacillus coagulans* E38–66 (denoted S1) are summarized here.

Methods for isolating S1 were similar to those described by Sleytr et al. (1986). S1 is a 100-kdal protein that forms an oblique S-layer lattice with base vector lengths of $a = 9.4$ nm, $b = 7.4$ nm, and $\gamma = 80°$. Carboxyl groups occur on both the inner and outer faces of S1, but only the inner face has a net negative charge (Pum et al., 1989). The S1 layer was adsorbed from solution onto mica surfaces on which a monolayer (1.0 nm) of polylysine had been previously deposited. A schematic representation of the deposited S-layer is shown in Fig. 5.16. This figure is also supported by electron micrographs of the deposited layers. The polylysine was used to obtain a uniform deposition of the S-layer. (Also shown for comparison is a monolayer of the lipid dipalmitoyl phosphatidy-lethanolamine.) The S1 surface that normally faces outward has been found to adsorb onto the polylysine layer. Hence, the measurements discussed here involve interactions between the inner surfaces of the S-layers. The response to hydration and time-dependent features are assumed at this time to be roughly similar for both the inner and outer surfaces.

Surface force measurements were carried out in a standard SFA in which the atmosphere could be controlled. Adhesion energies were determined from the adhesion (pull-off) forces, F_S, according to

$$\gamma = F_S/3\pi r, \tag{5.66}$$

or, when capillary forces dominate the adhesion,

$$\gamma = F_S/4\pi r. \tag{5.67}$$

Under dry conditions, the forces between the two S-layers were indicative of reasonably hard, but still compressible layers. This was ascertained by noting that on coming into adhesive contact, the layers could be compressed by about 2 Å under a pressure of 100 atm. This gives a value for their compressibility modulus of $Y = 1 \times 10^8$ N m^{-2}. This is similar to that of hydrated phospholipid bilayers in the gel state, that is, intermediate between the solid crystalline and

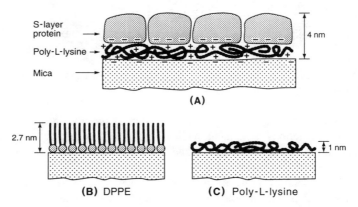

Figure 5.16. S-layer deposition on poly-L-lysine. The dimensions of the layers are indicated.

fluid states (Cevc and Marsh, 1987). However, the proteins clearly have some configurational or conformational flexibility, which was found to play a major role in determining how humidity and time affected their adhesion to each other and to other nonprotein surfaces.

Figure 5.17 shows how both the adhesion and thickness of two interacting S-layers increase with relative humidity (RH). The shaded band in the top figure gives the range of adhesion energy γ measured at each RH value. The vertical arrows indicate how γ increases with increasing contact time, t, from a minimum of 30 s to a maximum of 100 min. Up to a RH of 90%, the thickness increase could be accounted for by the polylysine substrate, so the protein thickness was not changing significantly in this range. Above approximately 90% RH, there is a sharp increase in thickness by about 7 Å that is attributed to water uptake or conformational rearrangements by the protein of two S-layers which increases with RH. The trend is qualitatively similar to that observed with other amphiphilic or polyelectrolyte surfaces. The results indicate that some hydrophilic (polar) groups replace the hydrophobic groups in the surfaces once the S-layers are exposed to humid air at a relative humidity above 90%. In addition, a small water meniscus was seen to capillary condense around the contact site for RH

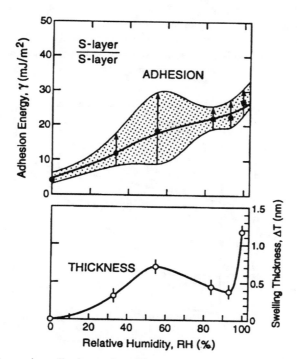

Figure 5.17. Increasing adhesion and swelling of interacting S-layers with increasing relative humidity.

greater than 55%. Whether the increased adhesion is due entirely to capillary forces or whether some of it also arises from molecular conformational rearrangements (nonpolar chain interdigitation and/or the exposure of more polar hydrophilic surface groups) is discussed below.

Unlike the simpler surfaces such as lipid bilayers or the polylysine layers, which equilibrate within 1 s of coming into contact, the interactions between S-layers exhibited pronounced time-dependent effects which provided additional insights into the molecular mechanisms and relaxation processes with their interactions. As shown in Fig. 5.18, the adhesion of two S-layers increases with the time the surfaces were allowed to remain in contact. It is interesting that the relaxation time (approach to equilibrium) is fast at low and high humidities, and slowest at intermediate humidities (cf. the curve 55%, which has the highest slope).

The results of Figs. 5.17 and 5.18, when taken together with previous findings on adhering layers, provide a fairly consistent picture as follows: In dry air, the S-layer surface is rigid and hydrophobic, but the surface groups are capable of slowly interdigitating with those of an opposing S-layer surface. This results in a slow but steady increase in the adhesion with contact time (see lower curve in Fig. 5.18). With increasing hydration, the adhesion steadily increases as the S-layer surfaces become more hydrophilic due to the reorientation and rearrangements of polar groups that now become exposed to the surfaces. These rearrangements are likely to be more extensive and should have higher activation energies. Finally, at even higher humidities, the increased hydration ensures that

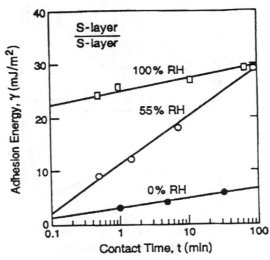

Figure 5.18.   Adhesion of two S-layers as a function of contact time *t* at different relative humidities.

the polar–hydrophilic groups are already at the surfaces before contact occurs, so that equilibrium (but higher) adhesion is now reached almost immediately upon contact of the surfaces.

Related time-dependent effects were observed in the rate of separation of the surfaces. If two S-layer surfaces are separated (or peeled apart) quickly, the adhesion is higher than if they are separated slowly. Previous adhesion studies on adhesion hysteresis of amphiphilic and organic surfaces (Chen et al., 1991) has shown that this arises because there is insufficient time for the entangled hydrocarbon chains or surface polymer groups to disentangle. This is indicated in Fig. 5.19. We found that at low and high humidities the adhesion did not depend much on the rate of separation. However, at intermediate humidities, the adhesion increased significantly if the separation was done rapidly. This correlation between the two time-dependent effects—contact time and separation rate—is not unexpected. Any adhesion that increases with contact due to interdigitation is also expected to increase if the separation rate is high (which does not give enough time for the interdigitated chains to disentangle).

Finally, it was found that when the surfaces were brought into contact at high humidity, and subsequently dried prior to separation, the resulting adhesion was extremely high ($\gamma > 100$ mJ m^{-2}). This effect is presumably due to the freezing or immobilization of the entangled chains across the interface on drying, which now resist disentanglement upon separation. This observation supports the idea that the interdigitation of molecular groups across an adhesive interface is a major contributor to enhancing the adhesion of S-layers (in addition to the exposure of polar groups). These and other effects noted above also emphasize the fact that bioaerosol–surface interactions are complex, time dependent, and

○ Non-polar groups (hydrophobic)
● Polar groups (hydrophilic)

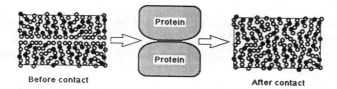

Before contact   After contact

Figure 5.19. Time-dependent interdigitation of molecules from opposed layers.

only poorly approximated by the Lifshitz or simple London force term, Eq. (5.35).

To summarize the time-dependent results with the S-layers, maximum adhesion is obtained between surfaces that are left in contact for a long time, then separated quickly, whereas the converse procedure yields the lowest adhesion. They also indicate that humidity (or hydration) can have a large effect on both the final adhesion value and on the rate at which the system approaches its equilibrium adhesion. It is interesting to note that maximum (though nonequilibrium) adhesion occurs at some intermediate humidity where slow molecular group interdigitations result in a strongly adhesive interface that cannot be as easily cleaved as totally hard but flat interface, or a totally fluid interface.

Acknowledgments

I would like to acknowledge a number of useful discussions with Jacob Israelach- vili, Madilyn Fletcher, and Debbie Leckband on the subject of adhesion and force measurement. I am especially indebted to Bruce Lightheart for introducing me to the area of bioaerosol research. I also wish to thank Professor Uwe B. Sleytr for very helpful assistance and comments concerning bacterial S-layers, their properties and isolation. Finally, it is a pleasure to thank my wife Sara Church for her helpful assistance with figure preparation.

References

Aylor, D. E. 1975. Force required to detach conida of *Helminthosporium maydis*. Plant Physiol. 55:99–101.

Aylor, D. E. 1978. Dispersal in time and space: Aerial pathogens. In J. G. Horsfall and E. B. Cowling (eds.), Plant disease. An advanced treatise. Vol. 2. Academic Press, New York.

Bagnold, R. A. 1941. The physics of blown sand and desert dunes. Methuen & Co. London.

Batchelor, G. K. 1976. Brownian diffusion of particles with hydrodynamic interaction. J. Fluid. Mech. 74:1–29.

Binnig, G., and H. Rohrer. 1986. Scanning tunneling microscopy. IBM J. Res. Develop. 30:355–369.

Binnig, G., C. F. Quate, and Ch. Gherber. 1986. Atomic force microscopy. Phys. Rev. Lett. 56:930–933.

Boehme, G., H. Krupp, H. Rabenhorst, and G. Sandstede. 1962. Adhesion measurements involving small particles. Trans. Inst. Chem. Engrs. 40:252–259.

Bongrand, P. (ed.). 1988. Physical basis of cell–cell adhesion. CRC Press, Boca Raton, FL.

Bowling, R. A. 1985. An analysis of particle adhesion on semiconductor surfaces. J. Electrochem. Soc. Solid-State Sci. Technol. 132:2209–2214.

Bradley, R. S. 1932. The cohesive force between solid surfaces and the surface energy of solids. Philos. Mag. 13:853–862.

Brenner, H., and M. E. O'Neil. 1972. Stokes resistance of multiparticle systems in a linear shear field. Chem. Eng. Sci. 27:1421–1439.

Brenner, S. S., H. A. Wriedt, and R. A. Oriani. 1981. Impact adhesion of iron at elevated temperatures. Wear 68:169–190.

Burris, R. H. 1989. Field testing genetically modified organisms. National Academy Press.

Cevc, G., and D. Marsh. 1987. Phospholipid bilayers. John Wiley & Sons, New York.

Chamberlain, A. C. 1967. Deposition of particles to natural surfaces. In P. H. Gregory and J. L. Monteith (eds.), Airborne microbes. Cambridge University Press, Cambridge, pp. 138–164.

Chan, P., and B. Dahneke. 1981. Free-molecule drag on straight chains of uniform spheres. J. Appl. Phys. 52:3106–3110.

Chaudhury, M. K., and G. M. Whitesides. 1991. Direct measurement of interfacial interactions between semispherical lenses and flat sheets of poly(dimethylsiloxane) and their chemical derivatives. Langmuir 7:1013–1025.

Chen, Y. L., C. A. Helm, and J. N. Israelachvili. 1991. Molecular mechanisms associated with adhesion and contact angle hysteresis of monolayer surfaces. J. Phys. Chem. 95:10736–10747.

Cleaver, J. W., and B. Yates. 1975. A sublayer model for the deposition of particles from a turbulent flow. Chem. Eng. Sci. 30:983–992.

Clough, W. S. 1975. The deposition of particles on moss and grass surfaces. Atmos. Environ. 9:1113–1119.

Cooper, D. W., and H. L. Wolfe. 1990. Electrostatic removal of particle singlets and doublets from conductive surfaces. Aerosol Sci Tech. 12:508–517.

Corino, E. R., and R. S. Brodkey. 1969. A visual investigation of the wall region in turbulent flow. J. Fluid Mech. 37 (Part I):1–30.

Corn, M. 1961a. The adhesion of solid particles to solid surfaces. I. A review. J. Air Pollut. Control Assoc. 11:523–528.

Corn, M. 1961b. The adhesion of solid particles to solid surfaces. II. J. Air Pollut. Control Assoc. 11:566–575.

Corn, M. 1966. Adhesion of particles. in C. N. Davies (ed.), *Aerosol science*. Academic Press, New York, pp. 359–392.

Corn, M. and F. Stein. 1965. Re-entrainment of particles from a plane surface. Amer. Ind. Hyg. Assoc. J. 26:325–336.

Dahneke, B. 1971. The capture of aerosol particles by surfaces. J. Colloid Interface Sci. 37:342–353.

Dahneke, B. 1973. Measurements of bouncing of small latex particles. J. Colloid Interface Sci. 45:584–590.

Davies, C. N. 1966. Deposition of aerosols from turbulent flow through pipes. Proc. Roy. Soc. (London) A 289:235–246.

Davis, M. H. 1969. Electrostatic field and force on a dielectric sphere near a conducting plane—A note on the application of electrostatic theory to water droplets. Amer. J. Phys. 37:26–29.

Derjaguin, B. V., I. I. Abrikosov, and E. M. Lifshitz. 1956. Direct measurement of molecular attraction between solids separated by a narrow gap. Quart. Rev. Chem. Soc. 10:295–329.

Derjaguin, B. V., V. M. Muller, and Yu. P. Toporov. 1975. Effect of contact deformations on the adhesion of particles. J. Colloid Interface Sci. 53:314–326.

Deryaguin, B. V., and A. D. Zimon. 1961. Adhesion of powder particles to plane surfaces. (Consultants Bureau translation). Kolloid Zh. 23:544–552.

Ducker, W. A., T. J. Senden, and R. M. Pashley. 1991. Direct measurement of colloidal forces using an atomic force microscope. Nature 353:239–241.

Dzyaloshinskii, I. E., E. M. Lifshitz, and L. P. Pitaevskii. 1961. The general theory of Van der Waals forces. Adv. Phys. 10:165–209.

Fernandez de la Mora, J., and D. E. Rosner. 1982. Effects of inertia on the diffusional deposition of small particles to spheres and cylinders at low Reynolds numbers. J. Fluid Mech. 125:379–395.

Fisher, L. R., and J. N. Israelachvili, 1980. Determination of the capillary pressure in menisci of molecular dimensions. Chem. Phys. Lett. 76:325–328.

Fisher, L. R., and J. N. Israelachvili. 1981a. Experimental studies of the applicability of the Kelvin equation to highly curved concave menisci. J. Colloid Interface Sci. 80:528–541.

Fisher, L. R., and J. N. Israelachvili. (1981b). Direct measurement of the effect of meniscus forces on adhesion: A study of the applicability of the macroscopic thermodynamics to microscopic liquid interfaces. Colloids Surfaces 3:303–319.

Friedlander, S. K. 1957. Mass and heat transfer to single spheres and cylinders at low Reynolds numbers. Amer. Inst. Chem. Engrs. 3:43–48.

Friedlander, S. K., and H. F. Johnstone. 1957. Deposition of suspended particles from turbulent gas streams. Ind. Eng. Chem. 49:1151–1156.

Fuchs, N. A. 1964. The mechanics of aerosols. Dover Publications, New York.

Garland, J. A., and L. C. Cox. 1982. Deposition of small particles to grass. Atmos. Environ. 16:2699–2702.

Giddings, L. V. (ed.). 1988. New developments in biotechnology–field-testing engineered organisms: Genetic and ecological issues. Vol. 3. (New developments in biotechnology). U.S. Government Printing Office, Washington, DC.

Gillespie, T., and W. J. Settineri. 1967. The effect of capillary liquid on the force of adhesion between spherical solid particles. J. Colloid Interface Sci. 24:199–202.

Gupta, D., and M. H. Peters. 1985. A Brownian dynamics simulation of aerosol deposition onto spherical collectors. J. Colloid Interface Sci. 104:375–389; erratum 110:301–303.

Gupta, D., and M. H. Peters. 1986. On the angular dependence of aerosol diffusional deposition onto spheres. J. Colloid Interface Sci. 110:286–291.

Gutfinger, C., and G. I. Tardos. 1979. Theoretical and experimental investigations on granular bed dust filters. Atmos. Environ. 13:853–867.

Hamaker, H. C. 1937. The London–Van der Waals attraction between spherical particles. Physica 4:1058–1072.

Hansma, H. G., J. Vesenka, C. Siegerist, G. Kelderman, H. Morrett, R. L. Sinsheimenr, V. Elings, C. Bustamante, and P. K. Hansma. 1992. Reproducible imaging and dissection of plasmid DNA under liquid with the atomic force microscope. Science 256:1180–1184.

Happel, J., and H. Brenner. 1986. Low Reynolds number hydrodynamics: with special applications to particulate media, 2nd ed. (paperback). Kluluner, Boston.

Hartmann, G. C., L. M. Marks, and C. C. Yang. 1976. Physical models for photoactive pigment electrophotography. J. Appl. Phys. 47:5409–5420.

Hays, D. A. 1988. Electric field detachment of charged particles. In K. N. Mittal, ed., Particles on Surfaces, Vol. 1. pp. 351–360. Plenum, New York.

Hertz, H. 1881. J. Reine Agew. Math. 92:156; English translation reprinted in collection: Miscellaneous Papers. MacMillan, London (1896), p. 146–181.

Hidy, G. M., and J. R. Brock. 1970. The dynamics of aerocolloidal systems. Pergamon Press, Oxford.

Hiemenz, P. C. 1986. Principles of colloid and surface chemistry. 2nd ed. Marcel Dekker, New York.

Hinds, W. C. 1982. Aerosol technology. John Wiley & Sons, New York.

Horn, R. G., J. N. Israelachvili, and F. Pribac. 1987. Measurement of the deformation and adhesion of solids in contact. J. Colloid Interface Sci. 115:480–492.

Hough, D. B., and L. R. White. 1980. The calculation of Hamaker constants from Lifshitz theory with applications to wetting phenomena. Adv. Colloid Interface Sci. 14:3–41.

Hughes, B. D., and L. R. White. 1979. "Soft" contact problems in linear elasticity. Quart J. Mech. Appl. Math. 33:445–471.

Hughes, B. D., and L. R. White. 1980. Implications of elastic deformation on the direct measurement of surface forces. J. Chem. Soc. Faraday Trans. I 76:963–978.

Ingold, C. T. 1960. Dispersal by air and water: the take-off. J. G. Horsfall and A. E. Diamond (eds.), In Plant pathology. Vol. III: The diseased population, epidemics and control. Academic Press, New York, pp. 137–168.

Israelachvili, J. N. 1972. Calculation of van der Waals dispersion forces between macroscopic bodies. Proc. Roy. Soc. (Lond.) A331:39–55.

Israelachvili, J. N. 1985. Intermolecular and surface forces. Academic Press, London.

Israelachvili, J. 1987. Solvation forces and liquid structure as probed by direct force measurements. Accounts. Chem. Res. 20:415–421.

Israelachvili, J. N., and G. E. Adams. 1978. Measurement of forces between two mica surfaces in aqueous electrolyte solutions in the range 0–100 nm. J. Chem. Soc. Trans. Faraday Soc. I 74:975–1001.

Israelachvili, J. N., and D. Tabor. 1973. Van der Waals forces: Theory and experiment. Prog. Surf. Memb. Sci. 7:1–55.

Johnson, K. L. 1985. Contact mechanics. Cambridge University Press, Cambridge.

Johnson, K. L., K. Kendall, and A. D. Roberts. 1971. Surface energy and the contact of elastic solids. Proc. Roy. Soc. (Lond.) A324:301–313.

Kline, S. J., W. C. Reynolds, F. A. Schraub, and P. W. Runstadler. 1967. The structure of turbulent boundary layers. J. Fluid Mech. 30 (Part 4): 741–773.

Kordecki, M. C., and C. Orr. 1960. Adhesion of solid particles to solid surfaces. Arch. Environ. Health 1:1–9.

Krupp, H. 1967. Particle adhesion. Adv. Colloid Interface Sci. 1:111–239.

Kraemer, H. F., and H. F. Johnston. 1955. Collection of aerosol particle in presence of electrostatic fields. Ind. Eng. Chem. 47:2426–2434.

Landman, U., W. D. Luedtke, N. A. Burnham, and R. J. Coulton. 1990. Atomistic mechanisms and dynamics of adhesion, nanoindentation, and fracture. Science 248:454–461.

Langmuir, I., and K. B. Blodgett. 1946. A mathematical investigation of water droplet trajectories. Army Air Forces Technical Report No. 5418.

Leckband, D. E., Y.-L. Chen, J. N. Israelachvili, H. H. Wickman, and M. Fletcher. 1992. Measurements of conformational changes during adhesion of lipid and protein (polylysine and S-layer) surfaces. Biotech. and Bioeng. 42:167–177.

Levich, V. G. 1962. Physicochemical hydrodynamics. Prentice-Hall, Englewood Cliffs, NJ.

Little, P. 1977. Deposition of 2.75, 5.0 and 8.5 μm particles to plant and soil surfaces. Environ. Pollut. 12:293–305.

Loeb, L. B. 1958. Static electrification, Springer-Verlag, Berlin.

London, F. 1937. The general theory of molecular forces. Trans. Faraday Soc. 33:8–26.

Mahanty, J., and B. W. Ninham. 1976. Dispersion Forces. Academic Press, New York.

Marshall, K. C. 1986. Adsorption and adhesion processes in microbial growth at interfaces. Adv. Colloid Interface Sci. 25:59–86.

Marra, J. 1986. Direct measurements of attractive van der Waals and adhesive forces between uncharged lipid balances in aqueous solutions. J. Colloid Interpace Sci. 109:11–20.

May, K. R., and R. Clifford. 1967. The impaction of aerosol particles on cylinders, spheres, ribbons and discs. Ann. Occup. Hygiene 10:83–95.

McFarlane, J. S., and D. Tabor. 1950. Adhesion of solids and the effect of surface films. Proc. Roy. Soc. (Lond.) 202:224–243.

McLachlan, A. D. 1963a. Retarded dispersion forces between molecules. Proc. Roy. Soc. (Lond.) A271:387–401.

McLachlan, A. D. 1963b. Retarded dispersion forces in dielectrics at finite temperatures. Proc. Roy. Soc. (Lond.) A274:80–90.

McLachlan, A. D. 1963c. Three-body dispersion forces. Mol. Phys. 6:423–427.

Mittal, K. L. (ed.). 1988, 1991. Particles on surfaces. Plenum Press, New York, Vols. 1 (1988), 2 (1989), Vol. 3 (1991).

Muller, V. M., V. S. Yuschenko, and B. V. Derjaguin. 1980. On the influence of molecular forces on the deformation of an elastic sphere and its sticking to a rigid plane. J. Colloid Interface Sci. 77:91–101.

Muller, V. M., V. S. Yushchenko, and B. V. Derjaguin. 1983. General theoretical considerations of the influence of surface forces on contact deformations and the reciprocal adhesion of elastic spherical particles. J. Colloid Interface Sci. 92:92–101.

Nazaroff, W. W., and G. R. Cass. 1989. Mathematical modeling of indoor aerosol dynamics. Environ. Sci. Technol. 23:157–166.

Nicholson, K. W. 1988a. A review of particle resuspension. Atmos. Environ. 22:2639–2651.

Nicholson, K. W. 1988b. The dry deposition of small particles: A review of experimental measurements. Atmos. Environ. 22:2653–2666.

Nielsen, K. A., and Hill, J. C. 1976. Collection of inertialess particles on spheres with electrical forces. Ind. Eng. Fund. 15:149–157.

O'Brien, W. J., and J. J. Hermann. 1973. The strength of liquid bridges between dissimilar materials. J. Adhesion 5:91–103.

Ounis, H., G. Ahmadi and J. B. McLaughlin. 1991. Brownian diffusion of submicrometer particles in the viscous sublayer. J. Colloid Interface Sci. 143:266–277.

Parsegian, V. A., and G. H. Weiss. 1981. Spectroscopic parameters of computation of van der Waals forces. J. Colloid Interface Sci. 81:285–289.

Paw, U, K. T. 1983. The rebound of particles from natural surfaces. J. Colloid Interface Sci. 93:442–452.

Pum, D., M. Sara, and U. B. Sleytr. 1989. Structure, surface charge, and self-assembly of the S-layer lattice from *Bacillus coagulens* E38-66. J. Bacteriol. 171:5296–5303.

Ranade, M. B. 1987. Adhesion and removal of fine particles on surfaces. Aerosol. Sci. Technol. 7:161–176.

Reist, P. C. 1984. Introduction to aerosol science. Macmillan Publishing Co., New York.

Robinson, K. S., and W. Y. Fowlkes. 1989. Multipolar interactions of dielectric spheres. J. Electrostatics 6:207–224.

Rogers, L. N., and J. Reed. 1984. The adhesion of particles undergoing an elastic-plastic impact with a surface. J. Phys. D. (Appl. Phys.) 17:677–689.

Ruelle, D. 1990. Chance and chaos. Princeton University Press, Princeton, NJ.

Sarid, D. 1991. Scanning force microscopy. Oxford University Press, New York.

Schlicting, H. 1960. Boundary layer theory. 4th ed. McGraw-Hill, New York.

Schröder, H. 1960. Dispersal by air and water—the flight and landing. In J. G. Horsfall and A. E. Dimond (eds.), Plant pathology. Vol. III. Academic Press, New York, pp. 169–227.

Sehmel, G. A. 1980. Particle resuspension: A review. Environment Internat. 4:107–127.

Shapiro, M., H. Brenner, and D. C. Guell. 1990. Accumulation and transport of Brownian particles at solid surfaces: Aerosol and hydrosol deposition processes. J. Colloid Interface Sci. 136:552–573.

Shimada, M., K. Okuyama, Y. Kousaka, Y. Okuyama, and J. H. Seinfeld. 1989. Enhancement of Brownian and turbulent deposition of charged aerosol particles in the presence of an electric field. J. Colloid Interface Sci. 128:157–168.

Sleytr, U. B., and A. M. Glauert. 1982. Bacterial cell walls and membranes. In J. R. Harris (ed.), Electron microscopy of proteins. Vol. 3. Academic Press, London, pp. 41–78.

Sleytr, U. B., P. Messner, D. Pum, and M. Sara. (eds.). 1988. Crystalline bacterial cell surface layers. Springer-Verlag. Berlin.

Sleytr, U. B., and P. Messner. 1988. Minireview: Crystalline surface layers in procaryotes. J. Bacteriol. 170:2891–2897.

Sleytr, U. B., M. Sara, and D. Pum. 1991. Application potentials of two-dimensional protein crystals. Phillips Electron Optics Bull. 126:9–14.

Sleytr, U. B., M. Sara, Z. Kupcu and P. Messner. 1986. Structural and chemical characterization of S-layers of selected strains of *Bacillus stearothermophilus* and *Desulfotomaculum nigrificans*. Archiv. Microbiol. 146:19–24.

Smith II, W. J., F. W. Whicker, and H. R. Meyer. 1982. Review and categorization of saltation, suspension, and resuspension models. Nuclear Safety 23:685–699.

Spurny, K., and J. Pich. 1965. Diffusion and impaction precipitation of aerosol particles by membrane filters. Collect. Czech. Chem. Comm. 30:2276–2287.

Tardos, G. I., N. Abuaf, and C. Gutfinger. 1978. Dust deposition in granular bed filters: Theories and experiments. J. Air Pollut. Control Assoc. 28:354–363.

Tsai, C-J., D. Y. H. Pui, and B. Y. H. Liu. 1991. Elastic flattening and particle adhesion. Aerosol Sci. Tech. 15:239–255.

Tsai, C-J., D. Y. H. Pui, and B. Y. H. Liu. 1990. Capture and rebound of small latex particles upon impact with solid surfaces. Aerosol Sci. Tech. 12:497–507.

Visser, J. 1972. On Hamaker constants: A comparison between Hamaker constants and Lifshitz–Van Der Waals constants. Adv. Colloid Interface Sci. 3:331–363.

Wall, S., W. John, and S. L. Goren. 1988. Impact adhesion theory applied to measurements of particle kinetic energy loss. J. Aerosol Sci. 19:789–792.

Wells, A. C., and A. C. Chamberlain. 1967. Transport of small particles to vertical surfaces. Brit. J. Appl. Phys. 18:1793–1799.

White, H. J. 1963. Industrial electrostatic precipitation. Addison-Wesley, Reading, MA.

Yeh, H. C., and B. Y. H. Liu. 1974. Aerosol filtration by fibrous filters. J. Aerosol Sci. 5:191–217.

Ying, R., and M. H. Peters. 1991. Interparticle and particle–surface gas dynamic interactions. Aerosol Sci. Technol. 14:418–433.

Zimon, A. D. 1982. Adhesion of dust and powder. 2nd ed. Consultants Bureau, New York.

6

Death Mechanisms in Microbial Bioaerosols with Special Reference to the Freeze-Dried Analog

Eitan Israeli,* Janina Gitelman, and Bruce Lighthart

6.1. Introduction

Bioaerosols may be generated from a liquid suspension of microorganisms, or upon drying from a dust or powder. Man makes use of both suspensions and powders of microorganisms in his biotechnological applications in manufacturing, agriculture, and forestry. Although, airborne, the deleterious effects of the stressful atmospheric conditions may damage the organisms, there is a point where resuscitation repair is necessary,[1] or a point beyond repair (i.e., death).

This chapter discusses a possible death mechanism in airborne bacteria as deduced from freeze-drying (FD) experiments. The discussion includes the freeze-drying process and the cellular damage involved and presents evidence suggesting that many functional and synthetic systems are still operative in the "dead" microorganism. A model is suggested for the death mechanism during FD and subsequent exposure to atmospheric oxygen. Similarities and differences between FD and spraying or freezing of microorganisms in suspensions are discussed, and a proposed common death mechanism is suggested.

6.2. Theoretical Background

For a DNA synthesis initiation event to occur, the following conditions are required: (a) protein and RNA synthesis (Lark et al., 1963; Smith, 1973; Messer, 1972). [Upon protein synthesis inhibition, DNA synthesis continues until comple-

---

*This chapter was written while completing a National Research Council and U.S. Environmental Protection Agency associateship

[1]The need for resuscitation may be interpreted to mean microorganisms that are viable but not culturable.

tion of DNA strands in process of synthesis (Bremer and Churchward, 1991; Pritchard and Zaritsky, 1970; Skarstad et al., 1986)]; (b) the enzymatic complex responsible for initiation and elongation must attach to the cytoplasmatic membrane (Jacob et al., 1952); (c) at least two proteins are involved with the initiation process (Smith, 1973); (d) after relief from thymine starvation or heat inactivation of thermolabile DNA A or DNA C protein, a burst of initiation occurs (Bremer and Churchward, 1985; Hanna and Carl, 1975; Zaritsky, 1975). This phenomenon, termed rate initiation, suggests that the initiation of replication requires the accumulation of a stable protein. Because initiation ceases as soon as protein synthesis ceases, the required protein must act stoichiometrically, rather than catalytically. The protein could be DNA A or alternatively a protein that is active before the DNA A step. When temperature-sensitive mutants for DNA initiation were grown in nonpermissive temperature, freeze-dried, and exposed to oxygen, the following results were obtained for DNA synthesis at the nonpermissive temperature (42°C) (Israeli et al., 1975). In FD cells, DNA synthesis continued until the completion of one round of replication. There was no synthesis in freeze-dried cells exposed to oxygen (FDO). On the basis of these results, a model for DNA elongation and initiation was proposed: Three factors are necessary for DNA initiation and two of them are required for DNA elongation. One factor is a membranal component, sensitive to oxygen in the dry state (and to colicin E); the second is a temperature-sensitive enzymatic factor (membranal?); and the third is protein-sensitive to chloramphenicol (DNA A product?). When all three are intact, both initiation and elongation occur. When one of the three is damaged, initiation stops, but elongations continues. If two are damaged, both initiation and elongation are stopped (Fig. 6.1).

6.3. Death Mechanisms

Many empiric studies documented the damages to microorganisms during freezing, freeze and thawing, freeze-drying, aerosolization, and desiccation. However, there are not many studies dealing with the mechanisms responsible for these damages. To know how to protect and preserve FD cultures, it is important to understand the mechanisms involved with death during these processes.

6.3.1. Vital Effects of Freeze-drying

Fry and Greaves (1951) concluded that oxygen is "probably harmful to lyophilized bacteria." However, the first to establish the sensitivity of FD bacteria to oxygen were Stark and Herrington (1931). Lion and Bergman (1961) showed that atmospheric oxygen is the lethal factor, attributing it to its paramagnetics, and they tried to explain its action by analyzing protectant properties which they screened. They also postulated that by reaching the dried cell components, oxygen would form free radicals that would cause the damage. Lion and Bergman (1961) proved

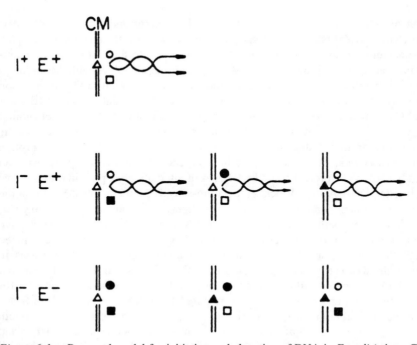

Figure 6.1. Proposed model for initiation and elogation of DNA in *E. coli* (where CM is the cell membrane, I is iniation, E is elongation, the square is a temperature step, the triangle is an oxygen and colicin E₁ step, the circle is a chloramphenical resistant protein; empty symbols are active compounds and the filled symbols are inactive compounds.

the formation of free radicals in oxygen-exposed dried cultures are formed by electron spin resonance signals. Similar results were reported by Heckley et al. (1963). Lion (1963) found a correlation between lyophilization protectants and their properties to prevent free radical formation. Lion (1963) argued that the protectants interfere with the reaction between oxygen and some cellular receptor and that its lethal effect is exerted in the dry state. On the basis of the Lion (1963) results, Gutman (1966) suggested that exposure damage is localized in a quinone group in the reduced state of a membranal NADH dehydrogenase. However, the loss in NADH dehydrogenase enzymatic activity could not account for the high decay of bacteria upon oxygen exposure. Cox and Heckley (1972) established a direct correlation between oxygen concentration and decay rates of FD *S. marcescens*. With respect to oxygen concentration, the toxicity was of first-order kinetic at low concentration and zero order at high concentrations, indicating that the site(s) for O_2 reaction may become saturated. In contrast to previous studies, Cox and Heckley (1972) excluded the involvement of free radicals in oxygen-induced death of bacteria. Israeli (1975) and Israeli et al. (1972, 1974, 1975) conducted extensive studies with FD and FDO *E. coli*, thus locating the site on which oxygen exerts its lethal effect in the cell and establishing the structural and synthetic systems involved with the injuries caused to bacteria during FD and exposure to oxygen.

6.3.1.1. DNA Synthesis and Initiation After FD and FDO

DNA synthesis in reconstituted FD and FDO bacteria progressed at the same rate as in control untreated cultures. In FDO samples, however, DNA synthesis stopped after 90–120 min at a level corresponding to completion of one cycle. Because a similar phenomenon was observed when blocking DNA initiation (by amino acid starvation, chloramphenicol treatment, and growing initiation mutants in nonpermissive temperature), it was postulated that the same mechanism was involved with the loss of colony-forming ability in FDO bacteria. Indeed, in cultures starved for amino acids, the DNA initiation complex was inactive before FD-DNA synthesis progressed normally for a few cycles even in the oxygen-exposure treatment. Additional information about this system was acquired from experiments with thymineless mutants: Upon thymine starvation, this mutant accumulates stable initiation proteins, which enabled a few initiation events even under protein synthesis inhibition. Under these conditions, FDO cells lost their initiation ability, pointing to the membranal site of DNA synthesis initiation complex as the target for oxygen effect. Under the same conditions, DNA synthesis initiation assay in FD cells (grown with thymine starvation) under protein synthesis inhibition, demonstrated no initiation at all, suggesting that the damage to the membranal site involved with DNA initiation is damaged upon freezing per se and can be repaired only in FD cells, a mechanism requiring protein synthesis. Upon exposure to oxygen, the damage becomes irreversible. The reversible damage observed in FD cells which becomes irreversible upon exposure to oxygen was demonstrated in phage productions, ONPG transport and β-galactosidase production.

6.3.1.2. Damage to Bacterial Envelopes During FD and FDO

6.3.1.2.1. Theoretical Background

[After Oseroff (1973) and Lindberg (1977)]. The outer layer of bacteria is the capsule, having some receptors for phages and immunological properties. Under this layer is the cell wall. In Gram-negative bacteria, this layer is built from lipopolysaccharide (LPS) protein and contains phospholipids (Fig. 6.2). The LPS is exposed to the outer surface and contains receptors to phages T_7, T_4, T_3; some of the protein residues are also exposed and absorb phages T_2 and T_6. This layer protects the base layer, built from polysaccharide chains cross-linked by N-acetyl-glucose-amine and N-acetal-muramine. Lysozyme can break the cross-linking bonds only if the LPS layer was damaged chemically by EDTA or physically by freezing and thawing. Another bond, between the lipoprotein and

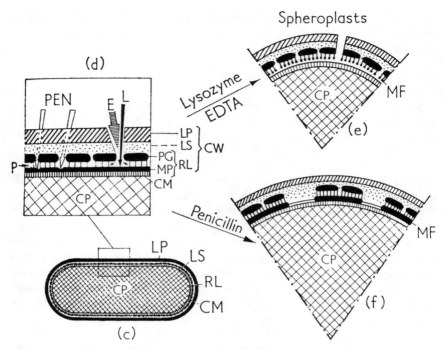

Figure 6.2. Schematic of bacterial envelope [after Martin (1965)] where LP is lipoprotein, LS is lipopolysacharide, BL is basement layer, CW is cell wall, CM is cytoplasmic
membrane, CP is cytoplasm, E is EDTA, L is lysozyme, P is penicillin, PG is protein
granule, and MP mucopeptide.

murein, can be broken by trypsin. The cytoplasmic membrane under the cell
wall, contains 40% lipids, 60% protein, and some carbohydrates. Sections of
the membrane reach the outside space through apertures in the cell wall and
contain some phage receptors (Bayer, 1968). The membrane includes specific
active transport systems for sugars, amino acids, and so on, and controls influx
and efflux of materials. Oxidative phosphorylation and electron transport systems
are also located in the cell membrane. They also have a crucial role in DNA
synthesis control, the septum formation, and cell division.

6.3.1.2.2. Structural Damages

Freezing and thawing render the cell sensitive to lysozyme due to breaks
in the LPS chain (Kohn, 1960), and similar phenomenon was observed after
aerosolization (Hambleton, 1971). In contrast, FD and FDO are stress factors
which act in different sites because they *do not* render the cells sensitive to
lysozyme. It was concluded that cell wall damage, which renders the cell sensitive

to lysozyme, occurs during thawing or dehydration from the fluid state, not during freezing (Israeli, 1975).

Other structural damages were observed in FDO bacteria by electron microscopy. The damages included membrane shrinkage, plasmolysis, and loss of ribosomal material. These damages could be avoided by colicin E_1 absorption before lyophilization (Israeli and Kohn, 1972).

Microscopic observations that demonstrated additional damages were observed in the FD cells, in the cell wall synthesis mechanism, septation, and cell division. Ninety percent of FD cells that lost colony-forming ability could form filaments and synthesize wall materials and DNA, but could not form septa. The same phenomenon was observed in *E. coli* upon depletion of thymine (Reeve et al., 1970) and magnesium (Webb, 1953). Although septation is not the only prerequisite for cell division, (Inouye, 1969), this damage in FD cells may have a predominant effect on the loss of reproduction capacity. Although the mesosome has a crucial role in the control of septation (Jacob et al., 1963; Rogers, 1970; Ellar et al., 1967), it was postulated that initial damage occurs during freeze-drying which intensifies upon exposure to oxygen.

6.3.1.2.3. Membrane Damage in FDO Cells Exposed to Colicin

Colicins, antibiotic proteins produced by colicinogenic strains, can adsorb sensitive strains and kill them. These molecules are attached to membranal receptors and move through the membrane, or induce movement of another component. Tolerant strains, which adsorb colicin molecules but do not die, lack membranal proteins (Fry and Greaves, 1951). A model for colicin action, presented by Changeux and Thiery (1968), included the following points: (a) Colicins are specific constituents of the colicinogenic cell membrane; they have a high affinity for the sensitive cell membranes and are actually structural membranal analogs. (b) Cell membranes have protomer structures, each having receptors for different molecules. Each protomer has two conformational states. (c) Attachment of colicin to the cell membrane changes the protomer conformational state. Due to the cooperative structure of the membrane, the change occurs throughout the membrane. (d) The conformational change causes dissociation of membranal associated systems, like DNA synthesis, oxidative phosphorylation, permeases, and so on. (Fig. 6.3). (e) The conformational change is reversible by trypsin treatment.

This model requires some membranal rigidity and is not compatible with the fluid membrane model suggested by Singer and Nicolson (1972). However, both models may account for the reversible conformational changes at different temperatures, which reversibly affect DNA synthesis initiation. This model may explain the colicin protective effect of FDO bacteria from oxygen damage: Colicin adsorption dissociates the initiation complex from the membrane; this membranal damage by oxygen will leave the initiation complex unharmed; for example, the

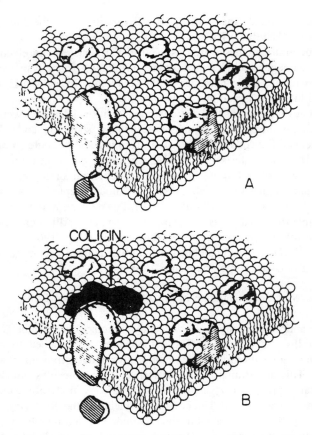

Figure 6.3. Membrane conformational change due to colicin adsorption (redrawn after Singer and Nicolson, copyright 1972 by the Amer. Assoc Adv. Sci.). Note the functional dissociation due to colicin adsorption in the lower figure.

same or close site for the action of both colicin and oxygen. Colicin adsorption will hinder the oxygen effect by blocking the active site, or by inducing a general physico-chemical change which renders the membrane less sensitive to oxygen. Experiments with the *E. coli* strain treated with as low as 100 colicin E_1 molecules per cell before freeze-drying showed the following results (Israeli and Kohn, 1972): the absorption of colicin E_1 to bacteria before lyophilization reduced their decay curves constants during the first hour from 3.70 ± 0.25 to 1.80 ± 0.23, and during the second hour of exposure—from 2.08 ± 0.35 to 0.85 ± 0.20. These results indicate that colicin E^1 probably protected a sensitive site in the cells from oxygen. This protective effect of colicin was specific. The specificity of protection by colicin is supported by the following data: (a) The protein concentration in the lyophilization medium did not exceed 0.1 μg/ml, which is too low for nonspecific protection. (b) After colicin treatment, the cells were washed three times, so only the adsorbed colicin was present in the freeze-drying mixture. (c) Colicin, biologically inactivated by heating, did not adsorb to bacteria, had no protective effect even when the titer of the inactive colicin was 100 times greater than that of native colicin with protective properties. (d) Adsorption of colicin E_1 to a tolerant strain of *E. coli* produced the same protective effect, without loss of viability (observed in the nontolerant strains). The protective effect of colicin E_1 was also demonstrated by minimizing and even eliminating the structural damages to cell's membranes upon exposure to oxygen as described earlier. These results support the prediction that the sensitive site for oxygen is a membranal one associated with the initiation of DNA synthesis.

6.3.1.2.4. Membrane Damage in FD Cells Exposed to Trehalose

Trehalose, an effective cryoprotectant, exerts its effect by stabilizing lipids, proteins, and phospholipid bilayers (see sec 6.3.2). In a recent study (Israeli, unpublished results), it was demonstrated that there was a protective effect of trehalose on FDO *E. coli*. *E. coli* cultures were suspended in 100 m*M* trehalose prior to FD, and death rates were measured upon exposure to oxygen and light. Although the nontreated cultures lost the colony-forming ability in a second-order pattern, practically no decay was observed in either the wild type or genetically engineered strain (GEM) in the trehalose-treated samples. This was true for the oxygen-exposed samples over a wide range of relative humidities (15, 45, 60, and 95%) for at least 3 h after exposure at 20°C. Even after 24 h exposure (at 60% RH and 20°C), a high fraction of the treated cultures survived, that is, 10% of the wild type and 5% of the GEM. No other chemical was found to have such a dramatic protective effect on FD *E. coli*. The trehalose concentration used (100 m*M*) was found to be optimal for protection from death upon exposure of FD cultures (both wild type and GEM). Partial protection was noted upon FD from solution of 10 m*M* and 1 m*M* trehalose in water.

To answer the question whether trehalose protected the FD cells from the inside or outside the cell membrane, the following experiment was performed. *E. coli* W311 DHRα was grown under conditions which prompt trehalose accumulation inside the cell (i.e., at high osmolarity (250 m*M* NaCl) or by adding 100 m*M* trehalose to the growth medium). Another culture was grown in Luria-Bertoli (LB) and resuspended in water with 100 m*M* trehalose prior to FD. Otherwise, all samples were washed and prepared for FD in the same way. Only the sample frozen from 100 m*M* trehalose exhibited the protection from death upon exposure to oxygen. The death rate of the cultures having trehalose inside the cell were not different from that of a control grown in LB from water and frozen. The results clearly indicated that trehalose is a very effective protective chemical for FD *E. coli* and probably other bacteria exposed to environmental conditions (oxygen, visible light at various RHs), suggesting that trehalose exerts its protective effect by stabilizing the liquid-crystalline phase of the outer bacterial membrane, probably by replacing water molecules among the phospholipids. The best protection is achieved when trehalose molecules saturate all the inter-phospholipid spaces.

6.3.1.2.5. Membrane Active Transport Systems Malfunction in FDO Cells

Because structural damage was demonstrated in membranes of FDO bacteria and damage to the membrane sites involved in DNA initiation was established, the biological activity of the membrane active transport of β-galactoside (ONPG), amino acid, and potassium ion was investigated more intensely.

6.3.1.2.5.1. ONPG transport. When ONPG is introduced to bacteria containing β-galactosidase (induced or constitutive), the compound is actively transported through the membrane by a permease. ONPG is degraded to galactose and ONP which passively diffuses out to the medium. The inward transport is the rate-limiting step for these chain reactions and reflects the integrity of the membrane, both from structural and biological activity points of view. Israeli and Kohn (1972) showed that in control bacteria, the rate of ONPG uptake was slow ($K = 0.2$), whereas in the FD and FDO culture immediately after reconstitution, the uptake rates were five to six times greater ($K = 1.0$ and $K = 1.2$), respectively. After 3 h of incubation of the reconstituted FD and FDO samples, significant changes in the rates of uptake of ONPG were observed. In the FD samples, the uptake rate had been decreased ($K = 0.6$). In the FDO bacteria, the uptake rate had risen ($K = 2.16$) The data indicated that some membrane damage had occurred during freeze-drying, but that this damage would be repaired upon incubation of the reconstituted bacteria. However, oxygen exposure intensified the injury, making it irreversible; thus there was no repair after incubation. In the FDO bacteria, the toluene treatment did not increase the ONPG uptake, as would be expected in a maximally damaged membrane.

6.3.1.2.5.2. Potassium transport. The membranal sodium-potassium pump in the cell membrane is dependent on energy-producing processes. By studying this system, additional information as to the role of the membrane in the damage occurring during freeze-drying and oxygen exposure was obtained. *E. coli* B cells were loaded with ^{42}K to a level of isotropic equilibrium and then freeze-dried and exposed to oxygen. After freezing alone, the ^{42}K content in the thawed cells dropped to a value corresponding to 25% of the untreated samples. Similar values (the same 25%) were also obtained for FD and FDO bacteria. It seems, therefore, that during all these procedures, probably during the first step (i.e., freezing), membrane damage had occurred, permitting leakage of ^{42}K from the cells.

In another type of experiment, the uptake of ^{42}K was measured in bacteria after freeze-drying. The rate of influx and intracellular concentration of ^{42}K was similar in the control bacteria and in frozen and thawed bacteria. This result indicated that if there was damage done by freezing and thawing, it was repaired very rapidly. In FD bacteria, however, the influx was reduced to about 50% of the control value. In FDO bacteria, the uptake was further reduced to 30% of the control. Incubation of the reconstituted FDO cultures in a growth medium at 37°C for 90 min before the addition of ^{42}K further diminished the influx and the potassium concentration in the cells, reaching a level of only 20% of the control.

6.3.1.4. Functional Complex Synthetic Systems in FDO Cells

To get a comprehensive picture of synthetic mechanisms in FDO cells, two systems were studied: the β-galactosidase system, which includes the following subsystems. One system was inducer transport through the membrane, induction of the β-galactosidase operon, transcription of DNA to mRNA, and translation of mRNA to protein. The other system was bacteriophage production, which includes a few mechanisms as well: adsorption of phage to membranal receptors, penetration of phage DNA into the cytoplasm, and again transcription of (phage) DNA, protein synthesis, and packaging.

6.3.1.4.1. Induction of β-Galactosidase in FDO Cells

Novick et al., (1972) were the first to study this system in FD *E. coli.* In these experiments, the inducer (IPTG) was added to samples at the beginning of the experiment, and enzymatic activity was determined in different time intervals. The activity measured reflected cumulative enzymatic concentration while the inducer was present throughout the experiment. The results indicated that lyophilized and FDO cells could be induced for β-galactosidase production. In another set of experiments, FD and FDO bacteria were preincubated for various time intervals and the inducer was added for 10 min, the reaction was stopped, and activity of β-galactosidase was measured. In the FD samples, a full repair was observed in this system after 90 min of preincubation. In FDO samples, however, only partial repair was noted at a level of 10% of the control cells.

This system provides another indication of the damage occurring during FD, which may be repaired upon preincubation, but becomes irreversible upon exposure to oxygen. The synthetic systems can function after FD or FDO and are not involved with the death mechanism. It was suggested that the impairment in this system is due to membranal damage, which impairs cell capability to concentrate the inducer, which effects the induction mechanism and enzyme production.

6.3.1.4.2. Bacteriophage Production

Freeze-dried *E. coli* was studied by Israeli and Shapira (1973) for its ability to produce phage. Although FD and FDO cultures lost their colony-forming ability with exposure time, they kept their phage production ability at a high level (10% of the control untreated cells) in all samples, regardless of exposure time. There was no difference in burst size between FD and FDO cells and the control. Again, a repair mechanism was active in FD cells (but not in FDO samples) after 60 min incubation in medium prior to phage production assay. It was concluded that the macromolecular synthetic systems (DNA, mRNA, and protein) were only partially damaged during freeze-drying or exposure to oxygen, which could not account for the loss of viability. Because the introduction of an external control mechanism over these systems resulted in a normal function, it was concluded that the damage incurred by oxygen is located in the control mechanism of DNA synthesis. This site is probably membranal, and the damage impairs membranal ability to control cell wall growth as well as influx and efflux of metabolites. The damaged site is involved with the initiation of DNA synthesis, but not macromolecules synthesis per se, which stops DNA synthesis after completion of one round of replication; hence, cell division and colony formation is impaired.

6.3.2. Freezing and Aerosolization

Freezing, aerosolization, freeze-drying, and desiccation are all processes in which water is removed from the system. These processes are all stress factors for microorganisms, causing damages to cellular systems, which can lead to the death of cells or can be partially or fully repaired under certain resuscitation conditions. Following the difference and similarity between freeze-drying, freezing and aerosolization are discussed.

6.3.2.1. Freezing

To study frozen cells, either for viability or metabolic follow-up, they must be thawed. Therefore, freezing and thawing (FT) are inseparable processes. However, freeze-drying is a process in which the thawing stage is eliminated but is replaced by rehydration upon resuspension. Therefore, although these two processes have a common stress factor—freezing—each includes additional different stress factors—thawing or rehydration—which may result in different cell damage.

Nei et al. (1968) studied the effects of freezing and thawing on *E. coli* and suggested that intracellular water crystal formation might cause cell death. This process was dependent on the ratio between dehydration rate upon freezing and membranal water permeability. High permeability and slow cooling rate resulted in high viability. In contrast, a high rate of cooling upon freezing for FD results in better survival of microorganisms in the dry state. Speck and Cowman (1968) demonstrated that freezing and thawing results in leakage of ultraviolet absorbing metabolites and changes membranal permeability. Some enzymes (proteinase) lost activity, probably due to aggregation changes from the monomer to the polymer state. FT damages lysozomic membranes, thus activating hydrolytic enzymes, which may confer further damage [reviewed by Tappel (1966)]. Bacteria are also rendered sensitive to lysozyme (Kohn 1960) after FT. Israeli (1975) indicated that FD and FDO bacteria were not sensitive to lysozyme. Another difference is potassium leakage, which can be repaired after FT but not after FD. These results suggest that the damage associated with membrane malfunction occurs during freezing, whereas damage to cell wall occurs only during thawing. Crow et al. (1990) is in agreement with Israeli's conclusion and concluded that "freezing and dehydration must be regarded as fundamentally different stress vectors." Crowe summarized the evidence as follows:

(a) Proteins, membranes, phospholipids, and living cells and organisms all contain about 0.25 g nonfreezable H_2O/g dry weight. By definition, this water is not removed by freezing; (b) Dehydration, by contrast with freezing, may remove the nonfreezable water. Removing this water results in profound changes in the physical properties of biomolecules, particularly phospholipids and proteins; (c) The specificity for solute requirements for stabilization of proteins during freezing is low; any solute that is preferentially excluded from the hydration shell of a protein is also a cryoprotectant; (d) By contrast, stabilization of proteins during drying requires direct interaction between the stabilizing molecule and the protein, probably involving hydrogen bonding between the stabilizer and polar residues in the protein. The specificity is very high in this case; only carbohydrates are effective, and of those that have been tested, trehalose is the most effective; and (e) Less is understood about the mechanism of stabilization of phospholipid bilayers during freezing, but it is clear that while many solutes will preserve liposomes during freezing, only a few (of which trehalose is the most effective) will preserve them during drying. Stabilization of bilayers during drying requires direct interaction between the sugar and polar head groups of the phospholipids."

The proposed mechanism by which trehalose stabilizes dry bilayers is by inhibiting fusion during drying (as do at least some other solutes). However, this effect may not be sufficient to effect maximal preservation. Trehalose and, to some extent, other disaccharides have the ability to depress T_m enormously in dry phospholipids. Crowe et al. (1990) have suggested that this effect permits dry phospholipids to exist in liquid-crystalline phase at temperatures at which they would normally be in gel phase. As a result, the gel to liquid-crystalline phase transition that would accompany rehydration is avoided (Fig. 6.4). It follows that a phospholipid that remains below T_m while it is hydrated and dry should be stabilized maximally by inhibiting fusion alone. The group extended this idea to intact cells (Crowe et al., 1989) which is in agreement with the

(Israeli, et al., 1974) suggestion that stabilization of cell membranes during FD and FDO results with higher survival rates (Fig. 6.4).

The same mechanism might be responsible for the protective effects of some disaccharides on desiccated and spray-dried cells after chilling of plants as well as of homeothropic animals, under conditions which cause, in the lipid components of the membrane, a phase change from a liquid-crystalline to a gel structure, also described by Lyons (1972). As mentioned above, oxygen was not involved in damage occurring during freezing or thawing. Survival of frozen bacteria was not influenced by exposure to oxygen (Cox and Heckly, 1972).

6.3.2.2. Aerosolization

Anderson and Cox (1967) discussed the similarity in the damage to bacteria upon FD or aerosolization. The results of Benbough et al. (1972) showed enhancement in the membrane permeability to ONP in spray-dried *E. coli* correlated with Israeli et al. (1974) described earlier. Another similarity in the damage to bacteria using FD or aerosolization already mentioned is demonstrated by the observed leakage of potassium in both processes (Anderson et al., 1968). However, there is a difference between these two processes when lysozyme sensitivity (cell wall damage) is present. Aerosolized *E. coli* becomes sensitive (same as freeze and thawed cells), whereas FD cells do not. The observed damage can be repaired (energy dependent) and can be mitigated by certain bivalent and trivalent cations (Hambleton, 1971). The damages to aerosolized bacteria were summarized by Cox (1987) as follows:

> *E. coli* strains following aerosolization into inert atmospheres suffer damage to their surface structures. This damage results in leakage of ions, reduced DNA, RNA and protein synthesis, impaired active transport and greatly decreased oxygen consumption. Surface damage arising through denaturation, possibly of proteins or lipoproteins occurs to the greatest extent at high RH. Of the biochemical failures-loss of ability to utilize oxygen seems to be the most marked and the most fundamental, because cells which cannot produce energy cannot possibly divide to form colonies.
>
> The ability of *E. coli* to synthesize β-galactosidase in the presence of a specific inducer is considerably impaired in the period immediately following collection from the aerosol. But the loss of this ability precedes loss of ability to form colonies. The active transport mechanism for the uptake of inducer is impaired by aerosolization rather than an impaired ability to synthesize β-galactosidase (Benbough et al., 1972). Loss of potassium also occurs, but is not in itself a lethal event. On the other hand, the greatest breakdown of RNA in *E. coli* K12 HfrC (collected from aerosols stored in nitrogen) is associated with lowest survival. This breakdown possibly results from the loss of moieties necessary for RNA stability as could the reduced ability to synthesize RNA and protein which parallels loss of viability. As well as these defects in metabo-

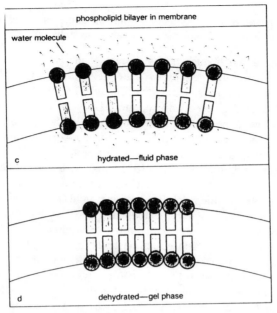

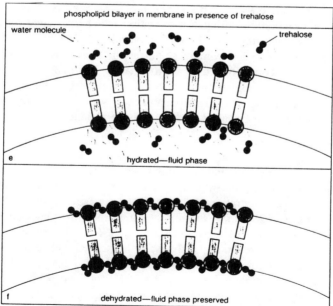

Figure 6.4. Diagram showing preservation of the membranal fluid phase of phospholipid bilayer by trehalose upon dehydration.

lism, DNA synthesis is slightly reduced while the ability of *E. coli* strains to utilize oxygen is markedly affected and follows the ability to form colonies.

A different type of biochemical study (Straat et al., 1977) found that under certain conditions *Serratia marcescens* can metabolize and divide in aerosol droplets. When aerosolized into air at 95% RH and 30°C, these bacteria produced $^{14}CO_2$ from ^{14}C-labeled glucose, incorporated ^3H-thymidine into DNA, and divided for at least two generations. Such behavior is likely also for other bacteria when aerosolized under conditions conducive for heir survival and cell division, for example, being sprayed from suspensions of protecting additives and stored at very high relative humidity. Cox (1987) suggested that these results, and those that demonstrated very significant enzyme activity in solutions of high concentration of solutes which stabilize proteins, indicated that airborne bacteria should not be considered inactive metabolically and may have implications for the aerosol dispersal of microorganisms in the environment.

The protective effect of trehalose, demonstrated in frozen, FD, and FDO cells was observed in aerosolized *E. coli* as well (Israeli, unpublished results). The effect was less dramatic than in FDO *E. coli* but was significant. When sprayed in 0.1 m*M* trehalose solution into a 15% RH and 20°C atmosphere, death rates of two strains of *E. coli* were two to three times slower than those of cultures sprayed from distilled water. This partial protection may be explained by the high rate of evaporation during aerosolization which prevented the trehalose molecules from replacing enough water molecules among the cell membrane phospholipids to achieve maximum stabilization. The trehalose concentration applied (0.1 m*M*) corresponded to approximately 100 m*M* in the evaporated aerosol droplet. Using higher concentrations of trehalose in the spray medium resulted in higher death rates. This phenomenon was attributed to cell damage due to crystalization of trehalose in the evaporate residue.

6.3.3. Oxygen

Some aerosolized microorganisms exhibit sensitivity to oxygen, same as FD bacteria. Oxygen susceptibility usually increases with degree of dehydration, increased oxygen concentration, and time. However, at RH above 70%, survival in air, in nitrogen +20% (v/v) oxygen, and in nitrogen was about the same for wet disseminated *E. coli* B (Cox and Baldwin, 1967). These results suggested that the cells have to lose most of the water associated with them in order for oxygen to exert its toxicity. Cox (1970) also showed that dry disseminated bacteria while susceptible to oxygen at low RH, survived better in air than in nitrogen atmospheres, at high RH; again suggesting that the death mechanism, at least at low RH, is due to oxygen effect both in spray-dried and freeze-dried bacteria, whereas at high RH, different mechanisms are involved. Cells, which lost their colony-forming ability, after spraying into air at low RH, retained their ability to support replication of phage T_6. This result is very similar to those obtained for FDO cells, as described earlier. Again, as in FDO cells, the most crucial metabolic pathways are not impaired (DNA, RNA, and protein synthesis) and can function when an external control mechanism (phage DNA) is introduced into them. Although it was not determined specifically, the most obvious sites for oxygen-induced damage (in *E. coli*) are DNA synthesis directed by *E. coli* DNA, cell wall synthesis, and cell division. A marked resemblance to these incurred in FDO cells. DNA synthesis was found indeed to be impaired in *E. coli* recovered from aerosols stored at low RH (Cox et. al., 1971; Israeli, 1973), however the residual synthesis could not be accounted for by the surviving fraction. Therefore, some DNA synthesis is occurring in the non-colony-forming cells. In contrast to FDO cells, in aerosolized cells the β-galactosidase induction system was damaged beyond repair (Israeli, 1973). The very low detectable enzyme activity was ascribed to a small percentage of colony formers present in the suspension. Benbough et al. (1972) attributed this observed damage to impairment of the active transport system of the inducer. Again, this result is in good correlation with those observed for FDO cells. Other damage to the envelopes of bacteria recovered from aerosols were reported by Hambleton and Benbough (1973) and Benbough and Hambleton (1973). They described the susceptibility of these cells to lysozyme, which correlated with their surviving fraction. They suggested that these damages, while not lethal, are related to the subsequent longer-term survival of airborne microbes. Aerosolized bacteria with wall damage may be rendered susceptible to other stresses which damage essential biochemical functions. For example, those associated with the cytoplasmic membrane or essential repair mechanisms. Also, a loss of wall components might affect the infectivity of pathogenic bacteria. Viable airborne pathogens with an altered wall structure may either be more susceptible to the hydrolytic enzymes of the host's defense mechanisms or have immunological characteristics significantly different from those of unaerosolized bacteria. Storage for 1 s at different relative humidities diminished the active transport of α-methyl glucoside and galactosides. The additional loss of control of potassium ions (Anderson and Dark, 1967), which are essential co-factors in the phosphoenol pyruvate-phospho-transferase system, may have contributed to the loss of other materials might be involved because transport activity could be partially restored by exudate from aerosolized bacteria, whereas storage of aerosols for a few seconds caused membranes to become leaky yet repairable. Storage for longer periods led to additional inactivation of membranes of dehydrated microorganisms. Mackey (1984) attributed such viability loss to amino-carbonyl reactions between cell membrane protein molecules and cell membrane reducing sugar molecules. Another similarity between aerosolized cells and FD ones can be pointed out with respect to repair mechanisms. Repair of damages in FD cells was observed in some systems (β-galactosidase production, cell-wall, and membranal damages and phage production), the same as in cells recovered from aerosols after a few seconds of storage. However, in both methods of drying, prolonged exposure to oxygen made the damages irreversible and more pronounced.

6.4. Conclusion

Freezing, freeze-drying, and spray-drying are common methods for preservation of biological materials for storage or as a preliminary step for application in an environmental setting. During these processes, water is removed from the materials, which transform living cells from active to resting state. Biomembranes, as phospholipid bilayers, in general go through conformational change from liquid-crystalline to the gel phase. These stress factors cause damages to cells and proteins, resulting in loss of activity or viability. The common denominator for all these processes is water loss. Although the outcome is also similar, the loss of viability, different mechanisms might be involved with the observed "death" of microorganisms. Yet, similar effects might be observed for either two or all three processes.

The future use of genetically engineered microorganisms will involve spray release, either from a dry state or from liquid suspensions. The forming bioaerosols are supposed to cover large areas of vegetation, soil, or water bodies. The effectiveness of these operations, as well as their safety, are dependent on the survival of these organisms in the environment. Higher survival rates will lead to coverage of large areas and may cause overspill from the intended site. In some cases, lower survival rates might be advantageous; in others, high survival and stability are required. In each case, understanding the mechanisms involved with bioaerosol decay is important, both for risk assessment and for developing methods to protect the cells from dying and/or to limit their multiplication and spread in the environment. The evidence presented in this chapter as to the death mechanism involved with FD and FDO aerosolized bacteria can be summarized as follows (see Table 6.1):

1. DNA synthesis proceeded normally in FD bacteria, but in cells exposed to oxygen this synthesis halted after a short time, at a level corresponding to completion of the ongoing cycle of DNA elongation process. (The cause of cessation of DNA synthesis in FD and oxygen-exposed cells was a damage to the initiation mechanism and not to the DNA elongation process.) Damage to DNA synthesis mechanism was observed in aerosolized cells as well.

2. Adsorption of colicin E_1 to bacteria before FD reduced the decay constants of oxygen exposed cells by a factor of 2, and minimized structural damages in the cells envelopes, caused by oxygen exposure demonstrated by electron microscopy. This led to the assumption that oxygen exerts its effect on a membranal site which is identical, or closely related, to the one to which colicin E_1 adsorbs, and which is linked with the control over DNA synthesis.

Table 6.1. *Damage comparison between freeze-dried (FD) and FD oxygen-exposed (FDO), frozen and thawed (FT), and spray-dried (SD) bacteria*

	Observed Damage				
	Freeze-dried			SD	
System Studied	FD	FDO	FT	Short Storage	Long Storage
1. Cell wall (lysozyme sensitivity)	−	−	+(R)[a]	+(R)	+(R)
2. Membrane					
A. K$^+$ leakage	±(R)	+(Rp)[b]	+	+(R)	+
B. Active transport	?	+	?	+	+
3. DNA synthesis	−	+[c]	?	?	±[c]
4. β-galactosidase production	−	−	?	?	+(R)
5. Phage production	±(R)	+	?	?	−
6. Viability					
A. Oxygen effect	NA	+	−	?	+[d]
B. Dissacharide effect	±	±	±		±

[a]Repairable after incubation.

[b]Partially repairable after incubation.

[c]Initiation.

[d]At low RH.

3. FD and oxygen exposure affected the performance of the cytoplasmic membrane: It injured the sodium-potassium pump, with the resulting leakage of potassium ions from the cells. This damage was found both in FT cells and in spray-dried ones. However, oxygen is not involved with the damages to frozen cells. The initial damage to bacteria after FD, which affected the ability of the cells to take up and to concentrate potassium, was intensified upon oxygen exposure.

4. Membrane performance in controlling uptake and efflux of materials of low molecular weight (amino acid and glucoside) was partially damaged in FD bacteria; the same was found in aerosolized cultures. Also, in this system, an initial reversible damage was found after FD intensified and became irreversible after exposure to oxygen.

5. Oxygen exposure did not render the bacteria sensitive to lysozyme and, hence, did not injure the lipopolysaccharide complex in the cell wall, damages that were observed after FT and in spray-dried cell. However, in light and electron microscopic examination, structural damages to the cell wall and membrane in some of the FD cells after exposure to oxygen were observed. Part of the FD bacteria lost the ability to septate and formed filamentous forms without dividing. In cells exposed to oxygen, a marked plasmolysis, shrinkage of the membrane, and loss of ribosomal material were observed.

6. Freeze-drying cells from trehalose solution enhanced survival upon exposure to environmental conditions of oxygen, light, and/or RH. The effect on these same cells sprayed from a solution only partially enhanced survival. Trehalose exerts its effect by substituting water molecules among the phospholipids in the bacterial outer membrane. Survival was best upon saturation of all phospholipid molecules, thus preserving the liquid-crystalline phase of the membrane upon rehydration.

7. Studying the induction of β-galactosidase in FD bacteria showed that the macromolecular synthetic mechanism was not damaged during FD or oxygen exposure. However, damage to the induction system was observed. This malfunction may be due to the inability of he membrane to concentrate the inducer. A similar phenomenon was observed in bacteria collected from aerosols.

A repair mechanism was active in this system upon incubation in growth medium. The repair was complete in FD bacteria and partial in oxygen-exposed ones.

By studying phage production by FD and FDO cells, it was demonstrated that the mechanisms of macromolecules synthesis in FD bacteria were not damaged during FD or oxygen exposure. The injury therefore, must be, in the control mechanisms over these systems: When an external control mechanism provided by the phage was introduced to the cells, these synthetic mechanisms proceeded normally.

The phenomenon was finding that the partial damage exhibited after FD was reversible and could be repaired upon incubation, whereas prolonged exposure to oxygen made the damage irreversible and more intense. This system was also functional in spray-dried bacteria.

The final conclusions of this chapter are as follows: During freeze-drying of bacteria there occurs an initial reversible damage; this damage is membrane-linked and connected with the control over macromolecular synthesis and cell division.

In the dried state, the bacterial membrane is rendered sensitive to oxygen. Oxygen exposure intensifies the damages caused during FD and makes them irreversible.

Upon FD or spray-drying, trehalose can substitute water molecules in the membranal interphospholipid spaces, thus preserving the liquid-crystalline phase of the membrane, which protects it from the deleterious effect of oxygen.

The membranal site on which oxygen exerts its action is linked with the control mechanism over the initiation of DNA synthesis and is probably related to the receptor to colicin E_1.

Following the arrest in initiation of DNA synthesis, the oxygen-exposed bacteria lose their ability to reproduce and to form colonies and are, therefore, considered to be "dead" although the synthetic mechanism in them may still be unharmed and active.

6.5. Appendix: Freeze-drying

6.5.1. Freeze-drying Concept

Freeze-drying is a process where biological materials are frozen at a very low temperature followed by a rapid dehydration by sublimation under a high vacuum.

There are many advantages of FD over other drying methods such as desiccation or aerosol spray-drying. For example, during desiccation, with the declining of water content, the cell surface is subjected to forces that cause shrinkage and deformation. These forces in FD are minimal. Also, during desiccation, a higher concentration of solutes, such as enzymes and electrolytes, form to induce changes in biochemical systems. This process in FD is much less pronounced due to lower temperatures slowing down the biochemical reactions. After desiccation, resuspension is problematic due to the smaller surface area. In contrast, the FD materials exhibit a larger surface area due to the spaces created in place of the evaporated water crystals facilitating resuspension.

In addition, the apparatus used in aerosol spray-drying is complicated, the quality of the product is hard to control, and the cell decay rate is higher. Again, in contrast, the FD process produces large amounts of biological materials while conserving the viability for long periods when stored under a high vacuum.

Further, the FD process is also used to assess environmental risks from the dispersal of genetically engineered microorganisms (GEMS) to the environment.

6.5.2. Freeze-drying Systems

FD systems require the following: (a) A **vacuum chamber**, which is usually stainless-steel walled, in or to which the frozen containers are connected under high vacuum. It is important to shorten the distance from the material to the condenser. (b) A **freezing apparatus**, where biological materials may be frozen by dipping in super-cooled liquids (e.g., liquid nitrogen, methanol–dry ice mixture, etc.). Alternatively, biological materials may be frozen in the vacuum chamber by means of circulating freezing fluid (Freon). Freezing should be fast for ensuring high viability of biological: (c) A **vacuum pump**, which creates a final pressure in the system of the order of 1 μHg. Two-stage vacuum pumps or diffusion pumps are required. Because it is sometimes necessary to control the vacuum, it is monitored throughout the FD process: (d) A **Condenser and desiccant** to trap water vapor in a chemical desiccant, such as P_2O_5, $Ca_3 (PO_4)_2$, or a very low-temperature surface. The condenser is cooled by a cooling mixture.

6.5.3. Factors Affecting the Viability of FD Microorganisms

There are five stages in the FD process: preparation, freezing, drying, storage, and resuspension. The microorganisms endure different stresses during each stage and are exposed to parameters that may cause damage, resulting in loss of viability.

6.5.3.1. Preparation

Before freezing, cultures are washed in different solutions, concentrated, usually by centrifugation, and additives may then be added. The cultures are then kept for different periods, usually at 4°C. The handling of cultures during this stage may affect viability after drying.

6.5.3.2. Freezing

Two parameters that may affect cell viability during freezing are the freezing rate and suspension additives. Damage during slow freezing due to concentrating solutes mimic the damages during desiccation. Although fast-freezing may cause intracellular crystals to damage cellular structures, it is usually the preferred method for microorganisms. Additives such as sugars (lactose, glucose, trehalose) and proteins protect cells from freezing and FD damages.

6.5.3.3. Drying

Usually airborne microorganisms (like Gram-positive micrococci) are more stable to FD than Gram-negative bacteria. The growth phase is also a factor. Log-phase cells are more sensitive to FD than stationary cultures. Additives protect microorganisms from decay during drying the process, desiccation, spraying, or FD. Some of the protectants are thiourea, mono- or di-saccharides and their derivatives, and simple inorganic salts. Ambient temperature may also affect survival. At the beginning of the FD process, water evaporation causes a drop in temperature under the freezing point, and heat is supplied to the material to enhance drying. Later in the process, the material reaches room temperature, causing higher decay rate.

6.5.3.4. Storage

A few parameters affecting survival upon storage are gaseous atmosphere, temperature, light, residual moisture, and relative humidity.

6.5.3.4.1. Atmosphere

Usually FD biological materials are kept in vacuuo for better preservation. But nitrogen or inert gases are also applicable. Heckly et al. (1963) demonstrated a correlation between microbial death and formation of free radicals by electron spin resonance (ESR). Cox and Heckly (1972) demonstrated correlation between oxygen concentration and survival of FD S. marcescens. Oxygen proved to be toxic and kinetic (with respect to O_2 concentration) were first order at low concentration and zero order at high O_2 concentration. These results indicated that the site for O_2 action may become saturated. Free radical studies showed that they were not involved in causing oxygen-induced loss of viability (Cox and Heckly, 1972). Cox et al. (1973) also reported a mathematical expression for oxygen-induced death in dehydrated bacteria. Both ESR and death rate were diminished upon storage *in vacuuo*. Lion et al. (1961) found strong ESR in *E.coli* FD from water upon exposure to oxygen. Additives (thiourea, NaI) protected the cells from the lethal effect of oxygen and diminished ESR.

6.5.3.4.2. Temperature

Usually low storage temperature is better for microorganism viability in the dried state. However, under high vacuum, FD cultures may survive for years at room temperature.

6.5.3.4.3. Light

Usually FD cultures are kept in the dark. Webb (1963) studied the light effects on *Serratia marcascens* FD from water and found that the lethal effect of light depends on the RH, as long as light intensity is not too high. Fry and Greaves (1951) showed that FD *E. coli* D 201H viability exposed in sun light was the same as in dark storage over a 24-h period. But after longer periods, light exposure enhanced decay. Recent results by Israeli et al. (1992) demonstrated that FD *E. coli* strains were sensitive to visible light. FD cultures were exposed *in vacuuo* to the visible light spectrum, whereas the UV effect was eliminated by filters. A first-order death rate was observed during a 3-h exposure period. Upon exposure to a combination of light and oxygen, death rates were higher than those observed for either oxygen *or* light exposure alone. These results suggest a cooperative damage exerted by the two parameters.

6.5.3.4.4. Residual Humidity

Cultures should not be overdried. Several studies indicated that a residual humidity of 0.2%–0.6% was the most beneficial to microorganism survival after FD (Hutton et al., 1951; Maister et al., 1958). The optimal residual humidity is dependent on the species and should be between 0.5% and 1.5%.

6.5.3.4.5. Relative Humidity (RH)

RH in the exposure atmosphere is a major factor affecting survival in spray-dried and FD cells. Different mechanisms are involved with the damages at low and high relative humidities. At low RH oxygen-induced damage is involved in the death mechanism of bacteria (Cox et al., 1973); however, death mechanisms at high RH is not completely established.

6.5.3.4.6. Resuspension

Addition of water to freeze-dried cultures may form concentration gradients, may have a higher rate of solutes transported through membranes, and may experience a temporary osmotic imbalance, thus ending up with a final solution completely different from the original one. For certain microorganisms, the slow addition of water results in higher viability (Leach et al., 1959). It is also essential for yeasts and phage T_4 (Shapira 1974). However, it is not an important parameter for *E. coli* or *S. marcescence*. The temperature of the suspending medium can also affect the survival of FD bacteria. Israeli (unpublished results) demonstrated a direct correlation between resuspension temperature and survival of FD *E. coli* strains. Survival was measured as percentage of CFU in unexposed FD samples from the counts before FD. Whereas survival percentage was around 3 at 4°C, it was twice as high at 10°C, and threefold at 22°C. Freezing from trehalose suspension enhanced survival at the above combinations 7- and 12-fold, respectively.

References

Anderson, J. D., and C. S. Cox. 1967. Microbial survival. Pp. 203–226. In Airborne microbes, Cambridge University Press, Cambridge.

Anderson, J. D., F. A. Dark, and S. Peta. 1968. The effect of aerosolization upon survival and potassium retention by various bacteria. J. Gen. Microbiol. 52:99–105.

Anderson, J. D., and F. A. Dark. 1967. Studies on the effects of aerosolization on the rates of efflux of ions from populations of *Escherichia coli* strain B. J. Gen. Microbiol. 46:95–105.

Bayer, M. E. 1968. Adsorption of bacteriophages to adhesions between wall and membrane of *Escherichia coli*. J. Virol. 2:346–356.

Benbough, J. E., and P. Hambleton. 1973. Structural organizational and functional changes associated with envelopes of bacteria sampled from aerosols. Pp. 134–137. In J. F. Ph. Hels and K. C. Winkler (eds.), Fourth International Symposium on Aerobiology. Oosthoek, Utrecht, The Netherlands.

Benbough, J. E., P. Thimbleton, K. L. Martin, and R. E. Strange. 1972. Effect of aerosolization on transport of alfa-methyl glucoside and galactosides in *E. coli*. J. Gen. Microbiol. 72:511–520.

Bremer, H., and G. Churchward. 1985. Initiation of chromosome replication in *Escherichia coli* after induction of dnaA gene expression from a lac promoter. J. Bacteriol. 164:922–924.

Bremer, H., and G. Churchward, 1991. Control of cyclic chromosome replication in *Escherichia coli*. Microbiol. Rev. 55:459–475.

Changeux, J. P., and J. Thiery, 1968. On the mode of action of colicins: A model of regulation at membrane level. J. Theoret. Biol. 17:315–318.

Cox, C. S. 1970. Aerosol survival of *E. Coli* B disseminated from the dry state. Appl. Microbiol. 19:604–606.

Cox, C. S. 1987. The aerobiological pathway of microorganisms. John Wiley & Sons. Chichester.

Cox, C. S., and F. Baldwin. 1967. The toxic effect of oxygen upon the aerosol survival of *Escherichia coli* B. J. Gen. Microbiol. 49:115–116.

Cox, C. S., J. Baxter, and B. J. Maidment. 1973. A mathematical expression of oxygen-induced death in dehydrated bacteria. J. Gen. Microbiol. 75:179–185.

Cox, C. S., and R. Heckly. 1972. Effects of oxygen upon freeze-dried and freeze-thawed bacteria: viability and free radical studies. Can. J. Microbiol. 19:189–194.

Cox, C. S., M. C. Bondurant, and M. T. Hatch. 1971. Effects of oxygen on aerosol survival of radiation sensitive and resistant strains of *E. coli* B. J. Hyg. (Camb.) 69:661–672.

Crowe, J. H., J. F. Carpenter, L. M. Crow, and T. J. Anchordoguy. 1990. Are freezing and dehydration similar stress vectors? A comparison of modes of interaction of stabilizing solutes with biomolecules. Cryobiology 27:219–231.

Crowe, J. H., F. A. Hoekstra, and L.M. Crowe. 1989. Membrane phase transitions are responsible for imbibitional damage in dry pollen. Proc. Nat. Acad. Sci. USA 86:520–523.

Ellar, O. J., O. G. Lundgren, and R.A. Slepecky. 1966. Fine structural bacillus megaterium during synchronous growth. 76:1189–1205.

Fry, R. M., and R. I. N. Greaves. 1951. The survival of bacteria during and after drying. J. Hyg. Camb. 220–243.

Guttman, M. 1966. Ph.D. Thesis. Hebrew University, Jerusalem.

Hambleton, P. 1971. Repair of wall damage in *Escherichia coli* recovered from aerosol. J. Gen. Microbiol. 69:81–88.

Hambleton, P. and J. E. Benbough. 1973. Damage to the envelopes of Gram-negative bacteria recovered from aerosols. Pp. 131–134. In J. F. Ph. Hels and K. C. Winkler (eds.), Fourth International Symposium on Aerobiology. Oosthoek, Utrecht, The Netherlands.

Hanna, M. H., and P. L. Carl. 1975. Reinitation of deoxyribonucleic acid synthesis by deoxyribonucleic acid initiation mutants of *Escherichia coli*: role of ribonucleic acid synthesis, protein synthesis and cell division. J. Bacteriol. 121:219–226.

Heckly, R. J., A. L. Dimmick, and J. J. Windle, 1963. Free radical formation and survival of lyophilized microorganisms. J. Bacteriol. 85:961–966.

Hutton, R. S., R. J. Hilmoe, and J. L. Roberts. 1951. Some physical factors that influence the survival of *Brucella abortus* during freeze-drying. J. Bacteriol. 61:309–319.

Israeli, E., E. Giberman, and A. Kohn, 1974. Membrane malfunction in freeze-dried *Escherichia coli*. Cryobiology 11:473–476.

Israeli, E., and A. Kohn. 1972. Protection of lyophilized *E. coli* from oxygen by colicin E treatment. FEBS Lett. 26:323–326.

Israeli, E., A. Kohn, and J. Gitleman. 1975. The molecular nature of damage by oxygen to freeze-dried *Escherichia coli*. Cryobiology 12:15–25.

Israeli, E., and A. Shapira, 1973. Production of bacteriophage by lyophilized and oxygen exposed *E. coli*. J. Gen. Microbiol. 79:159–161.

Israeli, E. 1975. Ph.D. Thesis, Tel Aviv University.

Israeli, E. 1973. Effects of aerosolization and lyophilization on macromolecular synthesis in *E. coli*. Pp. 110–113. *In* J. F. Ph.Hels and K. C. Winkler (eds.), Fourth International Symposium on Aerobiology. Oosthoek, Utrecht, The Netherlands.

Israeli, E., B. T. Shaffer, J. P. Hoyt, B. Lighthart, and L. M. Ganio. 1992. Survival of freeze-dried genetically engineered microorganisms in air and the effect of visible light. ASM 92nd General Meeting, New Orleans, LA.

Jacob, F., L. L. Simionovich, E. Wolman. 1952. Sur La Biosynothose D'une Colicin Et Sur Son Mode D'Action, Ann. Inst. Pasteur. 83:295–315.

Jacob, F., S. Brenner, & F. Cuzin. 1963. On The Regulation of DNA Replication in Bacteria. Cold Spring Harbor Symp. Quant. Biol. 28:329–348.

Kohn, A. 1960. Lysis of frozen and thawed cells of *E. coli* by lysosome, and their conversion into spheroplasts. J. Bacteriol. 79:697–706.

Lark, K. G., T. Redko, & E. J. Hoffman. 1963. The effect of amino acid deprivation on subsequent DNA replication. Biochem. Biophys. Beta. 76:9–24.

Lindberg, A. A. 1973. Bacteria Phage Recptors. Ann. Rev. Microbiol. 27:205–241.

Lion, M. B. 1963. Quantitative aspects of the protection of freeze-dried *E. coli* against the toxic effect of oxygen. J. Gen. Microbiol. 32:961–324.

Lion, M. B., and P. D. Bergman. 1961. Substances which protect lyophilized *Escherichia coli* against the lethal effect of oxygen. J. Gen. Microbiol. 25:291–296.

Lion, M. B., J. S. Kirby Smith, and M.L. Randolph. 1961. Electron spin resonance signals from lyophilized bacterial cells exposed to oxygen. Nature (Lond.) 192:34–36.

Lyons, J. M. 1972. Phase transitions and control of cellular metabolism at low temperatures. Cryobiology 9:361–350.

Mackey, E. Y. 1984. In M.H.E. Andrew and A. D. Russell (eds.) The revival of injured microbes, Academic Press, London, pp. 45–75.

Maister, H. G., V. F. Pfeiffer, W. M. Bogart, and E. N. Heger. 1958. Survival during storage of Serratia marcescens dried by continuous vacuum sublimation. Appl. Microbiol. 6:413–419.

Martin, H. H. 1963. Bacterial Protoplasts—A Review. J. Theoret. Biol. 5:1–34.

Messer, W. 1972. Initiation of DNA replication in *Escherichia coli* B/r: Chronology of events and transcriptional control of initiation. J. Bacteriol. 112:7–12.

Nei, T., T. Araki, and T. Matsusaka. 1968. Freezing injury to aerated and non-aerated cultures of *Escherichia coli*. Pp. 3–15. In T. Nei (ed.), Freezing and drying of microorganisms. University of Tokyo Press, Tokyo.

Novick, O., E. Israeli, and A. Kohn. 1972. Nucleic acid and protein synthesis in reconstituted lyophilized *Escherichia coli* exposed to air. J. Appl. Bacteriol. 35:185–191.

Death Mechanisms in Bioaerosols / 191

Oseroff, A. R. P. W. Robbins, and M. M. Burger. 1973. The cell surface membrane: biochemical aspects and biophysical probes. Ann. Rev. Biochem. 42:647–682.

Pritchard, R. H., and A. Zaritsky. 1970. Effect of thymine concentration on the replication velocity of DNA in a thymineless mutant of *Escherichia coli*. Nature (Lond.) 226:126–131.

Reeve, J. N., D. J. Groves, and D. J. Clark. 1970. Regulation of Cell Division Mutants. J. Bacteriol. 104:1051–1064.

Rogers, H. J. 1970. Bacterial growth and the cell envelope. Bacteriol. Rev. 34:196–214.

Shapira, A. 1974. Ph.D. Thesis. Tel Aviv University, Tel Aviv.

Singer, S. J., and G. L. Nicolson. 1972. The fluid mosaic model of the structure of cell membranes, Science 175:720–731.

Skarstad, K., E. Boye, and H.B. Steen. 1986. Timing of initiation of chromosome replication individual *Escherichia coli* cells EMBO J. 5:1711–1716.

Smith, D. W. 1973. DNA synthesis in prokaryotes: replication. Pp. 321–393. In Progress in biophysics and molecular biology. Vol. 26. Pergamon Press, New York.

Speck, M. L., and R. A. Cowman. 1968. Metabolic injury to bacteria resulting from freezing. Pp. 39–51. In T. Nei (ed.), Freezing and drying of microorganisms. University of Tokyo Press, Tokyo.

Stark, C. N., and B. L. Herrington. 1931. The drying of bacteria and the viability of dry bacterial cells. J. Bacteriol. 21:13–14.

Straat, P. A., H. Wolochow, R. L. Dimmick, and M. A. Chatigny. 1977. Evidence of incorporation of thymidine into deoxyribonucleic acid in airborne bacterial cells. Appl. Environ. Microbiol. 34:292–296.

Tappel, A. L. 1966 In: "Cryobiology." Ed. Meryman, H. T. Effects of Low Temperatures & Freezing on Enzymes & Enzyme Systems. Academic Press. London & New York.

Webb, S. J. 1963. The effect of relative humidity and light on air-dried organisms. J. Appl. Bacteriol. 26:307–313.

Webb, M. 1953. Effects of magnesium on cellular division in bacteria. Science 118:607–611.

Zaritsky, A. 1975. Rate stimulation of deoxyribonuclecic acid synthesis after inhibition. J. Bacteriol. 122:841–846.

7

Resuscitation of Microbial Bioaerosols

Balkumar Marthi

7.1. Introduction

The response of microorganisms to stressful conditions has long been an important and contentious issue among microbiologists. In any kind of laboratory experimentation, the endeavor has been to develop nutrient media and conditions that are "optimal" for the growth of microorganisms. Thus, most bacteriological media are very rich sources of all the possible nutrients that a microorganism may need to grow. However, these laboratory growth conditions are far removed from conditions in the "natural" environment. In the latter case, the microorganisms are subjected to consistent fluxes in very low ("suboptimal") levels of nutrients. Attempts to grow these microorganisms in a laboratory environment where nutrients are plentiful often stresses microorganisms and impairs their ability to grow.

Quite apart from the above stresses are the stresses caused by various inhibitory factors in the environment. These inhibitory factors (examples include temperature fluxes, desiccation, chemicals like hydrogen peroxide, chlorine, nutrient deprivation, etc.) are often lethal to bacteria. However, at sublethal levels, these "stressors" can cause stress or injury to microorganisms. This type of sublethal injury is the subject of this chapter.

The phenomenon of sublethal injury in bacteria has been recognized for a long time. However, this subject has gained a great deal of importance only in recent years due to an increased understanding of the stress response, its effect on bacterial physiology and growth, and its implications for microbiological monitoring of various products/processes.

Perhaps the most important effect of sublethal stress is the inability of stressed bacteria to grow on different kinds of enumeration media, both selective and nonselective. This has potentially very profound impacts on the design and use of bacterial enumeration methods in different areas of microbiology such as food microbiology, drinking water microbiology, clinical/diagnostic microbiology, and applied/environmental microbiology. In many of these areas, processing techniques create conditions that can cause sublethal stress. For example, the preservation of foods by the addition of antimicrobial compounds or by freezing can cause sublethal injury to microorganisms present in the particular food. Inadequate heat processing of foods has often led to the development of sublethal stress as well as increased heat resistance in *Listeria monocytogenes* and *Escherichia coli* in meat products with serious implications (Linton et al., 1992; Murano and Pierson, 1992).

This recognition that sublethally injured microorganisms exist also has implications in the regulation of monitoring regimes. Increasingly, it has been felt that conventional microbiological monitoring methods do not adequately enumerate sublethally injured microorganisms, leading to potentially serious underestimations of microbial numbers. For example, the process of chlorination used commonly to sanitize drinking water frequently leads to sublethal damage to coliforms. Coliforms are routinely used as indicators of water quality and are enumerated on selective media, such as m-Endo and m-Endo LES agars (Standard Methods, 1985). However, injured coliforms have been shown to be unable to grow on these selective media (Braswell and Hoadley, 1974; Camper and McFeters, 1979). This has often led to the underestimation of coliform numbers in drinking water.

Similar situations are encountered in the food industry. As mentioned earlier, inadequate heat processing of foods may lead to the development of sublethal heat stress. The presence of *Staphylococcus aureus* in baby and geriatric foods is an indication of poor sanitation practices. Heat processing of foods containing this organism can lead to sublethal heat stress. Stressed staphylococci are unable to grow on the medium commonly used to enumerate them, Trypticase Soy Broth containing 10% NaCl (Draughon and Nelson, 1981). Similarly, freeze-drying of *Lactobacillus acidophilus* and *Salmonella anatum* causes them to lose the ability to grow on media containing selective agents (Bozoglu and Gurakan, 1989; Ray et al., 1971).

In the light of the above and other observations, a need has been felt to develop better microbiological monitoring methods that will allow for the enumeration of stressed microorganisms. These methods frequently include a resuscitation step (which facilitates repair of the sublethal injury) prior to growth and enumeration. Some of these methods are discussed in detail in later sections of this chapter.

In aerobiology, the study of airborne microorganisms, sublethal stress phenomena are frequently encountered. The deleterious effects of various environmental factors such as temperature, relative humidity (RH), dehydration, rehydration, radiations, and so forth can cause sublethal injury to airborne microorganisms (Benbough, 1967; Cox, 1966, 1986; Cox and Baldwin, 1967; Stersky et al.,

1972; Webb, 1960; Zentner, 1966). The airborne route is an important route for the transfer and dissemination of microorganisms to different locations. Also, the spread of respiratory pathogens is facilitated by the airborne route (Anderson, 1966; Goodlow and Leonard, 1961). In food and dairy processing plants, the air quality (i.e., the numbers of airborne microorganisms) of packaging areas often influence postpasteurization contamination of food and dairy products (Kang and Frank, 1989). Airborne dissemination of microorganisms also occurs from evaporative cooling towers (Lighthart and Frisch, 1976) and sewage irrigation and treatment plants (Sorber et al., 1976; Teitsch et al., 1980).

The importance of studying the stress response in the airborne state lies in the fact that although injured microorganisms are metabolically deficient and may not be detected by conventional methods, they still have the potential to recover from or repair the injury and regain their disease-causing properties (Cox, 1987). To accurately enumerate these bacteria, it is necessary to facilitate repair of these bacteria. Also, because most stressed bacteria are unable to grow in selective and nonselective media, it is also necessary to develop appropriate media for the recovery of these bacteria.

Another area where sublethal injury will play an important role is in the development of methods to assess the risks associated with the release of genetically engineered microorganisms (GEMs) into natural environments. With the tremendous progress in genetic engineering and recombinant DNA technology, the push in the 1990s is expected to be toward practical applications of this technology. One such area of application is microbiological pest control. The potential use of microbiological pest control agents (MPCAs) as alternatives to chemical pesticides is gaining popularity. Genetically engineered microorganisms may also find application in other areas of environmental and agricultural microbiology [see, for example, Seidler and Hern (1988)].

It is expected that with an increased potential for the usage of GEMs in natural environments, there will be an increased need for the assessment of risks associated with GEM use and dispersal. A common form of GEM release is as an aerosol. Aerosolized GEMs pose the potential risk of dispersal to unwanted locations and persistance in the environment for long periods of time (Lighthart and Kim, 1989). It is, therefore, necessary to carefully monitor, enumerate, and predict GEM dispersal from aerosols. The additional effects of stress and resuscitation of airborne microorganisms also need to be considered to develop guidelines for more efficient and safe releases of GEMs.

This chapter is designed to provide a basic understanding of the bacterial stress response, as well as the repair of resuscitation process. It is divided into three parts. In the first part (Section 7.2), the general phenomenon of stress and injury is discussed in detail. In part two (Section 7.3), the effects of sublethal stress on airborne bacteria are discussed. In the third part (Sections 7.4 & 7.5), the enumeration and resuscitation of stressed bacteria is discussed. Various general resuscitation media and conditions are discussed, along with specific examples.

Based on experiments done in the author's laboratory, and on other previously published work, recommendations for development of appropriate conditions for the resuscitation of air-stressed bacteria are also discussed.

This chapter will be restricted to the study of sublethal injury in bacteria because the majority of published work has been done in bacterial systems. It is expected that the stress response is similar in both fungal and virus systems.

7.2. Sublethal Injury in Bacteria

7.2.1. Definition

Sublethal injury or stress refers to a physiological response that bacteria exhibit when exposed to low levels of antagonistic chemical or physical agents ("stressors"). Sublethally injured bacteria are unable to carry out functions that "normal" (uninjured) bacteria can (Busta, 1978). These include:

(a) loss of control of transport functions,

(b) loss of enzyme activity,

(c) a prolonged lag period before initiation of growth,

(e) loss of resistance to selective agents.

Figure 7.1 hypothetically demonstrates the response of sublethally injured cells. The "base plating agar" and the "test plating agar" described by Busta refer to a nonselective medium (e.g., Trypicase Soy agar, TSA, for heterotrophic microorganisms) and a medium to which a selective agent has been added (e.g., TSA with 10% NaCl for *Staphylococcus aureus*, or MacConkey's agar with bile salts for coliforms), respectively. Due to the effect of the stressing agent, the cells have lost their ability to grow on the selective medium. However, there is minimal loss of growth ability on the base agar. The difference in counts between the two plating media can be used to quantitate the extent of the injury caused. If the injured microorganisms are allowed to repair the damage ("resuscitate") under suitable conditions, they eventually regain their ability to grow on the selective medium. As can be seen from Fig. 7.1, after complete resuscitation, the growth responses on the base plating agar and the test plating agar are essentially identical.

7.2.2. Agents That Cause Stress ("Stressors")

Bacteria may be exposed to stressful conditions in a variety of ways. Many of the physical and chemical methods used to prevent bacterial growth in food products can cause stress in bacteria. These methods include freezing and thawing, use of preservatives, antibiotics, and so on. In water environments, bacteria are likely to be stressed due to lack of nutrients, changes in water temperature

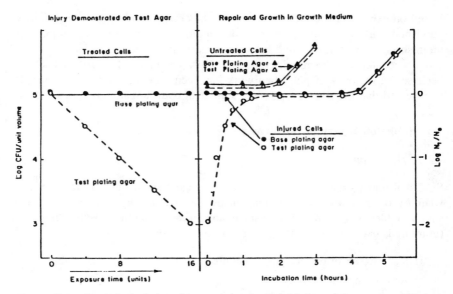

Figure 7.1. Hypothetical data demonstrating bacterial injury and recovery. The base plating agar and the test plating agar to a non-selective medium (e.g., Trypticase Soy Agar, TSA, for heterotrophic microorganisms) and a medium to which a selective agent has been added (e.g., TSA with 10% NaCl for *Staphylococcus aureus,* or MacConkey's agar with bile salts for coliforms), respectively. [From F. F. Busta (1978).]

and pH, and the use of disinfectants like chlorine. In other natural environments, bacteria may be subjected to temperature fluctuations, nutrient deprivation, desiccation, and osmotic shock (Hurst, 1977). Table 7.1 lists some common stressors that may affect bacteria. As a result of sublethal injury, bacteria may become considerably more sensitized to unrelated compounds ("secondary" stressors). For example, chlorine-injured bacteria are inhibited by bile salts commonly used in selective media for enumeration of enteric bacteria (Camper and McFeters, 1979; Herson et al., 1986; Lin, 1973), whereas heat-stressed *Staphylococcus*

Table 7.1. Stressors that cause sublethal injury.

Physical Agents	Chemical Agents
Temperature (heat- and cold-shock)	Antibiotics (penicillins, rifampicin, etc.)
Radiation (UV, X-rays)	Disinfectants (chlorine, phenols, etc.)
Osmotic pressure	Preservative systems
Hydrostatic pressure	Oxidative agents (chlorine, peroxides, etc.)
Desiccation	Media constituents (sodium chloride, potassium chloride, bile salts, etc.)
Nutrient depletion	Lack of nutrients

Table 7.2. Potential sites of stress-induced damage.

Site of Damage	Stressors	Effects
Outer membrane	Polynmyxins EDTA Heat	Increase permeability, sensitivity to surface-active agents
Cell wall	Penicillins Heat Freeze-thawing	Structural alterations, increased permeability, intracellular material
DNA	Acridines Radiation Heat	Breakage of DNA, mutations
RNA	Rifampicin Actinomycin D Heat	Degradation of RNA, inhibition of protein synthesis

aureus are inhibited by 10% NaCl used as selective agent in media (Brewer et al., 1977; Draughon and Nelson, 1981).

7.2.3. Sites of Damage in Stressed Bacteria

Depending on the type and degree of stress, the type of organism, and prevailing environmental conditions, stress-induced damage may be restricted to one or a few sites or may be generalized throughout the cell. However, even if the initial damage is only to a specific site (for example, the cell surface), its effects are translated to the rest of the bacterial cell and can affect numerous different sites and functions. The ultimate effects of all these events is the eventual death of the cell, providing that repair does not occur. Tables 7.2 and 7.3 list some of the potential cellular sites and activities that are the targets of stress induced damage. These are discussed in the following sections.

7.2.3.1. The Outer Membrane

The outer layers of the bacterial cell are very susceptible to stress-induced damage. In Gram-negative bacteria, the outermost layer is the outer membrane.

Table 7.3. Metabolic activities affected in injured bacteria.

Enzymes affected
  Catalase
  Peroxidase
  Deydrogenase
Metabolic/physiological activities affected
  Active transport of cations, sugars, amino acids
  ATP synthesis
  Mg^{2+} binding of cell-wall teichoic acids

The outer membrane of Gram-negative bacteria is contiguous with the cell wall. Chemically, it is composed of mucopeptide linked to lipopolysaccharide (Fig. 7.2). The Gram-negative envelope acts as a barrier to various compounds such as detergents, and surface-agents like bile salts. The ability of the envelope to exclude bile salts is utilized in the selective enumeration of coliforms and intestinal pathogens on media containing bile salts.

Heat stress and treatment with polymyxin causes damage to outer membranes of bacteria. In studies on sublethal heat treatment of *Escherichia coli*, Hitchener and Egan (1977) detected the release of lipopolysaccharide from the outer membrane. Treatment with lysozyme converts Gram-negative cells into spheroplasts which are now sensitive to bile salts (sodium deoxycholate) and ethylenediaminetetraacetic acid (EDTA). Treatments such as freeze-drying and chlorination also cause damage to the outer membrane, indicated by the release of periplasmic enzymes into the environment and increased sensitivity to bile salts (Braswell and Hoadley, 1974; Sinskey and Silverman, 1970). A characteristic feature of injured enteric bacteria is their inability to grow on selective media containing bile salts, indicating damage to the envelope.

7.2.3.2. Cell Wall

The cell wall of bacteria is composed of a class of macromolecules called peptidoglycans, which are heteropolymers of N-acetylglucosamine and N-actylmuramic acid. Figure 7.3 shows the composition of a peptidoglycan molecule. Peptidoglycan makes up about 80% of the Gram-positive cell wall. In Gram-

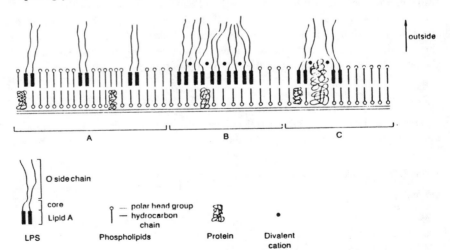

Figure 7.2. Schematic representation of the outer membrane of a Gram-negative bacterium. (Reprinted from H. Nakkaido. 1973. Biosynthesis and assembly of lipopolysaccharide. L. Lieve (ed., Marcel Dekker, New York.) In Bacterial membranes and walls.

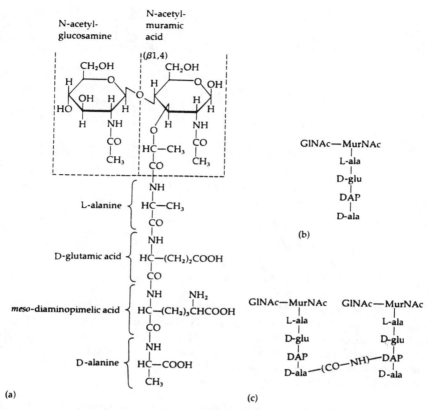

Figure 7.3. Structure of a peptidoglycan. (A) Complete structure of a single peptidogly-can subunit. (B) Schematic, simplified representation of structure shown in (A). (C) Cross-linking between the terminal carboxy group of D-alanine on one subunit and the free amino group of diaminopimelic acid on an adjacent subunit. (Reprinted from R. Y. Stanier et al. 1986. The microbial world, 5th ed., Prentice-Hall, Englewood Cliffs, N.J., p. 152.)

negative bacteria, only 5–10% of the wall weight is peptidoglycan and forms the thin, innermost layer of the outer envelope. The remainder of the envelope consists of the outer membrane which is predominantly made up of protein, phospholipid, and polysaccharide. Figure 7.4 is a schematic representation of the organization of the Gram-negative cell peptidoglycan.

A number of studies have shown that damage to the cell wall of Gram-positive bacteria may not significantly affect their survival after sublethal stress. Evidence for this is obtained from recovery studies done with heat-damaged *Staphylococcus aureus* (*S. aureus*). Because recovery was found to occur in the presence of penicillin and cycloserine, antibiotics that selectively inhibit cell-wall synthesis,

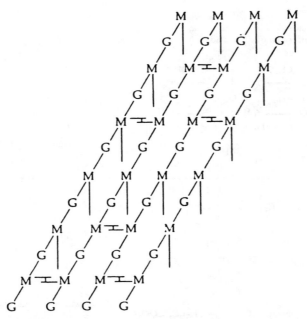

Figure 7.4. Peptidoglycan organization in the Gram-negative cell wall. G and M refer to N-acetylglucosamine and N-acetylmuramic acid, respectively, joined by beta-1,4-glycosidic bonds. (Reprinted from J. M. Ghuysen. 1968. Bacteriolytic enzymes in the determination of wall structure. Bacteriol. Rev. 32:425–437; with permission of the American Society of Microbiology, Washington, DC.)

it is concluded that sublethal damage to the cell wall is either very minor or does not affect cellular survival (Iandolo and Ordal, 1966).

However, it has been found that heat-stressed cells lose D-alanine from cell-wall teichoic acids (Hurst, 1977). Teichoic acid binds to magnesium ions (Mg^{2+} ions can prevent recovery and, thus, cause cell death.

7.2.3.3. Cytoplasmic Membrane

The cytoplasmic membrane consists of protein and lipid, with the proteins being embedded in a lipid bilayer (see Fig. 7.5), 60–80% of the membrane is made up of protein. The lipids are mainly phospholipids, such as phosphotidyl ethanolamine.

The cytoplasmic membrane is a semipermeable membrane, allowing selective influx and efflux of substances across it. Thus, its main function is the controlled transport of molecules into the cell. It also functions in energy production. Thus, damage to the cytoplasmic membrane can affect nutrient transport into the cell, leakage of intracellular constituents, and the activities of membrane-bound enzyme systems.

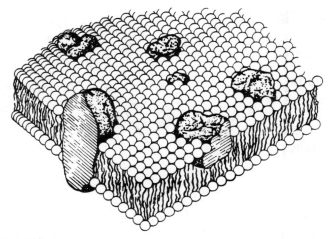

Figure 7.5. The cytoplasmic membrane. (Reprinted from S. J. Singer. 1972. The fluid mosaic model of the structure of cell membranes. Science 175:720; with permission of the American Academy for the Advancement of Science, Washington, DC.)

The cytoplasmic membrane can be damaged by the action of antibiotics such as valinomycins and nigericin, oxidative compounds like chlorine, disinfectants like phenols and quaternary ammonium compounds, by heat, and by the process of repeated freezing and thawing (Allwood and Russell, 1970; Alur and Grecz, 1975; Beuchat, 1978; Camper and McFeters, 1979; Ray et al., 1971; Sinskey and Silverman, 1970). The majority of research work has been done on the effects of temperature on the phospholipid profile of the membrane.

The phospholipid fatty acid content and composition is particularly affected by environmental conditions. The fatty acid composition influences the thermostability of the membrane. Thus, straight-chain saturated and isobranched fatty acids are found to be more thermostable than unsaturated fatty acids. Depending on the fatty acid composition, an organism is more or less susceptible to heat-induced damage. Upon exposure to elevated temperatures, bacteria have been shown to have increased leakage of intracellular components, such as DNA, RNA, and protein, along with impairment of membrane-transport functions. Figure 7.6 shows an example of heat-induced leakage in *Vibrio marinus* at 29.7°C.

Reduced temperatures cause injury by intracellular ice formation, dehydration, and due to the toxic effects of intracellular and extracellular solutes. At freezing temperatures, the release of water through cytoplasmic membrane channels is inhibited due to ice crystal formation. The freezing of this intramembranal water causes damage to the cytoplasmic membrane, changes its permeability characteristics, and causes leakage of intracellular material (Moss and Speck, 1966; Sinskey and Silverman, 1970). The formation of ice crystals also leads to an

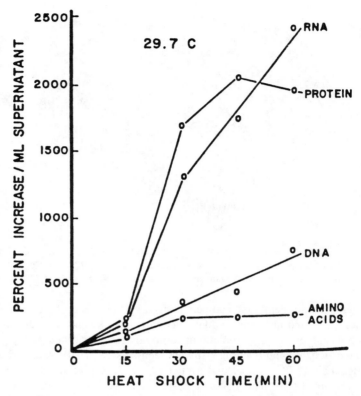

Figure 7.6. Leakage of DNA, RNA, amino acids, and proteins from *Vibrio marinus*, after heat-shock at 29.7°C. (Reprinted from R. D. Haight and R. Y. Morita. 1966. Thermally induced leakage from Vibrio marinus, an obligately psychrophillic marine bacterium, J. Bacteriol. 92:1388–1393; with permission of the American Society of Microbiology, Washington, DC.)

increase in the concentration of solutes within the cell. The topological character-istics of the cell membrane is altered due to the dissociation of proteins at or near the cell surface by the concentrated slats (Morichi, 1969). Other effects include changes in cellular pH, ionic strength, and consequent metabolic injury (Bozoglu et al., 1987). Frozen cells also lose cellular proteins, including permease systems and inhibitors of latent degradative enzymes.

Radiation-induced damage to the cytoplasmic membrane has also been re-ported. Co-factors for respiratory enzymes are particularly sensitive to ultraviolet light. Because the respiratory enzymes are located within the cytoplasmic mem-brane, any damage will cause impairment of activity. Visible and ultraviolet light have also been shown to inactivate active transport systems [e.g., phenylalanine transport in *E. coli* (Andrew and Russell, 1984)].

Sublethal injury by disinfectants and other chemical agents causes impairment of permeability functions, energy generation functions, and active transport functions of the cytoplasmic membrane.

7.2.3.4. RNA and Ribosomes

A variety of stresses (heat, freeze-thawing, chemical agents) cause release of 260 mn-absorbing material (Fig. 7.6). This material consists of ribonucleic acid (RNA) and RNA-derived nucleotides. This has been reported in *E. coli, S. aureus, Pseudomonas fluorescens, Vibrio parahaemolyticus,* and *Streptococcus faecalis* (Bozoglu et al., 1987; Moss and Speck, 1966; Pierson et al., 1978; Wagman, 1960; Witter and Ordal, 1977). The leakage of RNA is the combined effect of degradation of the RNA and weakening of the cytoplasmic membrane.

Ribosomal RNA (rRNA) constitutes about 80% of the total cellular RNA and is tightly associated with ribosomes. In any kind of sublethal stress, lesions are formed in the ribosomes, leading to their dissociation and the degradation of the rRNA (Hures, 1977; Iandolo and Ordal, 1966; Sogin and Ordal, 1967). Heat-induced damage causes denaturation of ribosomes and its unfolding. It is also postulated that destabilizing of ribosomes occurs due to leakage and loss of Mg^{2+} ions. This causes the ribosomes and the associated rRNa to be more susceptible to degradation by ribonuclease (RNase). Ribosomes denatured by sublethal heat treatment show intact 50S particles, whereas the 30S subunit is degraded. However, the 50S subunit, though intact, is also inactivated.

The 16S species in heat-stressed *S. aureus* is completely degraded. There is no effect on the primary structure of the 23S RNA species. However, the secondary and tertiary structures of 23S RNA in heat-stressed cells is extensively altered.

There is no evidence available to show any ribosome damage in freeze-injured cells. Likewise, the messenger RNA (mRNA) and the transfer RNA (tRNA) do not seem to show any damage caused by sublethal injury.

7.2.3.5. DNA

Evidence indicating damage to deoxyribonucleic acid (DNA) caused by stress comes from the finding that there is an increased rate of mutagenesis. For example, on heating or freeze-thawing, auxotrophic mutants of *Bacillus subtilis* are produced. Further evidence for the involvement of DNA damage in sublethal injury is the observation that repair-deficient mutants of *E. coli* are more sensitive to heat stress. Also, inhibition of DNA synthesis causes an inhibition of repair.

Sublethal levels of ionizing and ultraviolet radiations can cause breaks in the deoxyribonucleic acid (DNA) of bacteria, leading to mutations. Single-stranded and double-stranded breaks occur when bacterial cells are subjected to heat stress and to freezing and thawing. Bozoglu et al. (1987) showed that fragmentation of DNA occurs upon exposure of *E. coli* to freeze-drying. Breaks in DNA occur due to the induction of deoxribonucleases (DNases) by sublethal stress (for example, induction DNase H by thermal stress; Hures, 1977).

7.2.3.6. Metabolic Activities

In conjunction with damage caused by sublethal stress to specific cellular structures, there is also impairment of physiological and metabolic activities. Various enzyme systems are damaged. Among these are the inactivation of dehydrogenases by thermal stress in *S. aureus* and *Salmonella typhimurium,* the inactivation of catalase and peroxidase by thermal oxidative, and chemical stresses (Brewer et al., 1977; Flowers et al., 1977; Martin et al., 1976). The damage to cytoplasmic membranes causing disruption of transport systems and permeability controls has already been discussed. Many stresses also impair the ability of the cell to synthesize ATP by damaging energy-generating systems by inactivation of enzymes, transport systems, and so forth.

The damaging effects of elevated temperatures are dependent on a variety of factors (Andrew and Russell, 1984): (a) The **salinity** of the medium. In general, increasing the salt concentration causes bacteria to have an increased resistance to higher temperatures. (b) The **age** of the cell. Some bacterial species are more resistant to thermal stress in the stationary phase than in the logarithmic phase. However, this resistance may change, depending on the type of growth media used. (c) The **presence of solutes.** Thus, solutes such as yeast extract, sucrose, peptides, amino acids, and so forth can protect cells against thermal damage.

Damage to any site on the bacterial cell results in the death of the cell if the damaged structure/function is not repaired. The phenomenon of repair or resuscitation is discussed in detail in a subsequent section. Table 7.2 lists the sites of stress-induced damage in bacteria. For a detailed discussion of potential sites of stress-induced damage in bacteria, the reader is referred to Andrew and Russell (1984).

7.2.4. Sublethal Injury and the Growth Response

The most common manifestation of stress is the impaired growth response of stressed bacteria. Sublethally injured bacteria typically show a prolonged lag period prior to initiation of growth (see Fig. 7.1). Under favorable conditions, the damage caused by the stressor(s) is repaired during this lag period. The length of this lag period is dependent on the genus and species of the bacterium, type of stressing agent, degree of sublethal stress, and type of repair media/conditions used.

Stressed bacteria also show increased sensitivity to growth and enumeration media or to specific constituents of media, as already discussed. These bacteria are unable to grow even on those media that can support the growth of normal unstressed cells (Busta, 1978). Typically, stressed bacteria are inhibited by selective agents such as antibiotics, bile salts, and so forth. Other common media constituents like salts may also inhibit the growth of stressed bacteria. Some by-products of cellular activity, for example, hydrogen peroxide formed during respiration, may also be toxic to these bacteria. Although stress bacteria may be unable to form colonies on growth media, they are still viable. These "viable but nonculturable" cells have been shown to be metabolically active and, under favorable conditions, can repair stress-induced damage and regain their ability to grow.

7.3. Stress in the Airborne State

Cox (1989) has categorized the journey of airborne bacteria into three parts: (a) **take-off** (generation), (b) **aerial transport** (storage or dispersal); (c) **landing** (collection), The terms within the parenthesis refer to the equivalent processes that occur in an experimentally aerosolized bacterial population. As will be seen in the following pages, bacteria can encounter sublethal injury at any one or all of these three stages.

In addition, the extent of injury or stress that occurs in airborne bacteria is also dependent on a number of environmental factors. These include the relative humidity (RH) and temperature of the environment, the rates at which dehydration and rehydration of airborne cells can occur, the gaseous composition of the atmosphere in which the bacteria are airborne, and the types of radiation to which the cells are exposed (Kang and Frank, 1989).

In experimental bacterial aerosols, additional stress-causing factors are introduced by (a) the method of aerosol generation, (b) the method of aerosol collection, and (c) the method of enumeration of aerosolized bacteria (Cox, 1987). However, the response of bacteria to these "experimental" stresses is determined by the environmental conditions which affect them during aerosol storage. Thus, stress due to environmental conditions represents the "primary" stress to airborne bacteria. For more detailed information, the reader is referred to Cox (1987).

7.3.1. The "Primary" Stresses

7.3.1.1. Relative Humidity and Temperature

A change in cellular water content due to desiccation or dehydration represents probably the most fundamental stress to airborne bacteria (Cox, 1968). Dehydration effects are caused by evaporation of water from bacteria-carrying droplets, which can result in cellular water loss. The ambient relative humidity (RH) and the temperature of an environment determine the degree and rate at which evaporation occurs. In experimental aerosols, the nature and concentration of the spray fluids also affects the rate of evaporation.

It is very difficult to separate the effects of temperature from RH because RH is dependent on temperature. There is very limited information available on the effects of temperature on airborne bacteria. However, it is generally seen that as the temperature increases, the injury and death rates of airborne bacteria also increase.

The discussion that follows will be restricted to the effects of relative humidity on airborne bacteria. All the observations have been made with experimentally generated aerosols and, therefore, factors such as modes of aerosol generation, type of spray fluid used, and types of collection media and conditions will be crucial in determining the degree of stress.

The deleterious effects of RH are attributed to three variables (Cox, 1987):

(a) At low humidities (<29%), "air stress" plays an important role in determining survival. This stress is attributed to the toxic effects of oxygen present. However, it is not clear whether the toxic effects are due to oxygen alone or are also due to low (undetectable) levels of other lethal contaminants in air.

(b) At high humidities (>80%), survival is dependent on the "collection stress." The extent of sublethal damage is determined by the type of spray fluid used for aerosolization as well as the type and composition of the collection medium.

(c) Irrespective of the RH of the environment, bacteria-containing droplets undergo desiccation due to water loss. This is observed even at 100% humidities because the enhanced vapor pressure over an aerosol droplet surface causes evaporation through isothermal distillation.

A general observation of the effects of RH is that extremes in RH (<20% or >80%) are deleterious to bacterial survival (Webb, 1959). However, different species show differing responses to changes in RH. In fact, within the same species, different strains will differ in their response to RH changes. Some bacteria survive better at lower RH than at higher RH. An example is *Escherichia coli* Jepp, which shows impaired survival at higher relative humidities. On the other hand, *Escherichia coli* B is able to survive better at low RH than at high RH. The relationship between changes in RH levels and bacterial survival is not a smooth function. Rather, there are very narrow RH ranges where the survival patterns of the bacteria is drastically different from adjacent ranges (Cox, 1968). This observation is shown in Fig. 7.7.

The toxic effects of oxygen are probably due to free radical formation in the cell because it is found that metabolic inhibitors and free radical scavengers protect aerosolized bacteria (Benbough, 1967). Free radicals cause the occurrence of Maillard reactions in cells (Cox, 1989). These reactions are amino-carbonyl reactions between reducing sugars and amino groups of proteins, to form Schiffs' bases, which rearrange to give a number of products. Maillard reactions can cause changes in enzyme folding, structure, and activity and can have significant impact on cellular survival. In addition, Maillard reactions can occur in the

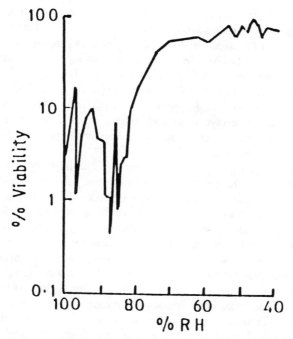

Figure 7.7. Effect of relative humidity on the aerosol survival of *Escherichia coli* B sprayed from a suspension of bacteria in distilled water and stored in nitrogen atmospheres. (Reprinted from C. S. Cox. 1989. Airborne bacteria and viruses. Sci. Prog., Oxford 73:469–500.

phospholipids, which can cause destabilization of membranes due to structural rearrangements.

Bacterial membranes (both the outer membrane of Gram-negative bacteria as well as the cytoplasmic membrane) are inherently unstable molecules. The mode of packing of the phospholipid component is dependent on the degree of hydration. Water loss due to desiccation can change hydration parameters and lead to structural changes that can cause destabilization of membranes. For a more detailed explanation of membrane changes due to desiccation, the reader is referred to Cox (1989).

Desiccation can also enhance the rate of Maillard reactions of phospholipids and other varieties within a cell. The loss or movement of water within a cell can have profound effects of cellular structures and functions. Structural changes in membranes have already been mentioned earlier. In addition, the structures of proteins, including enzymes, DNA, and RNA are also affected. These can lead to a series of secondary effects. Thus, damage to surface structures can affect permeability and transport of substances into and out of a cell. Aerosol-stressed bacteria have been shown to lose K^+, Na^+, and other ions by leakage through damaged membranes (Anderson et al., 1968; Benbough, 1967; Cox, 1969). Aerosolized Gram-negative bacteria also show an increased sensitivity to lysozyme, RNAse, DNAse, and trypsin, indicating possible damage to the cell surface (Hambleton, 1970, 1971). Further evidence of membrane damage during aerosolization is obtained from the observation that airborne *Salmonella newbrunswick* and *E. coli* are unable to grow on selective media containing bile salts, such as sodium deoxycholate and sodium taurocholate (Stersky and Hedrick, 1972). Metabolic processes, such as energy generation, are also adversely affected in the airborne state.

Another consequence of desiccation is an osmotic imbalance within the cell, causing osmotic stress to the bacteria. Proteins and other macromolecules can function only within certain ranges of water activities. Reduction in water activities by water loss can, thus, impair cellular functions. Desiccation also affects the concentration of various substances (e.g., metabolic by-products, intermediates, inhibitors, etc.), which can reach levels which are toxic to the cell.

The effects caused by desiccation are obviously enhanced at higher temperatures because the rate of desiccation increases with increasing temperatures. Due to the unique nature of bacterial aerosols, the deleterious effects of desiccation may either be compounded or mitigated by the types of processes used to generate and collect these aerosol droplets. Also, the types of media used to enumerate stressed bacteria will affect the final conclusions from any such study. These factors are discussed in a later section.

7.3.1.2. Radiation

In addition to causing mutagenic responses in bacteria, radiation (UV rays, X-rays) causes damage to proteins, lipids, and membranes (Cox, 1989). Radiation effects can induce the production of free radicals, which are toxic to cells. In addition to the stimulation of Maillard reactions, there is also extensive cross-linking of protein to DNa and protein to protein. The respiratory transport chain within the cytoplasmic membranes is a particularly vulnerable target. The effects of radiation are enhanced by dehydration and oxygen.

7.3.1.3. The "Open Air Factor"

The term "Open Air Factor" (OAF) collectively refer to the effects of reaction products of ozone with olefins (Cox, 1987). The OAF can cause damage to cellular nucleic acids. Gram-negative bacteria are more sensitive to the OAF than are Gram-positive bacteria. This is attributed to the higher lipid content in the former type of bacteria. The OAF can react with lipids to form highly reactive and toxic hydroperoxides.

Under natural conditions, the OAF is of significance in poorly ventilated environments rather than in open air environments. The OAF effect is enhanced at high RH levels, probably due to the formation of OH* radicals.

7.3.2. The "Secondary" Stressors

The severity of the effects of the "primary" stressors described above in experimental aerosols is dependent on the so-called "secondary" stressors. These include the types of procedures used to generate the aerosol (e.g., type of spray fluid used, additives, etc.), the types of collection processes and medium used, and the type of enumeration medium. The enumeration medium is of great importance because it affects the correct interpretation of the responses of air-stressed bacteria.

7.3.2.1. Aerosol Generation Procedures

The type of spray fluid used to generate the aerosol is of significant importance. Ideally, components of the spray fluid should have protective effects on the aerosolized bacteria. They should, thus, be able to prevent desiccation, stabilize cellular structures and functions, and so forth. Many compounds (additives) have been used in spray fluids (Cox, 1968, 1987; Stersky et al., 1972; Webb, 1960, 1967; Zimmerman, 1962). For example, raffinose has been shown to protect aerosolized bacteria at low RH levels. This is due to the fact that raffinose cannot enter the cell and forms a protective barrier outside it, decreasing water loss. Stersky and co-workers have shown increased survival of *Salmonella newbrunswick* when sprayed from skim milk as compared to distilled water. Other agents which prevent dehydration include glycerol, glucose, various sugars like sucrose, and mannose. However, reducing sugars may actually inhibit bacterial survival due to their participation in Maillard reactions. Incorporation of sugars such as glucose, compounds such as betaines, and so forth in spray fluids prevents osmotic damage (Marthi and Lighthart, 1990). Spent culture fluids, disaccharides and trisaccharides, and polyhydric alcohols (sorbitol, inositol) provide significant degrees of protection to airborne microorganisms.

Another factor that determines aerosol survival is the type of device or atomizer used to generate aerosols (Kang and Frank, 1989). If the process of atomizing or the device cause mechanical damage to cells, these cells are subsequently in a weakened state and can be damaged further during aerosol storage. Devices such as a spinning top or vibration-needle aerosol generator cause very low degrees of shear stress to microorganisms, thereby increasing viability. The Collison nebulizer, which has been developed to generate monodipersed aerosols, suffers from the disadvantage of producing high shear forces which may affect airborne microbial survival and viability.

7.3.2.2. Aerosol Collection Procedure

If the collection process imposes additional stress on the already stressed aerosolized bacteria, the survival of these cells will be significantly affected. The most important factor involved here is the rate of rehydration of the cells once they are collected. This situation is analogous to that seen in freeze-drying situations. When freeze-dried bacterial cells are rehydrated, it is generally seen that very rapid rehydration can cause injury to bacteria. The rate of rehydration can be controlled by the composition of the collection fluid and will depend on the solute concentration, the water activity, and the temperature of collection. Various types of collection fluids have been utilized to most efficiently collect and enumerate aerosolized bacteria: buffered gelatin (Tyler and Stupe, 1959), phosphate buffer (Lundholm, 1982; Marthi and Lighthart, 1990; Marthi et al., 1991), 2% peptone water (Delmore and Thompson, 1981), nutrient broth (duBuy et al., 1945), and gelatin milk broth (Lembke et al., 1981).

Another factor that plays an important role is the degree of osmotic stress that is imposed on bacteria during collection. This phenomenon is particularly significant when the collection menstruum is a liquid or fluid. Addition of osmo-protectants such as betaine, prolines, and so forth can prevent osmotic damage.

The type of collection device used also affects survival. There are two main types of collection devices: (a) the "impinger" type, such as the All-Glass Impinger (AGI) and (b) the "impactor" type, such as the Anderson or the Reynier samplers. In the first type, aerosol droplets are collected by impingement on to a liquid medium, frequently through a very small orifice at very high flow rates. In addition to causing physical loss due to dispersion, this method can also stress bacteria due to powerful mechanical shear forces and may not be ideal for many purposes.

The impactor type of samplers work by the principle of direct impaction onto the surface of a solid nutrient medium. The mechanical shear forces during collecting are greatly reduced. No further processing of the bacteria is required in order to enumerate them.

7.3.3. "Microbial" Factors

Only environmental and experimental factors which affect bacterial survival in the airborne state have been discussed so far. However, the ultimate behavior of airborne bacteria under any sorts of environmental conditions is dependent on certain factors which can be classified as "microbial" factors. These factors are summarized in Table 7.4.

The **type, species,** or **strain** of the organism under study will affect its survival. Bacteria are inherently more susceptible or resistant to difference stresses. This response is controlled at the genetic level. Even within a species, different strains

Table 7.4. "Microbial" factors affecting aerosol-induced stress.

Genotypic and Physiological factors
   Type/species/strain of organism
   Gram positive versus Gram negative
   Presence/absence of outer membrane
   Composition of cell surface structures
Growth/cultural factors
   Type of growth medium and conditions used
   Age of culture
   Type of enumeration medium used

---

will show divergent responses. As has been shown earlier for the effects of the relative humidity, *E. coli* B, Jepp and MRE162 show very different responses (Cox, 1966, 1987).

The **Gram nature** of the bacterium will influence its survival after stress. The basis for the Gram-staining characteristic lies in the difference in the cell wall composition and structure. Due to the higher levels of phospholipids present in Gram-negative bacteria, they are more susceptible to the effects of Maillard reactions, dehydration, osmotic shock, and so forth. The Gram-positive bacteria are generally more resistant to such stresses due to the relatively tough and rigid cell wall that they possess.

The **growth conditions** used also play an important role in determining the response of bacteria to stress. Bacteria grown in nutrient-rich media may carry over media components to the aerosolized state. This carryover may protect these bacteria. On the other hand, carryover of selective agents from media may cause secondary stresses on airborne bacteria. The survival response may be difference when bacteria are grown on solid versus liquid media. The toxic effects of oxygen at low RH values has been known to reduce when cells are grown on an agar (solid) medium prior to aerosolization. Finally, the presence of protective agents in the growth medium will affect survival.

The **physiological age** of the bacterial population determines its response to stress-inducing conditions. In general, "older" cells (cells in the stationary phase of growth) are more resistant to deleterious stresses than are "younger," logarithmic phase cells. This is probably due to the fact that log phase cells are very active physiologically and metabolically, and their cellular systems are more vulnerable to stresses than those of stationary phase cells.

The **type of enumeration medium** will affect the final growth of collected aerosolized bacteria. If the enumeration medium is very complex or contains selective agents, it can cause secondary stresses to the metabolically and structurally impaired bacteria. This can lead to an overestimating of the effects of aerosol-induced stress on the bacteria. The types of enumeration media are discussed in further detail in later sections on resuscitation.

The "microbial" factors discussed here play significant roles not only in aerosol-induced stresses, but also in all other types of stresses induced whether naturally or deliberately.

7.4. Enumeration of Stressed Bacteria

Development of enumeration methods for stressed bacteria is an important element of quality control in a variety of situations. Sublethally injured bacteria in processed foods may not be detectable on commonly used media. However, these organisms may recover and pose a health hazard. In drinking water systems, the presence of chlorine-injured organisms has frequently passed undetected due to their inability to grow on conventional media.

The conventional method of enumerating stressed bacteria uses the observation that many sublethally injured bacteria grow normally on nonselective media but are impaired on selective media (Busta, 1978). Stressed bacteria are plated on both a selective (e.g., m-Endo agar for coliforms) and a nonselective medium (e.g., Trypticase Soy agar). The difference in counts between the two media is used to determine the number of stressed bacteria in the population. Thus, the ratio of the colony-forming units (CFU) on selective media to the CFU on the nonselective medium gives the percentage of stressed bacteria. However, some studies have shown that even nonselective media may impair the growth of sublethally injured bacteria. Recognizing this fact, alternative methods need to be developed.

Methods to enumerate stressed organisms are divided into three categories: Those based on modified media, metabolic or physiological activity, and resuscitation.

7.4.1. Methods Based on Modified Media

These modifications include (a) the addition of "protective" agents or removal of "harmful" (often selective) ingredients; and (b) the utilization of diluted complex and minimal media.

The **addition of protective agents** to conventional media has been widely reported. Compounds such as pyruvate and catalase have been added to media to protect against stress-induced (oxidative) damage (Baird-Parker and Davenport, 1965; Brewer et al., 1977; Flowers et al., 1977; Martin et al., 1976). Catalase mediates the breakdown of hydrogen peroxide (H_2O_2) and other peroxides which are generated by bacterial metabolism and which are toxic to cells. Because cellular catalase activity of aerobic microorganisms is impaired by injury, the addition of exogenous catalase mitigates this effect. Pyruvate also protects against oxidative stress.

Betaines have been used in situations where osmotic stresses are encountered. Osmotic stress can be caused by the accumulation of various osmolytes, such as K^+ ions, glutamate, and trehalose. Such an intracellular increase can adversely affect nucleic acid–protein interactions and the activities of various enzymes (Yancy et al., 1982). Osmoprotectants such as betaines and proline reduce the cytoplasm levels of accumulated osmolytes, thereby balancing the osmolarity of the cell with respect to the osmolarity of the surrounding menstruum (Caley et al., 1992; Sutherland et al., 1986). Betaines are more effective osmoprotectants than proline.

Many enumeration media contain **selective agents** that are used to specifically detect certain types of bacteria. Frequently, however, these selective agents are toxic to injured bacteria, that is, they act as secondary stressors. For example, bile salts used to selectively grow coliform organisms like *E. coli* in m-Endo and MacConkey agars prevent the growth of coliform bacteria that have been injured by heat or sublethal chlorination. In an uninjured cell, the bile salts are excluded from the interior of the cell by the outer membrane. As we have seen in the foregoing discussion, this membrane can be damaged by stress and can allow the entry of the bile salts into the cells where they are toxic. Similarly, thermally injured *S. aureus* is incapable of growing on media containing 10% NaCl, which is used as a specific selective agent for this organism. To facilitate growth of injured organisms, therefore, the selective agent needs to be excluded from the medium formulation. This is the procedure that is followed when resuscitation of injured bacteria is carried out and is discussed in detail in a later section.

In addition to media constituents, the composition of the diluents used in viable counting procedures also affect colony-forming activity. Diluents containing low quantities of organic material such as peptone, tryptone, and milk solids are superior to plain buffers or saline (McFeters et al., 1982).

Researchers have developed media formulations which are specifically designed to grow injured bacteria. Besides containing protective agents like pyruvate, peptone, catalase, and so forth, these media have lower levels of a wide variety of nutrients and metabolic intermediates that facilitate growth and recovery of stressed bacteria. Two examples of such media are the R2A medium developed by Reasoner and Geldreich (1985) which has been used to enumerate heterotrophic bacteria in drinking water and Baird-Parker agar (Baird-Parker and Davenport, 1965) containing, in addition to nutrients, sodium pyruvate and blood for the enumeration of thermally injured *S. aureus*. The pyruvate in the medium degrades accumulated peroxides, whereas blood is added to neutralize any antimicrobial substances that may be present in the medium.

A survey of the literature on stress and injury in microorganisms indicates the importance of oxidative damage [see above and also Archibald and Fridovich (1981), Farr et al., (1988), and Imlay and Linn (1987)]. One of the most important effectors of the stress response is hydrogen peroxide, and consequently the superoxide radial (O_2^-). In an uninjured cell, the peroxides generated during metabolic activities are degraded by the enzymes catalase and superoxide dismutase (SOD) and are, thus, not lethal. However, the activities of these enzymes are impaired significantly by stress and, therefore, the cells are exposed to the toxic actins of peroxides. Consequently, exogenous incorporation of compounds such as catalase and pyruvate lead to better recovery of stressed microorganisms. In addition, incubation of stressed organisms under anaerobic conditions also enhances recovery (Knabel et al., 1990; Smith and Archer, 1988).

In the design of media for enumeration of stressed organisms, the type of stress and the effects that it has on the bacterial cell are important determinants. There is no "universal" enumeration (or, for that matter, resuscitation) medium for growing injured bacteria. In many instances, sublethally injured bacteria may grow better on minimal media than on complex, rich media (Gomez et al., 1973). However, the reverse may also be true. Hence, the medium to be used needs to be chosen taking into consideration the type of organism, the type of stress, and the response of the organism to that stress. This will determine the types of additives, conditions of growth, and so forth which should be used. As a general rule, it has been found that sublethally stressed bacteria are more sensitive to spread (surface) plating methods than to pure plating methods. Thus, the latter should be the method of choice. Also, a period of resuscitation is usually needed to allow repair of stress-induced damage. Generally, resuscitation should be carried out in the absence of any selective agents.

7.4.2. Methods Based on Metabolic or Physiological Activity

Although sublethal injury may inhibit the colony-forming ability of bacteria, these bacteria can be metabolically and physiologically active. It is, therefore, possible to utilize this property to enumerate these bacteria. Many different metabolic parameters can be used to assess viability. These include assays of enzyme activity (e.g., β-galactosidase, dehydrogenase), assessment of respiratory activity, determination of ATP production, and so forth.

The measurement of **respiratory activity** has been used as an index for the detection of stressed bacteria. The activity of key enzymes of the respiratory chain has been used to quantify respiration. This method, called the electron transport system (ETC) method, uses the electron acceptor 1–2 (p-iodophenyl)-3-(p-nitrophenyl)-5-phenyl tetrazolium chloride (INT), along with reduced nicotinamide adenine dinucleotide (NADH), to detect total respiratory activity. This method has the advantage of being very rapid (3–5 min) and reproducible.

A **dehydrogenase assay** has also been developed to measure bacterial activity. This INT-dehydrogenase assay measures the rate of reduction of the INT dye. This reduction is a reflection of microbial dehydrogenase activity and can be utilized to quantitate microbial populations.

A microscopic method based on INT reduction has also been utilized for the enumeration of stressed bacteria (Herson et al., 1986). This method is based on the ability of respiring bacteria to reduce INT and cause its deposition as a refractile crystal within the bacterial cell. After staining with acridine orange (AO), these AOINT-positive cells can be enumerated by epifluorescent microscopy.

The rate of **ATP production** can also be used to determine metabolic activity in stressed and recovering bacteria. This assay uses the luciferin–luciferase system to estimate the amount of ATP produced. The ATP-dependent catalysis of luciferin by luciferase emits light. The bioluminescence can then be quantitated, and the amount of ATP generated can be determined.

7.4.3. Methods Based on Resuscitation

As discussed above, modifications of conventional growth media for detecting stressed bacteria include incorporation of protective agents such as betaine, catalase, and pyruvate and incubation under conditions favorable for growth. In resuscitation methods, injured cells are pre-incubated with these protective agents, which allow the repair of stress-induced damage. The process of recovery or resuscitation and various methods to facilitate resuscitation are discussed in detail in the next section.

7.4.4. Enumeration of Air-Stressed Bacteria

There have not been any media that have been specifically designed to enumerate air-stressed bacteria. Most of the studies have concentrated on improving collection media to prevent collection stress. It is also in collection media that resuscitation can occur.

Work in this author's laboratory at the U.S. Environmental Protection Agency in Corvallis, Oregon has demonstrated the protective effects of betaine and catalase on air-stressed bacteria (Marthi and Lighthart, 1990; Marthi et al., 1991). Incorporation of 2–5 mM betaine or 1000 U catalase into enumeration media (Trypticase Soy agar, TSA) increased the colony-forming abilities of airborne bacteria by 60–110%. The protective effect of betaine is probably due to its function as an osmoprotectant to prevent osmotic stress in desiccated aerosol droplets. Catalase acts by removing H_2O_2 which is toxic to cells. Studies to determine the effects of catalase on resuscitation are described later.

7.5. Resuscitation of Injured Bacteria

In previous sections, we have shown that there are a variety of conditions/compounds which are stressful to bacterial cells. These include heat, freeze-drying, aerosolization, chemical treatments, and so forth. These sublethal stresses impair the growth and metabolic activity of cells. One striking feature of sublethally injured bacteria is there inability to form colonies on media on which their "normal" counterparts grow freely (i.e., the so-called "viable but nonculturable" state). It is also abundantly clear that to regain their "normal" properties, bacteria need to repair the damage caused by stress. The process of resuscitation is the process by which repair of the damage caused by sublethal stress takes place. Resuscitation has been also referred to as repair or recovery (Andrew and Russell, 1984).

Figure 7.1 showed the type of response seen in a resuscitating population. At the population level, recovery is manifested by a restoration of the ability to grow on the "test" (selective) plating medium. It is only when this ability it completely restored that doubling of the population occurs. The "lag" period that is very characteristic of injured populations is actually a period of intense activity, where cellular processes are repaired and restored prior to growth. This lag period is analogous to the one seen in typical bacterial growth curves when a culture is transferred to a different medium. Indeed, a change in the medium composition from a "normal" to an "abnormal" one is itself a form of stress.

A variety of factors influence the rate of resuscitation of bacterial population. These are listed in Table 7.5. These factors do not act individually but have a synergistic action which will affect the rate of resuscitation. Resuscitation can theoretically occur in any nutrient medium which is free of inhibitory or selective ingredients (for example, the "base plating medium" of Fig. 7.1). But, the rates of recovery in such an unmodified medium may be slow. This rate can be increased by adding specific protective agents to the medium.

The resuscitation process also has a requirement for protein synthesis because, in many cases, it does not occur in the absence of any protein synthesis. Indeed, many stressed bacteria synthesize specific "stress-related" proteins. Many such proteins have been identified: the heat-shock proteins (HSPs), oxidative stress proteins, chlorine stress-induced proteins, and so on. It is hypothesized that some of these proteins may be involved in the resuscitation process. These proteins are found to disappear once recovery is complete and the population has started to grow. Further detailed discussion on stress-induced proteins is beyond the scope of this chapter.

Table 7.5. Factors affecting stress and resuscitation.

Nutrient composition
Redox potential
Osmolarity
Ionic Strength
Water activity
Salt composition
Temperature
pH
Surface tension
Type (strain/species) of culture
Age/physiological state of culture

7.5.1. Repair of Stress Damage

Much of the evidence for repair of stress-induced damage comes from studies with specific inhibitors of bacterial metabolic and biosynthetic activities. Any form of injury can cause a variety of damaging lesions in a cell. There have been no extensive studies to determine the exact time course of repair of these lesions. It is also not known what specific role stress-induced proteins play in the recovery process.

7.5.1.1. Repair of Membranes

As indicated in earlier sections, the cell surface of bacteria is vulnerable to stress-induced damage, more so in Gram-negative bacteria than in Gram-positive bacteria. The latter organisms have more rigid cell walls that are more resistant to attack by stressors.

The lipid content of the outer and cytoplasmic membranes of bacteria has been shown to be an important determinant of injury and repair. Thermally injured *Salmonella typhimurium* showed incorporation of glucose-U^{-14}C into lipids during the recovery process. This newly synthesized lipid is probably used to synthesize damage membranes. It is also postulated that this lipid biosynthesis occurs very early during the recovery.

The presence of Na^+ ions during recovery helps stabilize the cell envelope by providing mechanical strength and by preventing lysis. Similar effects are seen with Mg^{2+} ions. Many solutes, such as sucrose and egg whites, say, help recovery by stabilizing the membrane.

Repair of active transport systems which may be impaired by membrane damage is affected by the pH of the recovery medium. In general, pH levels above neutral (i.e., 7.5–8.5) are more protective that pH values below neutral.

The synthesis of periplasmic enzymes in recovering cells is an energy requiring process (Ray et al., 1971). Because ATP synthesis is a through oxidative phosphorylation in such organisms (e.g., *E. coli*, *Salmonella* sp.), an appropriate energy source needs to be supplied during the recovery process.

7.5.1.2. Repair of Nucleic Acids

Thermal injury causes damage to rRNA and ribosomes, resulting in their degradation and the leakage of 260-nm absorbing material (see above). To repair this damage, RNA synthesis is required, as demonstrated by the fact that recovery does not occur in the presence of Actinomycin D and an RNA synthesis inhibitor (Gomez et al., 1976).

RNA synthesis in recovering cells is initiated immediately after recovery begins (Hurst, 1977). All species of rRna-5S, 16S, and 23S are synthesized during recovery. Protein synthesis does not begin until much later. The rate-limiting step in the recovery process is the maturation of precursor rRNA molecules or the ribosomes, rather that the synthesis of rRNa itself.

The 30S and 50S subunits of ribosomes are also regenerated during the recovery process.

Repair of **DNA** after stress-induced damage is inhibited more on complex media than on minimal media (Gomez et al., 1973). For repair of damaged DNA, DNA synthesis has been shown to be necessary.

7.5.2. Repair in Airborne Bacteria

Cell envelope damage in air-stressed bacteria is repaired in the presence of Mg^{2+}, Fe^{3+}, and Zn^{2+} (Hambleton, 1971). These ions stabilize membrane structures damaged by stress. For complete repair to occur, an energy source and ATP generation are required. Recovery of cell envelope damage can occur in the absence of protein and RNA synthesis. Repair of dehydration stress in many species of bacteria occurs only in minimal media. However, some bacteria (e.g., *Streptococcus* sp.) require certain peptides for recovery (Cox, 1987).

UV-radiation-induced damage to DNA in airborne bacteria, as in bacteria subjected to other forms of stress, is repaired by photoreactivation, which excises and repairs double-stranded breaks. A similar excision–repair system exists for repairing single-stranded breaks.

Some of the other aspects or repair in air-stressed bacteria are similar to those discussed above.

7.5.3. Minimal Medium Recovery

Certain kinds of stresses can be resuscitated only in minimal media (called "Minimal Medium Recovery"; (MMR). In these instances, the ingredients of complex media can be inhibitory to resuscitation and to survival. MMR is characteristic of injured cells which require repair of DNA in order to fully resuscitate. MMR has also been shown to occur in both cold- and heat-injured *Salmonella typhimurium* (Gomez et al., 1973). In this bacterium, the highest recovery was seen on a glucose–salts medium (M-9 medium). The MMR effect is also related to peroxide sensitivity. On complex media, catalase activity of cold-shocked cells is reduced as compared to that on M-9. Addition of organic supplements like yeast extract decreased recovery.

7.5.4. Resuscitation Methods

To efficiently resuscitate a population of injured bacteria, it is necessary to provide conditions that are most favorable for growth, free from inhibitors and factors that may cause secondary stresses. In addition, additives may be needed to provide either (a) **protection** from deleterious effects of stress or (b) **intermediates** for the de novo synthesis of damaged cellular structures. Examples of the former are betaines, solutes such as sucrose used as osmoprotectants, raffinose, dextrans to protect from desiccation, and catalase to prevent the toxic action of peroxides. Examples of added intermediates are peptones, which provide peptides for new protein synthesis, compounds, such as pyruvate, succinate, and so forth, which can serve as intermediates of energy-generating processes, and so on.

The type of recovery medium to be used will depend on some or all of factors listed in Table 7.5. The type of stress that is involved will determine which of the various cellular structures and activities are damaged. Based on this information, the appropriate recovery medium can be formulated. In general, a resuscitation medium, like any growth medium, should be complete and contain all the essential requirements—carbon, energy sources, vitamins, growth factors, nitrogen source, slats—needed for growth. In addition, the medium should also contain the additives and protective agents mentioned above.

Many complex media suffer from the disadvantage that they inhibit cellular catalase activity. Because peroxides are important effectors of the stress response, catalase captivity is required for recovery and survival. In these cases, it is preferable to carry out recovery in minimal media (see MMR above) or to supplement the media with exogenously added catalase. In addition, resuscitation can be carried out under anaerobic conditions to prevent accumulation of peroxides. Also, in many instances, autoclave sterilizing of complex media leads to the production of toxic products, which can inhibit and kill the already stressed bacteria that are grown on these media. Thus, it is preferable to filter sterilize complex media used for resuscitation.

Irrespective of the type of medium used, there are two basic methodologies that are followed for resuscitation of stressed bacteria. These methodologies are based on the principle that a preincubation step is necessary for resuscitation to occur before actual enumeration of cells is done. These methods are discussed below [see Ray (1979)].

7.5.4.1. Liquid Medium Resuscitation (Repair)

Liquid medium repair simply indicates the fact that the pre-incubation step to facilitate repair is done in a liquid menstruum. The procedure involves two steps:

(a) addition of samples to repair medium and incubation;

(b) enumeration of resuscitated organisms on (selective) medium.

The composition of the repair medium and the time of preincubation will depend on the several factors that were discussed earlier. The liquid medium repair method has applicability where bacteria, such as water-indicator bacteria and pathogens, are detected and enumerated by the most probable number (MPN) technique. A disadvantage with this medium is that during the preincubation, uninjured cells may grow and, thus, overestimate bacterial numbers. This problem could be overcome by incorporating an antibiotic such as penicillin which attacks dividing cells only and will, thus, kill only the uninjured population.

Due to problems with growth of uninjured organisms during resuscitation, this method has not won approval for regulatory monitoring. However, the liquid medium repair technique has been successfully used to repair populations of injured coliforms from water and from semipreserved foods. *Vibrio parahaemolyticus* from seafood, and so on and facilitate their detection.

The liquid medium repair is especially suited for resuscitation of air-stressed bacteria because some of the collection procedures employ liquid media. By collecting directly in a resuscitation medium, an extra step in processing is avoided. In our laboratory, we have used this method to study the effects of catalase addition on the resuscitation of aerosol-stressed bacteria. We have found that incubation of air-stressed cells in phosphate buffer containing 1000 U catalase increased viability by $> 100\%$ as compared to resuscitation on buffer alone (Marthi et al., 1991).

7.5.4.2. *Solid Medium Resuscitation (Repair)*

The main difference between this method and liquid medium repair is the fact that resuscitation is allowed to occur in a solid agar medium. Subsequent enumeration is done directly on the same medium, by overlaying with an appropriate, usually selective medium.

In practice, a population of stressed cells is added to the molten agar medium and pour-plated. After incubation for appropriate periods of time, the selective medium is overlayed, and after a further period of incubation, the cells are enumerated. The basic advantage of this system is that because pour-plating immobilizes the cells in the agar matrix, there is no effect of cell multiplication on the final count of the population.

This method has been validated by testing in a variety of situations. For resuscitation and enumeration of air-stressed bacteria, the solid medium repair method can be easily adapted for use with the impactor types of samplers, where cells are collected directly onto an agar medium.

7.6. Summary

The phenomenon of bacterial stress is of critical importance in a variety of public health, environmental, and regulatory monitoring scenarios [see, for example, Read (1979)]. One consequence of injury is the inability of these bacteria to grow on media used for their enumeration. They can, thus, escape detection. The second, far more serious consequence is the ability of stressed bacteria to repair the damage caused by sublethal injury. This latter property has serious implications. For example, sublethally injured pathogens in preserved foods can escape deletion by conventional techniques but will still recover and cause disease in humans. Similar situations also occur in drinking water systems.

A recent and more intriguing "application" of stressed bacteria is in the field of agricultural biotechnology, specifically dealing with the release of genetically engineered microorganisms (GEMs) into natural environments as aerosols. Because aerosolization can cause stress in bacteria, the interest here is not only related to whether injured GEMs can carry out their functions efficiently but is also related to regulatory issues. Release of GEMs into natural environments is a particularly sensitive issue given the fears (mostly unfounded) associated with the creation of "new" genetic material, and their possible effects on ecosystems. Thus, regulatory compliance has to be strict and should account for the *entire* released population, whether injured or not. Thus, the phenomenon of stress and resuscitation gains a great deal of importance.

In this chapter, we have shown how bacteria, including airborne bacteria, react to deleterious stresses and described the cellular structures and the physiological processes that are damaged by stresses. We have also discussed methods to more efficiently enumerate these stressed bacteria and also various methods and conditions that facilitate resuscitation. It is hoped that this information will be of significant use in developing more efficient methods for enumerating microorganisms. This is important from the standpoint of applied/environmental and public health microbiology, as well as the newly emergent science of agricultural biotechnology. Moreover, at the basic level, an understanding of the microbial stress response will lead to an improved understanding of the metabolic and physiological processes of microorganisms.

References

Allwood, M. C., and A. D. Russell. 1970. Mechanisms of thermal injury in non-sporulating bacteria. Adv. Appl. Microbiol. 12:89–118.

Alur, M. D., and N. Grecz. 1975. Mechanism of injury of *Escherichia coli* by freezing and thawing. Biochem. Biophys. Res., Commun. 62:308–312.

Anderson, J. D. 1966. Biochemical studies of lethal processes in aerosols of *Escherichia coli*. J. Gen. Microbiol. 45:303–313.

Anderson, J. D., F. A. Dark, and S. Peto. 1968. The effect of aerosolization upon survival and potassium retention by various bacteria. J. Gen. Microbiol. 52:99–105.

Andrew, M. H. E., and A. D. Russell. 1984. The revival of injured microbes. Academic Press, New York.

Archibald, F. S., and I. Fridovich. 1981. Maganese and defenses against oxygen toxicity in *Lactobacillus plantarum*. J. Bacteriol. 145:442–451.

Baird-Parker, A., and E. Davenport. 1965. The effect of recovery medium on the isolation of *Staphylococcus aureus* after heat treatment and after storage of frozen or dried cells. J. Appl. Bacteriol. 28:390–396.

Benbough, J. E. 1967. Death mechanisms in airborne *Escherichi aoli*. 1. J. Gen. Microbiol. 47:325–333.

Beuchat, L. R. 1978. Injury and repair of Gram-negative bacteria, with special consideration of the involvement of the cytoplasmic membrane. Adv. Appl. Microbiol. 23:219–243.

Bozoglu, T. F., and G. C. Gurakan. 1989. Freeze-drying injury of *Lactobacillus acidophilus*. J. Food Protect. 52:259–260.

Bozoglu, T. F., M. Ozilgen, and U. Bakir. 1987. Survival kinetics of lactic acid starter cultures during and after freeze-drying. Enzyme Microb. Technol. 9:531–537.

Braswell, J. R., and A. W. Hoadley. 1974. Recovery of *Escherichia coli* from chlorinated secondary sewage. Appl. Microbiol. 28:328–329.

Brewer, D. G., S. E. Martin, and Z. J. Ordal. 1977. Beneficial effects of catalase or ppyruvate in a most-probable-number technique for the detection of *Staphylococcus aureus*. Appl. Environ. Microbiol. 34:797–800.

Busta, F. F. 1978. Introduction to injury and repair of microbial cells. Adv. Appl. Microbiol. 23:195–201.

Camper, A. K., and G. A. McFeters. 1979. Chlorine injury and the enumeration of waterborne coliform bacteria. Appl. Environ. Microbiol. 37:633–641.

Caley, S., B. A. Lewis, and M. T. Record, Jr. 1992. Origins of the osmoprotective properties of betaine and proline in *Escherichia coli*. J. Bacteriol. 174:1586–1595.

Cox, C. S. 1986. The survival of *Escherichia coli* atomized into air and into nitrogen from distilled water and from solutions of protecting agents as a function of relative humidity. J. Gen. Microbiol. 43:383–399.

Cox, C. S. 1966. The aerosol survival and cause of death of *Escherichia coli* K-12. J. Gen. Microbiol. 54:169–175.

Cox, C. S. 1969. The cause of loss of viability of airborne *Escherichia coli* K12. J. Gen. Microbiol. 57:77–80.

Cox, C. S. 1987. The aerobiological pathway of microorganisms. John Wiley & Sons, New York.

Cox, C. S. 1989. Airborne bacteria and viruses. Sci. Prog. Oxford 73:469–500.

Cox, C. S., and F. Baldwin. 1967. The toxic effect of oxygen upon the aerosol survival of *escherichia coli* B. J. Gen. Microbiol. 49:15–21.

Delmore, R. P., and W. N. Thompson. 1981. A comparison of air sample efficiencies. Med. Device Diagn. Ind. 3:45–48.

Draughon, F. A., and P. J. Nelson. 1981. Comparison of modified direct plating procedures for recovery of injured *Escherichia coli*. 1. J. Food Sci. 46:1188–1191.

duBuy, H. G., A. Hollaender, and M. D. Lacky. 1945. A comparative study of sampling devices for airborne microorganisms. Publ. Health Rept. Suppl. No. 184.

Farr, S. B., D. Touati, and T. Kogoma. 1988. Effects of oxygen stress on membrane functions in *Escherichia coli:* role of HPI catalase, J. Bacteriol. 170:1837–1842.

Flowers, R. S., S. E. Martin, D. G. Brewer, and Z. J. Ordal. 1977. Catalase and enumeration of stressed *Staphylococcus aureus*. Appl. Environ. Microbiol. 33:1112–1117.

Goodlow, R. J., and F. A. Leonard. 1961. Viability and infectivity of microorganisms in experimental airborne infection. Bacteriol. Rev. 25:182–187.

Gomez, R. F., K. D. Blais, A. Herrero, and A. J. Sinskey. 1976. Effects of inhibitors of protein, RNA, and DNA synthesis on heat-injured *Salmonella typhimurium* LT2. J. Gen. Microbiol. 97:19–27.

Gomez, R. F., A. J. Sinskey, R. Davies, and T. P. Labuza. 1973. Minimal medium recovery of heated *Salmonella typhimurium* LT2. J. Gen. Microbiol. 74:267–274.

Hambleton, P. 1970. The sensitivity of Gram-negative bacteria, recovered from aerosols, to lysozyme and other hydrolytic enzymes. J. Gen. Microbiol. 61:197–204.

Hambleton, P. 1971. Repair of cell wall damage in *Escherichia coli* recovered from an aerosol. J. Gen. Microbiol. 69:81–88.

Herson, D. S., B. Marthi, M. A. Payer, and K. H. Baker. 1986. Enumeration of chlorine-stressed organisms with acridine orange 2-(*p*-iodophenyl)-3-(*p*-nitrophenyl-5-phenyl tetrazolium chloride (AOINT). Curr. Microbiol. 13:77–80.

Hitchener, B. J., and A. F. Egan. 1977. Outer membrane damage in sublethally heated *Escherichia coli*. Can. J. Microbiol. 23:311–318.

Hurst, A. 1977. Bacterial injury: a review. Can. J. Microbiol. 23:936–942.

Iandolo, J. J., and Z. J. Ordal. 1966. Repair of thermal injury of *Staphylococcus aureus*. J. Bacteriol. 91:131–142.

Imlay, J. A., and S. Linn. 1987. Mutagenesis and stress responses induced in *Escherichia coli* by hydrogen peroxide. J. Bacteriol. 169:2967–2976.

Kang, Y. F., and J. H. Frank. 1989. Biological aerosols: A review of airborne contamination and its measurement in dairy processing plants. J. Food Protect. 52:512–524.

Knabel, S. J., H. W. Walker, P. A. Hartman, and A. F. Mendonca. 1990. Effects of growth temperatures and strict anaerobic recovery on the survival of *Listeria manacyto-genes* during pasturization. Appl. Environ. Microbiol. 56:370–376.

Lembke, L. L., R. N. Niseley, R. C. V. Nostrand, and M. D. Hale. 1981. Precision of the All-Glass Impinger and Anderson microbial impactor for air sampling in solid waste handling facilities. Appl. Environ. Microbiol. 42:222–225.

Lighthart, B., and A. S. Frisch. 1976. Estimation of viable airborne microbes downwind from a point source. Appl. Environ. Microbiol. 31:700–704.

Lighthart, B., and J. Kim. 1989. Simulation of airborne microbial droplet transport. Appl. Environ. Microbiol. 55:2349–2355.

Lin, S. 1973. Evaluation of coliform tests for chlorinated secondary effluents. J. Water Pollut. Control Fed. 45:498–506.

Linton, R. H., J. B. Webster, M. D. Pierson, J. R. Bishop, and C. R. Hackney. 1992. The effect of sublethal heat shock and growth atmosphere on heat resistance of *Listeria monacytogenes* Scott A. J. Food Protect. 55:84–87.

Lundholm, I. M. 1982. Comparison of methods for quantitative determination of airborne bacteria and evaluation of total viable counts. Appl. Environ. Microbiol. 44:179–183.

Marthi, B., and B. Lighthart. 1990. Effects of betaine on the enumeration of airborne bacteria. Appl. Environ. Microbiol. 56:1286–1289.

Marthi, B., B. T. Shaffer, B. Lighthart, and L. Ganio. 1991 Resuscitation effects of catalase on airborne bacteria. Appl. Environ. Microbiol. 57:1775–1776.

Martin, S. E., R. S. Flowers, and Z. J. Ordal. 1976. Catalase: its effect on microbial enumeration. Appl. Environ. Microbiol. 32:731–734.

McFeters, G. A., S. C. Cameron, and M. W. LeChevallier. 1982. Influence of diluents, media, and membrane filters on detection of injured waterborne coliform bacteria. Appl. Environ. Microbiol. 43:97–103.

Morichi, T. 1966. In T. Nei, ed., Freeze drying of microorganisms. University of Tokyo Press, Tokyo, pp. 53–58.

Moss, C. W., and M. L. Speck. 1966. Release of biologically active peptides from *Escherichia coli* at subzero temperatures. J. Bacteriol. 91:1105–1111.

Murano, E. A., D. Merle, and M. D. Pierson. 1992. Effect of heat shock and growth atmosphere on the heat resistance of *Escherichia coli* 0157:h7. J. Food Protect. 55:171–175.

Pierson, M. D., R. F. Gomez, and S. E. Martin. 1978. The involvement of nucleic acids in bacterial injury. Adv. Appl. Microbiol. 23:263–285.

Ray, B. 1979. Methods to detect stressed microorganisms. J. Food Protect. 42:346–355.

Ray, B., J. J. Jezeski, and F. F. Busta. 1971. Repair of injury in freeze-dried *Salmonella anatum*. Appl. Microbiol. 22:401–407.

Read, R. B., Jr. 1979. Detection of stressed microorganisms—Implications for regulatory monitoring. J. Food Protect. 42:368–369.

Reasoner, D. J., and E. E. Geldreich. 1985. A new medium for the enumeration and subculture of bacteria from potable water. Appl. Environ. Microbiol. 46:1–7.

Seidler, R. J., and S. Hern. 1988. Special report: release of ice-minus recombinant bacteria. Environmental Research Laboratory, U.S. Environmental Protection Agency, Corvallis, OR, p. 83.

Sinskey, T. J., and G. J. Silverman. 1970. Characterization of injury incurred by *Escherichia coli* upon freeze-drying. J. Bacteriol. 101:429–437.

Smith, J. L., and D. L. Archer. 1988. Heat-induced injury in *Listeria monocytogenes*. J. Ind. Microbiol. 3:105–110.

Sogin, S. J., and Z. J. Ordal. 1967. Regeneration of ribosomes and ribosomal ribonucleic acid during repair of thermal injury to *Staphylococcus aureus*. J. Bacteriol. 94:1082–1087.

Sorber, C. A., H. T. Bausum, S. A. Schaub, and M. J. Small. 1976. A study of bacterial aerosols at a wastewater irrigation site. J. Water Pollut. Control Fed. 48:2367–2379.

Standard methods for the examination of water and wastewater. 1985. American Public Health Association Inc., New York.

Stersky, A. K., and T. I. Hedrick. 1972. Inhibition of growth of airborne coliforms and other bacteria on selective media. J. Milk Food Technol. 35:156–162.

Stersky, A. K, D. R. Heldman, and T. I. Hedrick. 1972. Viability of airborne *Salmonella newbrunswich* under various conditions. J. Dairy Sci. 55:14–18.

Sutherland, L., J. Cairney, M. J. Elmore, II. Boothe, and C. F. Higgins. 1986. Osmotic regulation of transcription: induction of the proU betaine transport gene is dependent in accumulation of intracellular potassium. J. Bacteriol. 168:805–814.

Teltsch, B., H. I. Shuval, and J. Tador. 1980. Die-away kinetics of aerosolized bacteria from sprinkler application of wastewater. Appl. Environ. Microbiol. 39:1191–1197.

Tyler, M. E., and E. L. Stupe. 1959. Bacterial aerosol samplers. I: Development and evaluation of the All-Glass Impinger. Appl. Microbiol. 7:377–349.

Wagman, J. 1960. Evidence of cytoplasmic membrane injury in the drying of bacteria. J. Bacteriol. 80:558–564.

Webb, S. J. 1959. Factors affecting the viability of airborne bacteria. I. Bacteria aerolozied from distilled water. Can. J. Microbiol. 5:649–669.

Webb, S. J. 1960. Factors affecting the viability of airborne bacteria. II. The effect of chemical additives on the behavior of airborne cells. Can. J. Microbiol. 6:71–105.

Webb, S. J. 1967. The influence of oxygen and inositol on the survival of semi-dried microorganisms. Can. J. Microbiol. 13:737–745.

Witter, L. D., and Z. J. Ordal. 1977. Stress effects and food microbiology. In M. Woodbine (ed.) Antibiotics and antibiosis in agriculture. Butterworths, Reading, MA, pp. 102–112.

Yancey, P. H., M. E. Clark, S. C. Hand, R. D. Bowlus, and G. N. Somero. 1982. Living with water stress: evolution of osmolyte systems. Science 217:1214–1222.

Zentner, R. J. 1966. Physical and chemical stresses of aerosolization. Bacteriol. Rev. 30:551–557.

Zimmerman, L. 1962. Survival of *Serratia mercescens* after freeze-drying or aerosolization at unfavorable humidity. I. Effects of sugars. J. Bacteriol. 84:1297–1302.

8

Instrumentation Used with Microbial Bioaerosols*

Paul A. Jensen, B. Lighthart, A. J. Mohr, and Brenda T. Shaffer

8.0. Introduction

8.0.1. General

Bioaerosol monitoring is a rapidly emerging area of industrial hygiene that is finding increased use and overuse. It is often used in conjunction with indoor environment quality investigations, infectious disease outbreaks, and agricultural health investigations. Bioaerosol monitoring includes the measurement of viable (culturable and nonculturable) and nonviable microorganisms in both indoor (e.g., industrial, office, or residential) and outdoor (agricultural and general air quality) environments. In general, indoor bioaerosol sampling need not be performed if visible growth is observed. Contamination (microbial growth on floors, walls, or ceilings, or in the HVAC system) should be remediated. If personnel remain symptomatic after remediation, air sampling may be appropriate, keeping in mind that negative results are quite possible and they should be interpreted with caution. Other exceptions for which bioaerosol sampling may be appropriate include epidemiological investigations, research studies, or if indicated after consultation with an occupational physician and an immunologist.

Sampling for fungi and bacteria, including *Actinomycetes,* are included in this chapter. Less developed methods for bioaerosols such as viruses, protozoa, antigenic fragments, algae, arthropods, and mycoplasmas are not addressed at this time.

---

*We wish to acknowledge Millie P. Schafer, Ph.D., of NIOSH for her assistance in the preparation of the portions of this chapter pertaining to immunochemial and molecular analytical techniques and for her critical review of the chapter.

8.0.2. Indoor Versus Outdoor Bioaerosol Sampling

In general, indoor microflora concentrations of a healthy work environment are lower than outdoor concentrations at the same location. If one or more taxa are found indoors in concentrations greater than outdoor concentrations, then the source of amplification must be found and remediated. Bioaerosol sampling is often performed out-of-doors for pollen and fungi to assist allergists in their treatment of patients by identifying taxa distribution and concentration in air over time. On occasion, outdoor bioaerosol sampling is conducted in an occupational environment (e.g., agricultural investigations and sewage treatment plants). Indoor bioaerosol sampling is often conducted in occupational (industrial and office environments) and nonoccupational (residential and educational buildings) settings.

8.0.3. Viable Versus Nonviable Bioaerosol Sampling

Viable microorganisms are living organisms which have the potential to reproduce. Viable microorganisms may be defined in two subgroups: culturable and nonculturable. Culturable organisms reproduce under selected conditions. On the contrary, nonculturable organisms do not reproduce due to intracellular stress or because the conditions (e.g., medium or temperature) are not conducive to growth. Nonviable microorganisms are not living organisms; as such, they are not capable of reproduction. As the name implies, viable bioaerosol sampling involves collecting a bioaerosol and culturing the collected particulate. Only culturable microorganisms are enumerated and identified, thus leading to an underestimation of bioaerosol concentration.

As a general rule, sampling for aeroallergens (pollen and fungi) is performed without culturing the microorganism (i.e., nonviable bioaerosol sampling). The bioaerosol is collected on a "greased" surface or membrane filter. The microorganisms are then enumerated and identified using microscopy, classical microbiology, molecular biological, or immunochemical techniques. When sampling for culturable bacteria and fungi, the bioaerosol is generally collected by impaction onto the surface of a broad spectrum solid medium (agar), filtration through a membrane filter, or impingement into an isotonic liquid medium (water based). After impaction onto an agar surface and incubation for a short period of time, the organisms may be replica-plated (transferred) onto selective or differential media and incubated at different temperatures for identification and enumeration of microorganisms (Tortora et al., 1989). Impingement collection fluids are plated directly on agar, serially diluted and plated, or the entire volume of fluid is filtered through a membrane filter. The membrane filter is then placed on an agar surface and all samples may be replica-plated as previously described. Culturable microorganisms may be identified or classified by using microscopy, classical microbiology, or molecular biology techniques such as restriction fragment length polymorphic (RFLP) analysis. Classical microbiology techniques include observation of growth characteristics, cellular or spore morphology, simple and differential staining, and biochemical, physiological, and nutritional tests for bacteria. Analytical techniques which may be applied to both nonviable and viable microorganisms include polymerase chain reaction (PCR) and enzyme-linked immunosorbent assay (ELISA). Such methods may be used to identify specific microorganisms and may be used to locate areas of contamination. Though these methods are generally qualitative, current research efforts involve modifying the methods to obtain semiquantitative or quantitative results. Concentrations of endotoxin from Gram-negative bacteria determined using the *Limulus* amebocyte lysate (LAL) assay method have been correlated with patient symptoms in very few studies (Rylander and Vesterlund, 1982; Milton et al., 1990).

8.1. Principles of Operation

8.1.1. General Principles

Most aerosol analyses require sampling methods to separate and collect particles from the airstream. Impaction, filtration, and impingement are three typical principles of accomplishing this separation in bioaerosol sampling.

8.1.2. Impaction

Impaction is used to separate a particle from a gas stream based on the inertia of the particle. An impactor consists of a series of nozzles and a target. Ideal impactors have a "sharp cutoff" or step-function efficiency curve, in which all particles greater than a certain aerodynamic size are collected and all particles less than that size pass through. As a practical matter, an impactor may be assumed to be ideal and the efficiency curves characterized by a single number, Stk_{50}, the Stokes number, which gives 50% collection efficiency. This is equivalent to assuming that the mass of the particles larger than the cut diameter (d_{50}) that pass through the impactor equals the mass of the particles smaller than the d_{50} that are collected. Hence, the d_{50} is the aerodynamic diameter above which the collection efficiency of the impactor approaches 100% (Hinds 1982). Aerodynamic diameter (d_{ae}) is the diameter of a unit density ($\rho_p = 1$ g cm^{-3}) sphere that has the same settling velocity as the particle. Thus, impactors are selected so that the minimum size particle expected to be present will be collected. The distribution of mass and the distribution of count for the same sample of particles have different means, medians, geometric means, graphical representations, and probability density functions. The median of the distribution of mass is called the mass median aerodynamic diameter (MMAD) to distinguish it from the count median diameter (CMAD). The MMAD is defined as the diameter for which half the mass is contributed by particles larger than the MMAD and half by particles smaller than the MMAD. It is the diameter that divides the graphical representation of the distribution of mass into two equal segments.

Industrial hygienists are likely to employ stationary cascade impactors or individual impactors used in survey instruments, either as the primary collection mechanism or as a preclassifier (for example, to remove nonrespirable particles from the sampled airstream). Cascade impactors consist of a stack of impaction stages: Each stage consists of a one or more nozzles and a target or substrate. The nozzles may take the form of holes or slots. The target may consist of a greased plate, filter material, or growth medium (agar) contained in Petri dishes. Each succeeding stage collects smaller particles than the one preceding it. A filter may be used as the final stage so that all particles not impacted on the previous stages are collected. The target may be weighed to determine the collected mass, or it may be washed and the wash solution analyzed. Filters may induce more particle bounce than greased or oiled plates. Although personal cascade impactors are available, these devices are not as widely used in personal sampling for bioaerosols as are filters (Macher and Hansson, 1987).

Impactors used for the collection of airborne microorganisms may range from a single slit to more than 400 holes through which the air jets impact onto growth medium with one or more bacterial or fungal colonies forming at some of the impaction sites. Multiple particles passing through a single hole may be inaccurately counted as a single particle. As the number of particles deposited onto growth medium increases, the probability that the next particle will impact a "clean" hole decreases. The basic formula for the positive-hole correction adjustment follows:

$$P_r = N \left(\frac{1}{N} + \frac{1}{(N-1)} + \frac{1}{(N-2)} + \ldots + \frac{1}{(N-r+1)} \right);$$

P_r is the expected number of culturable particles to produce r positive holes (i.e., the number of colonies observed on the specimen plate) and N is the total number of holes per impactor stage. Andersen (1958) stated that in the actual case of bioaerosol sampling, the flow of particles present if r positive holes are observed would be equal to or greater than P_r but less than P_{r+1} and the average would be $(P_r + P_{r+1} - 1)/2$. For example, if 75% of the holes have received one particle, the chance that the next particle will impact a "clean" hole is one in four (25%) (Andersen, 1958; Leopold, 1988; Macher, 1989).

8.1.3. Filtration

Collection of particles from an aerosol sample is most commonly achieved by filtration. Filter media are available in both fibrous (typically glass) and membranous forms. A common misconception is that aerosol filters work like microscopic sieves in which only particles larger than the pore size are retained. In fact, particle removal occurs by collision and attachment of particles to the surface of the fibers or membrane. Thus, particles smaller than the pore size may be efficiently collected. Sampling filter media may have pore sizes of 0.01 to 10 μm. The efficiency of fibrous filters is a function of the face velocity (i.e., the air velocity across the surface of the filter). For particles less than 1 μm, the overall efficiency decreases with increasing face velocity. For particles greater than approximately 1 μm, the filter collection efficiency is greater than 99%. The overall efficiency of membrane filters is approximately 100% for particles larger than the pore size (Lippmann, 1989). Membrane filters are manufactured in a variety of pore sizes from polymers such as cellulose ester, polyvinyl chloride, and polycarbonate. Polymeric membrane filters lack rigidity and must be used with a support pad. The choice of a filter medium depends on the contaminant of interest and the requirements of the analytical technique. For gravimetric analysis, nonhygroscopic materials such as glass fibers, silver, or polyvinyl chloride membranes are selected. For analysis by microscopy, cellulose ester or polycarbonate membranes are the usual choices. Filters are often held in disposable plastic filter cassettes during bioaerosol sampling. The three-piece cassette may be used either open face or closed face. Open-face sampling is performed by removing the end plug and the plastic cover from the three-piece cassette and is used when the particulate must be uniformly deposited (i.e., for microscopic analysis). If a three-piece cassette is used in the open-face mode, the plastic cover is retained to protect the filter after sampling is concluded. All plastic cassettes are securely assembled and sealed with a cellulose shrink band or tape around the seams of the cassette to prevent leakage past the filter.

Membrane filters for use in industrial hygiene sampling are usually supplied as disks of 37 or 47 mm diameter. Because the pressure drop across a filter increases with the air velocity through the filter, the use of a 47-mm filter for a given volumetric flow rate results in a lower pressure drop than a 37-mm filter. The use of the smaller (37 mm) filter also concentrates the deposit of the contaminant into a smaller total area, thus increasing the density of particles per unit area of filter. This may be helpful for direct microscopic examination of low concentrations of organisms. In areas of high concentration, the microorganisms may have to be eluted, diluted, and then refiltered for microscopic analysis.

Filtration techniques are used for the collection of certain fungi and endospore-forming bacteria. These desiccation-resistant culturable organisms are washed from the surface of smooth-surface polycarbonate filters. The microorganisms in the wash solution are either cultured or refiltered to uniformly distribute the microorganisms on the membrane filter. In the latter case, the microorganisms are stained and examined microscopically (Wolf et al., 1959; Fields et al., 1974; Lundholm, 1982; Palmgren et al., 1986]. To culture the organisms, the membrane filter from each sampling cassette is washed with a 0.02% Tween™ 20 (J. T. Baker Chemical Co., Phillipsburg, NJ) in aqueous solution (three 2-ml washes) with agitation. A portion of the recovered wash volume is serially diluted from full strength (1 : 10, 1 : 100, 1 : 1000) and 0.1 ml of each dilution is inoculated onto duplicate 100-mm × 15-mm Petri dishes containing the appropriate medium. Residual culturable microorganisms on the membrane filter from each sampling cassette are counted by placing the filter on a medium in a Petri dish to allow the microorganisms to colonize. The Petri dishes are incubated and the colonies are identified and enumerated (Muilenberg et al., 1992). This method allows flexibility in dealing with high levels of spores by permitting a count of colonies on the filter by serial dilution of the wash solution. An inherent weakness in this procedure is that high analytical dilutions can statistically exclude taxa present in the air sample at low concentrations. This dilution technique favors the predominant fungi populations at the expense of minor populations.

8.1.4. Impingement

The liquid impingers are a special type of impactor. Impingers are useful for the collection of culturable aerosols (White et al., 1975; Lembke et al., 1981; Henningson et al., 1988). Impingers such as the Greenberg–Smith impinger or the AGI-30 use a liquid (simple salt solutions) as the collection medium. Additives to the collection medium such as proteins, antifoam, or antifreeze aid in resuscitation of bacterial cells, prevention of foaming and loss of the collection fluid, and minimize injury to the cells. The jet is positioned a set distance above the impinger base and consists of a short piece of capillary tube designed to reduce cell injury when the air is dispersed through the liquid and the particles entrapped. The Greenberg–Smith and AGI-30 samplers operate by drawing aerosols at nominal flow rates of 28.3 and 12.5 L min^{-1} through an inlet tube. The d_{50} of these samplers is approximately 0.3 μm (Wolf et al., 1959; Cown et al., 1957). The AGI-30 inlet tube is curved to simulate particle collection in the nasal passage (Cox, 1987). This makes it especially useful for studying the respiratory infection potential of airborne microorganisms. When the AGI-30 is used to recover total airborne organisms from the environment, the curved inlet tube is washed with a known amount of collecting fluid after sampling because larger particles (i.e., over 15 μm) are collected on the tube wall by inertial force. After sampling for the appropriate amount of time, 10 ml of the full-strength collection fluid is filtered through a 0.45-μm pore-size membrane filter. In addition, serial dilutions of the remaining collection fluid are handled similarly (APHA, 1989). The membrane filter is placed in a 100-mm × 15-mm sterile plastic Petri plate filled with the appropriate medium and incubated for later identification and enumeration.

8.1.5. Characteristics of Several Bioaerosol Samplers

Once the purpose or the goal of bioaerosol sampling is determined, the appropriate sampling method(s) may be chosen. The sampler used to collect the bioaerosol must be capable of collecting the aerosol with high efficiency and in the physical and biological conditions required by the bioaerosol sampling method. Experimental, theoretical, and physical characteristics of several commonly used bioaer-

Table 8.1. Experimental, theoretical, and physical characteristics of several commonly used bioaerosol samplers.

Sampler	d_{50} True (μm)	d_{50} Theoretical (μm)	No. of holes	Q (L min^{-1})	D_j or W_j (mm)	A_j (mm^2)	U_j (m s^{-1})
Andersen 6-Stage (Andersen, 1958, 1984)							
Stage 1	7.0	6.24	400	28.3	1.18	1.10	1.08
Stage 2	4.7	4.21	400	28.3	0.914	0.656	1.80
Stage 3	3.3	2.86	400	28.3	0.711	0.397	2.97
Stage 4	2.1	1.84	400	28.3	0.533	0.223	5.28
Stage 5	1.1	0.94	400	28.3	0.343	0.092	12.8
Stage 6	0.65	0.58	400	28.3	0.254	0.051	23.3
Andersen 2-Stage (Phillips, 1990)							
Stage 0	8.0	6.28	200	28.3	1.50	1.77	1.33
Stage 1	0.95	0.83	200	28.3	0.400	0.126	18.8
Andersen 1-Stage (Andersen, 1958; Phillips, 1990)							
Stage N6	0.65	0.58	400	28.3	0.254	0.051	23.3
Mattson–Garvin Slit-to-Agar		0.53	1	28.3	0.152	6.23	75.7
Ace Glass All-Glass Impinger-30		0.30	1	12.5	1.00	0.785	265.
PBI Surface Air Sampler (Lach, 1985)							
Compact	2.0	1.97	219	90	1.00	0.785	8.72
Standard	2.0	1.52	260	180	1.00	0.785	14.7

Sampler	d_{50}	d_{50}	N	Q	D_j or W_j	A_j	U_j
BIOTEST Reuter Centrifugal Sampler (BIOTEST n.d.; Macher and First, 1983)							
Standard RCS	3.8	7.5		280			
RCS-Plus				100			
Membrane Filter Samplers							
Burkard Spore Trap (1,7-day)							
Standard Nozzle	3.70		1	10	2.00	28.0	5.95
High-efficiency nozzle	2.17		1	10	2.00	10.0	16.7
Burkard (Personal) Sampler							
Slit			1	10			
Seive	4.18		100	20	1.00	0.785	1.94
Burkard May-Type Multi-Stage Impinger							
Stage 1	10		1	20			
Stage 2	4		1	20			
Stage 3			1	20			
Allergenco MK-2			1				

Note: d_{50} = cut diameter or aerodynamic diameter above which the collection efficiency of the impactor approaches 100%, both the true and the theoretical d_{50}'s are shown; Q = airflow rate; D_j or W_j = diameter of seive or hole j or width of slit j; A_j = area of hole j or slit j; U_j = velocity of air through hole j or slit j.

osol samplers are shown in Table 8.1. The physical characteristics (airflow rate, diameter of hole or width of slit, area of hole or slit, and velocity of air through the hole or slit) were used to calculate the theoretical cut diameters of the listed samplers. The theoretical characteristics were discussed in the preceding subsections. See also Jensen et al. (1992) for a comparison of bioaerosol samples in a wind tunnel.

In choosing a sampler, one would not collect *Escherichia coli* on a membrane filter for culturable counting because the cells would desiccate and become either nonviable or viable but not culturable under these conditions (Jensen et al., 1992). In another example, one would not try to collect *Aspergillus niger* spores ($d_{ae} \approx 1-3$ μm) using an impactor that has a d_{50} of 4 μm because most spores would remain entrained in the air passing through the instrument. General guidelines for matching the appropriate sampler with the bioaerosol of interest are shown in Table 8.2. The bioaerosol of interest categories include culturable bioaerosol sampling and nonviable bioaerosol sampling. Subcategories of include free bacteria (i.e., mostly single cells), free fungi (i.e., mostly single spores), and clumped bacteria and fungi with MMAD $\geq$ 4 μm. Culturable bioaerosol sampling instruments must minimize injury during the collection process and maintain the culturability of the collected microorganisms. In particular environmental settings, free bacteria and fungi are the bioaerosols of interest, and the sampler must collect these small aerosols. In many cases, however, the bioaerosols will be clumps of microorganisms or microorganisms attached to another particle such as a skin scale or piece of lint. In using any culturable bioaerosol sampler, the sampling time must be selected, relative to estimated concentration, such that 30–100 colonies (up to 300 in some situations) develop per plate (Tortora et al., 1989). The lower limit (30 colonies) is necessary to obtain sufficient statistical power for comparison purposes. The upper limit (100–300 colonies) is the maximum range in which one could easily count and differentiate colonies. When sampling for nonviable microorganisms or when viability (culturability) are not issues of concern, collection efficiency is the overriding concern. Table 8.2 is not an all inclusive listing of bioaerosol samplers. Take special note of the limitations listed at the bottom of the table.

8.2. Sampling Considerations

8.2.1. Safety

Investigators should use appropriate personal protective equipment (PPE) and practice good personal hygiene when investigating an outbreak of infectious disease or investigating environmental conditions that have resulted in medically diagnosed symptoms. PPE may include respiratory protection to prevent inhalation of microbes, and microorganism-resistant clothing to prevent the transmission to humans. Good personal hygiene practices include washing exposed skin

Table 8.2. General guidelines for matching the appropriate sampler with the bioaerosol of interest.

Sampler	Culturable Bioaerosol Sampling			Nonviable Bioaerosol Sampling	
	Free Bacteria MMAD < 4 μm	Free Fungi MMAD < 4 μm	Clumped MMAD ≥ 4 μm	Bioaerosols MMAD < 4 μm	Bioaerosols MMAD ≥ 4 μm
Andersen 6-Stage	√[a]	√[a]	√[a]		
Andersen 2-Stage	√[a]	√[a]	√[a]		
Andersen 1-Stage	√[a]	√[a]	√[a, b]		
Mattson-Garvin Slit-to-Agar	√[a]	√[a]	√[a]		
Ace Glass All-Glass Impinger-30	√[c, d]	√[c, d]	d, e		
PBI Surface Air System					
Compact	f	f	√[g]		
Standard	f	f	√[g]		
BIOTEST Reuter Centrifugal Sampler					
Standard			√[g]		
RCS-Plus			√[g]		
Membrane-Filter Samplers	√[c, h]	√[c]	√[c]	√[c]	√[c]
Burkard Spore Trap (1,7-Day)					
Standard nozzle					√[f]
High-efficiency nozzle				√	√
Burkard (Personal) Sampler					
Slit					√
Seive			√[g]		
Burkard May-Type Multi-Stage					
Impinger	√[c, d]	√[c, d]	e		
Allergenco MK-2					√

Note: √ denotes satisfactory for specfied application. [a]denotes concentrations greater than 5000–7000 cfu m^{-3} will overload sample. [b]denotes May underestimate concentration of large bioaerosols (MMAD > 10 μm) due to impactor entry losses. [c]denotes Good for very low to very high concentrations. [d]denotes Bioaerosols may be reaerosolized and drawn out of the impingers during sampling, resulting in an underestimation of concentration and a decrease in precision. [e]denotes May overestimate concentration due to breaking up of clumps. [f]denotes May be acceptable with new sampling head being evaluated by PBI/Spiral Biotech. [g]denotes Concentrations greater than 1000–2000 CFU m^{-3} will overload sample. [h]denotes For desiccation-resistant bacteria only. [i]denotes May underestimate concentration of small bioaerosols (MMAD ≤ 5 μm).

and clothing thoroughly and refraining from eating, drinking, and smoking in a contaminated area. These simple steps will help minimize the ingestion, inhalation, or assimilation of microorganisms.

All samplers, plates, equipment, and so forth, should be handled aseptically to prevent contamination of the samplers and, more importantly, prevent spread of potential human pathogens to the worker or the work environment (CDC/ NIH, 1993; Block, 1991). All surfaces, including washed hands, contain microorganisms or spores unless they are specifically sterilized. Practically speaking, however, not all objects may be sterilized. Therefore, disinfection with an oxidizing chemical or alcohol destroys most vegetative cells, but these agents do not destroy all spores. Samplers should be disinfected or sterilized after each sample collection. Special care should be given to samplers with convoluted inlets or air pathways where microorganisms may accumulate.

8.2.2. Environmental Conditions

Airborne bacteria are subject to dehydration caused by the evaporation of water from bacterium-carrying droplets and of cellular water (Marthi and Lighthart, 1990). The degree and rate of evaporation are dependent on ambient relative humidity and temperature. In field experiments (greenhouse), survival of certain bacteria was 35- to 65-fold higher at 80% RH than at 40% (Walter et al., 1990). In laboratory experiments, survival of certain bacteria was virtually complete at low RH but was reduced at RH values above 80% (Cox, 1968). Cox (1987) believes the potential for the movement of the solvent water is the most important environmental criteria in assessing survivability of bacteria, viruses, and phages. Limited studies have been made of temperature effects. Temperature also induces morphological changes in dimorphic fungi. *Histoplasma capsulatum,* a pathogen, exists as a spore or mycelial form below 25°C. However, higher temperatures have been shown to induce a transition from the mycelial form to the yeast form (Salvin, 1949). Like most particles, freshly generated microbial aerosols are nearly always electrostatically charged unless steps are taken to neutralize them. There appears to be very little published information about electric charges on actual workplace aerosols, and even less on bioaerosols (Johnston et al., 1985). In general, the effect of electrical charge has been overlooked, resulting in the possible bias of sampling results. At RH values above 70% RH, electrostatic phenomena are minimal (Hinds, 1982; Cox, 1987).

8.2.3. Flow Calibration

Accurate airflow rates are very important in calculating the concentration of microorganisms in the air. All samplers should be calibrated before *and* after sampling to ensure the flow rate was within the manufacturer's specifications and did not change from the initial calibration. Where it is not possible to calibrate using a primary standard, a secondary standard such as a dry gas meter may be used. The calibration of such a secondary standard should be traceable to a primary standard.

8.2.4. Culture Media

General detection and enumeration media are normally used in the collection of fungi, bacteria, and thermophilic *Actinomycetes*. As previously mentioned, plates can be replicated on differential or selective media for identification after the organisms have been collected. The following subsections are some general guidelines for media.

8.2.4.1. Fungi

Traditionally, malt extract agar (MEA) and rose bengal agar (RBA) have been recommended as broad spectrum media for the collection and enumeration of fungi (Morring et al., 1983; Burge et al., 1977; Smid et al., 1989). Worthy of note is the fact that MEA and RBA are generic terms and formulations will vary from supplier to supplier and laboratory to laboratory. One MEA recipe is a less nutritious, unamended 2% MEA which is reported to promote better sporulation than MEA amended with glucose and/or peptone (Hunter et al., 1988; Strachan et al., 1990). With RBA, the colonies remain small; however, the light may make the medium toxic to some fungi. In addition, the pigmentation of the fungal growth on RBA complicates the identification process. Based on the work of researchers in Holland, Dichloran Glycerol 18 agar (DG-18) is recommended for identification and enumeration of fungi (Hocking and Pit, 1980; Pit et al., 1983). This medium is adequate for most fungi, including xerophilic fungi. DG-18 does not have the disadvantages of RBA. To inhibit the growth of bacteria, antibiotics such as streptomycin may be added to the medium (DIFCO, 1984).

8.2.4.2. Bacteria

Tryptic soy agar (TSA), casein soy peptone agar (CSPA) and nutrient agar (NA) are broad-spectrum media for the collection and enumeration of bacteria. Special purpose media (i.e., selective media) are often used to select for specific microorganisms of interest. As with the media for fungi, growth-restricting chemicals may be added to the media.

8.2.4.3. Thermophilic Actinomycetes

Thermophilic *Actinomycetes* are a special class of bacteria which have been associated with indoor environment quality problems. Casein soy peptone agar (CSPA) and tryptone glucose yeast agar [also known as standard methods agar (SMA) and standard plate count agar (SPCA)] are broad-spectrum media for the collection and enumeration of thermophilic *Actinomycetes* (DIFCO, 1984; Amner et al., 1989; Burge et al., 1989).

8.2.4.4. Additional Media

Other media for the detection and enumeration of fungi and bacteria may be used. To discriminate for a general class of microorganisms by inhibiting or eliminating other microorganisms, a selective medium containing an antibiotic or other growth-restricting chemical may be used. To distinguish among species, a differential media may be used. Differential media contain indicators that permit the recognition of microorganisms with particular metabolic activities. Different incubation times or temperatures can also be used to get differential growth on the same medium. When a specific medium (i.e., selective or differential) is needed, the investigator(s) should refer to the most recent *Manual of Clinical Microbiology* published by the American Society for Microbiology, *DIFCO Manual: Dehydrated Culture Media and Reagents for Microbiology* published by DIFCO Laboratories, or the various catalogues published by the American *Type Culture* Collection (ATCC) (DIFCO, 1984; Lennette et al., 1985; Gherna et al., 1989; Jong and Edwards, 1991).

8.2.5. Blanks

8.2.5.1. Laboratory Media Blanks

Laboratory media blanks are unexposed, fresh media samples. These samples are generally not taken into the field. Before using any batch of media, incubate at least three culture plates under the same conditions as the field samples will be treated to check for sterility. Approximately five media blanks should be included with each sample set. If the samples are to be analyzed by an outside laboratory, consult the specific laboratory procedure for the number of blanks to be submitted.

8.2.5.2. Field Blanks

Field blanks are simply unopened, fresh media samples that are handled in every way the same as field samples, including labeling, except that no air is drawn through the sampler. The recommended practice for the number of field blanks is two field blanks for each 10 samples with a maximum of 10 field blanks for each sample set.

8.3. Chambers and Generators for Bioaerosols

Most microorganisms are of little interest to aerobiologists because of their poor aerosol stability and subsequent rapid die-off rate. When evaluating an aerosolized microorganism for possible health effects or infectivity, it is most important to determine its survival characteristics under various environmental conditions.

The results for many aerosol stability tests performed on the same microorganism often differ because testing techniques have not been standardized. Harper (1963) noted that variations in results can be caused by the following:

1. Method of aerosol generation, storage, and sampling procedure
2. Method of assay
3. Differentiation of total, physical, and viability decay
4. Presence or absence of light
5. Methods and extent to which relative humidity and temperature are controlled
6. Method of data presentation

Additionally, variations can occur because of choice of suspending fluid, collection fluid, and content of atmosphere (presence or absence of oxygen and other gases) used for testing.

As laboratory methods have developed, the effects of many variables such as chamber design, generation method, sampler variety, the composition of the fluids used for generation and sampling, physical tracer types, and assay methods have also matured. This is where the science of applied bioaerosols enters. The following subsections will introduce the various methods used for the generation, collection, containment, and general characterization of bioaerosols.

8.3.1. Chambers Used for the Characterization of Bioaerosols

Several types of chambers, usually variations of static or dynamic designs, have been found to be of use for the study of bioaerosols. A static chamber is one where the aerosol is generated for short periods of time for a short-duration experiment. A dynamic chamber is usually used to study the aerosol stability of microorganisms and the generation period can be either short or relatively long. Both static and dynamic aerosol test chambers come in all shapes and dimensions, but the operational principals are generally similar. Aerosols are usually generated into a mixing chamber or tube into which secondary air is also fed which has been conditioned to a predetermined temperature and relative humidity. The mixing chamber should be designed so that the aerosol is thoroughly blended with the secondary air before it reaches the actual test chamber. To achieve thorough mixing, the length of the duct should be 10 times the diameter of the mixing tube. If this is not possible, small-diameter metal tubing can be stacked until the inside of the mixing tube is filled. This assembly of small tubing should be twice the diameter of the mixing tube and will act as a flow straightener. A screen can be placed on the downwind end of the flow straightener to induce turbulence and enhance mixing (Baines and Peterson, 1951).

Physical loss of aerosols is directly related to the surface area of the test chamber. Physical loss is also related to the number of angles and bends in the test system plumbing. Test chambers should be designed to minimize the number of bends and angles in the system and, if possible, the test chamber should be placed in a vertical position so that the influence of gravity is minimized. Particles less than 5 μm tend to follow the streamlines of the airflow if turbulence is low. For particles larger than 5 μm, it is essential that bends and angles are kept to a minimum. Chambers should be constructed of stainless steel for decontamination and to control electric charge effects. The chamber should be grounded so that electrical charges do not build up on surfaces and increase physical losses. If stainless-steel construction is not practical, other metals can be used. Materials which could absorb water vapor, such as wood, should be avoided. Some materials such as plexiglass or plastics are sensible materials for chamber construction because they allow observation of the test system or aerosol. Special care must be taken to minimize electrical charge effects. This can be accomplished with the use of antistatic materials, but anytime a foreign substance comes in contact with a bioaerosol, its effect on viability must be evaluated. This includes materials such as steam, paraformaldehyde, hydrogen peroxide, and ethylene oxide, which are used for the decontamination of chambers.

Air systems for bioaerosol testing should be designed so that the chamber can be purged in a short period of time. Between repetitive tests, with the same organism, the chamber should be purged with an amount of air that is equal to 10 times the volume of the chamber. The air supply should be oil-free (Teflon-ringed or water-ringed compressors) and run through HEPA filters and air dryers before being introduced into the test chamber.

8.3.1.1. Test Chambers

The properties of aerosolized microorganisms are often a function of the time spent in the airborne state as well as temperature, relative humidity, oxygen concentration, and so forth. This aging can be accomplished using stirred settling chambers; however, Goldberg (1958) showed that a less complicated and more effective method for aerosol aging was that of generating an aerosol in a drum rotating around a central axis. This type of chamber is termed a dynamic aerosol toroid (DAT). The reported optimal rate of rotation reported by Goldberg was around three revolutions per minute. At this rate, it was reported that aerosol particles 6.0 μm and smaller could be kept suspended for 48 h with minimal physical loss. Gruel (1987) approached the question of drum rotation from a theoretical point of view and found that the optimum rotation rate depended on the density and viscosity of the suspension fluid (typically air), the drum radius, the aging period, and the suspended aerosol particle density and diameter. However, the optimum rotation rate was essentially independent of the particle size if the particles were smaller than 10.0 μm. For 2.0-μm particles, aged for 1, 6, 48, and 96 h, the optimum rotation rate was calculated to be 1.33, 0.73, 0.37, and 0.31 revolutions per minute (rpm), respectively. For these aging periods, the maximum physical loss of aerosols was calculated to be less than 3.0%. Most DATs are around 2 ft wide and 4–6 ft. in diameter. Asgharian (1992) expanded the theoretical expressions to include drum rotation at very low rates of rotation. The goal of this study was to determine the potential fate of particles in the drum after several days (up to 8 days) for use in inhalation studies. They found that for particles between 0.5 and 1 μm in diameter and for suspension times less than 96 hs there was no loss of suspended particles for drum rotation rates from 0.1 to 10 rpm. For 20-μm- and 5.0-μm-diameter particles, greater than 98% and 91%, respectively, remain suspended after 96 h under optimal rotation (optimal rotation for 2.0 μm and 5.0 μm was 0.3 rpm) of the drum chamber. Additionally, they found that optimal rotation rates were independent of particle diameter when they were smaller than 5 μm for periods less than 1 day.

Static aerosol test chambers have been found to be of limited use because short-duration studies have less application for typical bioaerosol investigations. In practice, a static aerosol chamber does not differ from an aerosol mixing tube. Static chambers are valuable devices for testing the collection efficiency of aerosol samplers.

Several chambers have been designed to study the effects of infection transmission to animals. Basically, there are three types of animal exposure chambers: whole body, head only, and nose only. Whole-body experiments normally involve the least complex test systems. Test chambers are either static (single, short generation time) or dynamic (continuous generation of test material). Currently, the most often performed investigation involves a dynamic dose–response study (Salem, 1987). The Henderson (1952) apparatus is essentially a long narrow dynamic chamber with breathing ports located along the side where animals are placed for inhalation of the test material. The aerosol concentration was difficult to control, and often the animals subjected to the aerosol first, received the highest dose. Hood (1971) designed an hermetically sealed 22-ft sphere into which the test aerosol was injected. The aerosol was mixed by three fans placed inside the chamber. The unit was designed so that aerosol could be withdrawn and utilized for whole-body inhalation studies. Druett (1969) added the toroid aging chamber concept and attached a dynamic chamber to a rotating drum to determine the influence of aerosol inactivation and infectivity. Cannon (1983) designed a nose-only system for mice which proved to deliver a consistent amount of aerosol to the test animals. His system provided separate manifolds to deliver the test aerosol to contain and expel exhaled air. A complete discussion of inhalation procedures and instrumentation will be addressed in the section on health aspects of bioaerosols.

8.3.2. Bioaerosol Generation

The physical properties of a test bioaerosol are dependent on the choice of an appropriate generation method. Two of the most important variables, particle size distribution and concentration, are directly related to the aerosol generator's performance. Initially the physical characteristics of a test aerosol are selected, which define the type of generator required, then the generator must be characterized with various types of suspending fluids and concentrations of biologics to evaluate generator efficiency.

Biological aerosols are usually produced from a liquid suspension. Most often small particles in the 1.0–10.0μm size range are required. Most bioaerosols generated from liquids are polydispersed with the exception of aerosols produced by two monodispersed aerosol generators, namely, the Spinning Top Aerosol Generator (STAG), BGI, Inc. and the Vibrating Orifice Aerosol Generator (VOAG), TSI, Inc. Both of these monodispersed generators have limited application for producing bioaerosols and will be discussed later.

The energy required to produce these small particles can be supplied by a pressurized airstream, electricity, shearing forces, centrifugal forces, impaction, or heat. Many of these forces are so violent that inactivation of microorganisms will result from the process. Other parameters which influence the particle size and viability are the flow of secondary dilution air and the makeup of the suspending fluid. After generation, the liquid associated with a newly aerosolized particles will instantaneously start to evaporate, except at high relative humidities, and an equilibrium with the surrounding water vapor will be established leaving a solid particle. The diameter of the resultant particle will undoubtedly be smaller than the originally manufactured particle.

Bioaerosols may be disseminated from the dry state (Cox, 1970; Cox, 1971), but most microorganisms are inactivated when dried. The energy applied to aerosolize dry particles may come in the form of compressed air, scraping a dried cake and then applying air, or by employing an explosive device. Dry disseminated aerosols may inherit excessive electric charge through the dissemination process causing problems with particle loss in test fixtures.

The following discussion pertains to specific aerosol generators and will be presented according to the basic way energy is applied to a liquid or powder to produce aerosol droplets.

8.3.2.1. Two-Fluid Aerosol Generators

Many bioaerosols are produced by the application of compressed air to a small liquid jet. These two fluids, air and water, give this class of atomizers their name. In application, these types of generators are simple and popular and most often used when a biological aerosol is required.

Collison Nebulizer. Compressed air is applied to this nebulizer (Fig. 8.1) in the range between 15 and 30 psi (May, 1973). A high-velocity air jet is created inside the unit which forms a low-pressure area above the liquid. This venturi action causes the suspending fluid to be drawn up into the inner tube, and

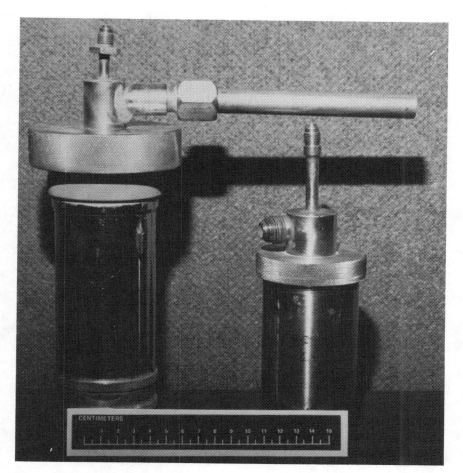

Figure 8.1. Collison nebulizer.

atomization occurs when the fluid is mixed with the air jet and forced against the inner side of the fluid reservoir. Small particles are carried out of the atomizer with a gentle reflux action created by the positive pressure inside the body of the instrument. The three-jet unit puts out approximately 0.25 ml/min but may be altered to increase the fluid output. A mass median aerodynamic diameter (MMAD) around 2.5 μm is produced by the Collison nebulizer. The concentration of solids in the suspending fluid may increase due to larger droplets rapidly settling out and carrying a high proportion of the particulate matter back into the suspending fluid. The relatively gentle reflux action and atomization process of this instrument make it suitable for aerosolizing many fragile vegetative microorganisms and complex phages. Additionally, the Collison nebulizer has been used for inhalation studies where small particles are desired (Young et al., 1977).

Wells Atomizer. This atomizer produces very small particles (0.5–3.0 μm) (Faber et al., 1944; Rosebury, 1947) and is usually used for inhalation studies when the required site of deposition is deep in the respiratory system. This device consists of a flask with an inlet and outlet set at a 90° angle. The inlet is formed by a tube which passes through the liquid suspension located at the bottom of the flask and discharges into the air space of the flask. A second tube surrounds the air inlet at the point of discharge and draws the fluid up into the air space of the flask, where it impacts the top of the flask. The resulting aerosol is discharged through the outlet tube. Concentration of solids in the fluid also occurs in this device because of the recirculation of fluid. The output of the Wells atomizer is quite low, around 0.1 ml min^{-1}.

DeVilbiss Nebulizer. This unit (Larson et al., 1976) is similar to Wells atomizer except the air line and the liquid feed tube are nearly in line with the outlet tube. Again, a low-pressure area is formed by the presence of a high-velocity jet which draws fluid upward where it is atomized by the air jet. The particles impact on the side of the vessel; the large ones fall back into solution, whereas the small ones are carried out with the reflux of the outgoing air. Most of the particles produced are in the 5.0-μm range. The fluid output is in the 0.2–2.0-ml min^{-1} range, and concentration of solids in the suspending fluid may occur. The pressure drop across the nozzle is high enough to inactivate many fragile microorganisms but not enough to inactivate spores and many other viruses or phages.

Chicago Atomizer. The Chicago atomizer (Fig. 8.2) is an all-glass two-fluid atomizer. Compressed air (15–30 psi) surrounds the liquid jet and breaks up the fluid as it exits a narrow orifice which is slightly recessed inside the tip of the atomizer. The generator is quite versatile in that the liquid feed rate is controlled with a syringe

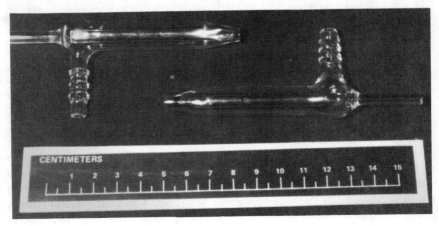

Figure 8.2.   Chicago nebulizer.

pump (from very low fluid rates up to 20.0 ml/min). The particle size produced (MMAD around 2.5–3.0 μm) is quite constant from low fluid outputs up to 18–20 ml min^{-1}. When a high-volume, small-particle-size-producing atomizer is required, the Chicago atomizer may be the instrument of choice. The pressure drop across the nozzle is high enough to inactivate many fragile microorganisms but not enough to inactivate spores and many other viruses or phages.

Babington Aerosol Generator. The Babington aerosol generator (Fig. 8.3) was originally designed to produce a very fine aerosol to improve the combustion of flammable liquids. Compressed air is introduced into a hollow sphere into which a small hole or slit has been manufactured. Liquid is introduced onto the outer shell of the sphere with a bubble pump, syringe pump, or a peristaltic pump in a very thin layer. As the liquid flows over the hole, it is atomized by escaping compressed air. Depending on the size of the orifice in the sphere and the air pressure, the unit will produce an aerosol with an MMAD from submicrometer sizes to the 10-μm size range. The fluid output of the generator can be adjusted from low liquid flows to over 1.0 ml min^{-1}. The atomization method is very gentle, making this instrument valuable when aerosolizing vegetative bacteria, phages with complex structures, or other fragile microorganisms.

8.3.2.2. Electrical Generators

Vibrating-Orifice Aerosol Generator. The VOAG (Bergland and Liu, 1973; Leong, 1986; Liu et al. 1974; Ramiarz, 1982) has limited application for the

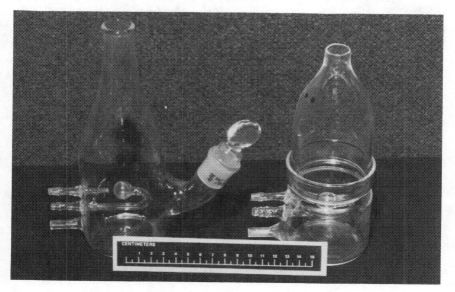

Figure 8.3. Babington aeriosol generator.

production of bioaerosols and is quite difficult to operate (Fig. 8.4). Liquids are fed (syringe pump) to a very small orifice which is oscillating at a specific frequency for the particular size of droplet to be generated. The problem arises because the liquids must be filtered (0.5-μm pore size) before they reach the orifice or it may become obstructed. Microorganisms larger than the pore size of the filter cannot pass through and will not be aerosolized. Solid particle aerosols will result, after evaporation of the liquid, if they are soluble in the original solution. The unit is useful for producing monodispersed aerosols (0.5–50.0 μm) which are necessary for the calibration of aerosol impactors and other collection devices.

8.3.2.3. Centrifugal Generators

Spinning Top Aerosol Generator. The STAG (Fig. 8.5) uses the centrifugal force of a spinning disk to created the energy required to produce droplets (1.0–50.0 μm). The standard deviation of the resultant particles is quite low (σ_g = 1.01–1.03) approaching a monodispersed aerosol. Liquid is fed (0.2–0.8 ml min^{-1}) (Mitchell, 1984) through a fine needle to the center of a rotating disk (2.54 cm in diameter) which rotates (particle size depends on the liquid and the

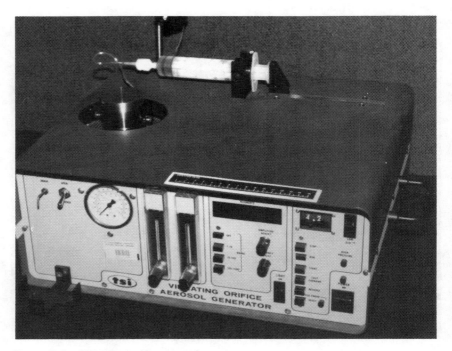

Figure 8.4. Vibrating-orifice aerosol generator.

Figure 8.5. Spinning top aerosol generator and controls.

concentration of solute in the solvent) at a high rate of speed, between 400 and 1000 revolutions per second. The size of particles can be varied by altering the rotor speed, by changing the rate of liquid feed, and by varying the concentration of solute or solids in the fluid. Modifications in the latest models allow for precise control over the rate of rotation of the disk (tachometer), control over the needle height above the disk, as well as control of small satellites (exhaust air) that are formed during the aerosolization process. The generator is not supplied with an appropriate containment system, making fabrication of a secondary vessel necessary.

8.4. Preparation, Identification, and Enumeration Procedures for Culturable Bioaerosols

8.4.1. Sample Preparation

Inoculated agar plates are incubated at the appropriate temperature for times ranging from hours for a fast-growing bacterium to develop a microcolony, to days for a fungus to develop into a visible colony and perhaps sporulated, to weeks for an organism such as multidrug-resistant *Mycobacterium tuberculosis* to produce visible colonies. As a general rule, plates are incubated at the temperatures shown in Table 8.3 (Burge et al., 1989; Baron and Finegold, 1990).

Table 8.3. *Incubation temperatures and conditions for viable (culturable)*
microorganisms.

Fungi	25°C or room temperature with natural light
Bacteria, environmental	25–30°C
Bacteria, human source	35–37°C
Bacteria, thermophilic *Actinomycetes*	50–56°C

Laboratory media blanks and field media blanks are be handled in the same manner as samples.

8.4.2. Enumeration

8.4.2.1. Total Concentration

Total concentration (colony-forming units per cubic meter) of culturable microorganisms is calculated by dividing the volume of air sampled into the total number of colonies observed on the plate. It is often necessary to use a dissecting-type microscope (10X–100X) to observe more than one colony at the same impaction point. Concentrations of culturable bioaerosols are normally reported as colony-forming units (cfu) per unit volume of air. The colony-forming unit is the number of colonies that replicate from individual or groups of bacterial cells, bacterial endospores, or fungal spores.

8.4.2.2. Adjusted Concentration

Often, it is difficult to identify multiple colonies at one location because of the lack of differential colony morphology or because the chemicals secreted by one microorganism might inhibit the growth of other microorganisms at that same location (Burge et al., 1977). In addition, some organisms produce large, spreading colonies, whereas others produce microcolonies. Also, the morphology of the colony of one microorganism may completely obscure that of another and a fast-grower might obscure a slow-grower. In that case, a statistical adjustment of the observed number of colonies is needed to account for the probability that more than one particle impacted the same site (Andersen, 1958; Leopold, 1988; Macher, 1989]. The adjusted concentration (cfu m^{-3}) of culturable microorganisms is calculated by dividing the volume of air sampled into the adjusted number of colonies observed on the plate (see Section 8.1.2).

8.4.2.3. Limitations

The methods thus far pertain to culturable microorganisms. Microorganisms that are stressed or injured either by environmental conditions or bioaerosol sampling procedure may be viable, but they may not be culturable (McFeters et al., 1982). Certain species may be too fastidious to grow in laboratory culture.

For instance, some bioaerosols (e.g., *Legionella pneumophila, Histoplasma capsulatum,* or *Pneumocystis carinii*) are very difficult, if not impossible, to collect and culture (Ibach et al., 1954; Dennis, 1990).

8.4.3. Identification of Culturable Bioaerosols

The science of classification, especially the classification of living forms, is called taxonomy. The objective of taxonomy is to classify living organism to establish the relationship between one group of organisms and another and to differentiate between them. Several criteria and methods for the classification of culturable microorganism and the routine identification of some are discussed in the subsections that follow. In addition to using these methods, nonviable and viable (nonculturable) methods of identification as discussed in Sec. 8.5 may be used with culturable microorganisms.

8.4.3.1. Classical or General Microbiology

Classical microbiology includes general methods for classifying or identifying microorganisms. The least specific of these is the observation of growth characteristics. Growth characteristics include the appearance of the microorganisms in liquid medium, colonial morphology on solid medium, and pigmentation. On the cellular level of bacteria, cell shape, cell size, arrangement of cells, arrangement of flagella, capsule, and endospores are characteristic of general classes of microorganisms. On the other hand, fungi are very difficult to classify. In general, fungi are classified by spore morphology or colonial morphology. Simple and differential staining may be performed on bacteria. Simple staining is a method of staining microorganisms with a single basic dye which highlights microorganisms to determine cellular shapes and arrangements using a microscope. Stains such as methylene blue, carbolfuchsin, crystal violet, or safranin may be used for bacteria; and stains such as lactophenol cotton blue, periodic acid–Schiff stain, or potassium hydroxide (10% KOH) may be used for fungi. A stain that distinguishes among structures or microorganisms on the basis of reactions to the staining procedure is called a differential stain. Two examples of differential stains are the Gram stain and the acid-fast stain. The mechanism of the Grain stain may be explained on the basis of physical differences in the cell walls of these two general groups of bacteria (Gram positive and Gram negative). The Gram-positive bacteria possess a cell wall composed of a relatively thick peptidoglycan layer and teichoic acids (Tortora et al., 1989). Gram-negative bacteria possess a cell wall composed of a thin peptidoglycan layer and an outer membrane which consists of lipoproteins, lipopolysaccharides, and phopholipids (Tortora et al., 1989). A few of the commercially available identification kits require a Gram-stain prescreening to assure that the correct reagents are used. Some species of bacteria, particularly those of the genus *Mycobacterium*, do not stain readily. In the acid-fast staining process, the application of heat facilitates the staining of the microorganism. Biochemical, physiological, and nutritional tests for bacteria evaluate cell-wall constituents, pigment biochemicals, storage inclusions, antigens, temperature range and optimum, oxygen relationships, pH tolerance, osmotic tolerance, salt requirement and tolerance, antibiotic sensitivity, energy sources, carbon sources, nitrogen sources, fermentation products, and modes of metabolism (autotrophic, heterotrophic, fermentative, respiratory). As a general rule, batteries of such tests, rather than any one individual test, are used to identify or classify microorganisms. A few commercially available test batteries are discussed in the following subsection.

8.4.3.2. Clinical and Environmental Microbiology

Historically, clinical microbiological techniques are being used for analysis of environmental samples. However, clinical strains and environmental isolates may differ, requiring modification of clinically based techniques.

8.4.3.2.1. Biochemical Analyses

All identification systems should permit the efficient and reliable distinguishment of microorganisms. A number of modifications of classical biochemical procedures have been used in recent years to facilitate inoculation of media, to decrease the incubation time, to automate the procedure, and to systematize the determination of species based on reaction patterns. Several commercial multitest systems have been developed for identification of members of the family *Enterobacteriaceae* and other pathogenic microorganisms because of the high frequency of isolation of Gram-negative rods in clinical settings. These microorganisms are indistinguishable except for characteristics determined by detailed biochemical testing. These systems require that a pure culture be examined and characterized. Following is a listing of commercially available identification kits: API® 20E (Analytab Products, Plainview, NY); Enterotube II and R/B Enteric (Roche Diagnostics Systems, Nutley, NJ; Hoffmann-La Roche & Co., AG, Basel, Switzerland); Micro-ID (Organon Teknika-Cappel, Durham, NC); Minitek and Sceptor (BBL Microbiology Systems, Cockeysville, MD); and MicroScan (American Microscan, Inc., Sacramento, CA). Automated identification systems include Quantum II (Abbott Laboratories, North Chicago, IL), Autobac IDX (General Diagnostics, Warner-Lambert Co., Morris Plains, NJ), and AutoMicrobic System (Vitek Systems, Inc., Hazelwood, MO). All 10 of these multitest systems have documented accuracies greater than 90% in clinical settings (Baron and Finegold, 1990; Konenman et al., 1988). Biolog (Biolog, Inc., Hayward, CA) is one of the newest multitest systems on the market, but its application to environmental and clinical settings is limited (Amy et al., 1992; Miller and Rhoden, 1991).

8.4.3.2.2. Cellular Fatty Acid Analysis

Cellular fatty acids (CFA) of bacteria are structural in nature, occurring in the cell membrane or cell wall of all bacteria. When the bacteria are grown under standardized growth conditions, the CFA profiles are reproducible within a taxon, down to the subspecies or strain level in some microorganisms. The Microbial Identification System (MIS), developed by MIDI (Newark, DE), provides a chromatographic technique capable of identifying various microorganisms based on their CFA composition (Sasser 1990a, 1990b). MIS has a database containing the analysis libraries for culturable Gram-negative and Gram-positive bacteria, and yeasts. Methods for extracting and analyzing fungi are under development. In a comparison study, Amy et al. (1992) found that of 18 isolates identified by either API or MIS, only 8 of the isolates were identified accurately for Biolog.

8.4.4. Interpretation of Data

Generally speaking, the literature is divided on whether identification is necessary or recommended. If clinical or research aspects of the investigation would benefit by identification of the source of an etiologic agent, the following general guidelines are suggested:

Dose–response data are not available for most microorganism exposures. Indoor bioaerosol levels must be compared with outdoor levels or with an asymptomatic control area. In general, indoor levels are lower than outdoor levels, and the taxa are similar (Adams and Hyde, 1965; Solomon et al., 1980). The Bioaerosol Committee of the American Conference of Governmental Industrial Hygienists (ACGIH) states that all interpretations of health risk due to saprophytic fungi are made with the understanding that the outdoor fungi concentration routinely exceed 1000 cfu m^{-3} and may average near 10,000 cfu m^{-3}. No occupational exposure limit for bioaerosols has been promulgated by the Occupational Safety and Health Administration (OSHA). Burge et al. (1989) indicate that except where immunosupressed people are routinely present, levels less than 100 cfu m^{-3} of nonpathogenic microorganisms are not of concern. However, the population of microorganisms must be evaluated for potential toxigenic microorganisms or microorganisms which emit microbial volatile organic compounds MVOCs. Thus, a low airborne concentration of microorganisms, in and of itself, does not indicate a clean and healthful environment.

Where local amplification and dissemination of bacteria have not occurred in an occupied, indoor environment, Gram-positive cocci (e.g., *Micrococcus* and *Staphylococcus*) are normally dominant (Morey et al., 1986). ACGIH states that high levels of Gram-positive cocci, primarily shed from human skin and respiratory secretions, may indicate overcrowding and inadequate ventilation. Dominance of Gram-negative bacteria in the indoor air is highly suggestive of the need to identify and remediate the source of contamination. Concentrations ranging from 4500 to 10,000 cfu m^{-3} have been suggested as the upper limit for ubiquitous bacterial aerosols (Nevalainen, 1989; Burge et al., 1989). These exposure limits, however do not apply to pathogenic microorganisms.

Actinomycetes (mesophilic and thermophilic) are unusual in nonfarm, indoor environments and their presence indicates contamination (Banaszak et al., 1970; Lacey and Crook, 1988). Thermophilic *Actinomycetes* at concentrations above 70 cfu m^{-3} in an affected person's work area have been regarded as the threshold for triggering remedial action (Otten et al., 1986).

8.5. Additional Identification and Enumeration Procedures for Nonviable or Viable Bioaerosols

Classifying nonviable and viable (nonculturable) microorganisms cannot be performed using the methods described in the previous section. Identification of nonviable and viable (nonculturable) microorganisms or components of microorganisms without culturing can be performed using microscopy and molecular biology techniques. In addition, microscopy techniques may be used for the enumeration of suspensions of viable and nonviable microorganisms.

8.5.1. Microscopy

8.5.1.1. Bright Field or Light

In bright-field or light microscopy, an ordinary microscope is used for simple observation or sizing. Ordinary light is used as a source of illumination and the specimen appears against a bright backfield. Objects smaller than 0.2 μm cannot be resolved. The image contrast (visibility) decreases as the refractive index of the substance/microorganism under observation and the mounting medium become similar. To maximize the contrast, the mounting medium should have the same refractive index as glass or the immersion oil. Membrane filters are often "cleared" by using the appropriate immersion oil. This method is commonly used to observe various stained (killed) specimens and to count microorganisms. In addition, pollen grains and fungi spores are often identified and enumerated in this manner (Eduard et al., 1990).

8.5.1.2. Phase Contrast

Phase-contrast microscopy is used when the microorganism under observation (e.g., *Escherichia coli*) is nearly invisible and an alternative mounting medium is not possible or permissible. A phase-contrast microscope uses a special condenser and diffraction plate to diffract light rays so that they are out of phase with one another. The specimen appears as different degrees of brightness and contrast. One cannot see an object *exactly* matching the refractive index of the mounting liquid; however, very slight differences produce visible images (Donham et al., 1986). This type of microscope is commonly used to provide detailed examination of the internal structures of living specimens; no staining is required.

8.5.1.3. Fluorescence

Fluorescence microscopy uses ultraviolet or near-ultraviolet source of illumination that causes fluorescent compounds in a specimen to emit light. Fluorescence microscopy for the direct count of microorganisms has been described in a number of studies. Direct count methods using fluorochromes, such as acridine orange, were originally developed for the total enumeration of microorganisms in soil samples (Strugger, 1948). This method has also been used to estimate total aquatic bacteria and for the rapid determination of bacteria in food products (Zimmerman and Meier-Reil, 1974). More recently, Palmgren applied this method to airborne microorganisms and concluded that it is of the utmost importance to combine viable counts with total count enumeration in the study of microorganisms in work-related situations (Palmgren et al., 1986).

8.5.1.4. Electron

Electron microscopy uses a beam of electrons instead of light. Because of the shorter wavelength of electrons, structures smaller than 0.2 μm can be resolved. Scanning electron microscopy (SEM) is used to study the surface features of cells and viruses (usually magnified $1000\times-10,000\times$); and the image produced is three dimensional. Also, SEM allows direct characterization of how airborne microorganisms appear (i.e., as single spores, in aggregates, or as spores bound to other particles). The size of the particles and the shape of individual microorganisms can also be determined. Taxonomic characterization is possible to a certain degree with morphological criteria. All organisms can be counted regardless of viability or aggregation (Karlsson and Malmberg, 1989). Transmission electron microscopy is used to examine viruses or the internal ultrastructure in thin sections of cells (usually magnified $10,000\times-100,000\times$); and the image produced is not three dimensional.

8.5.2. Endotoxin Assay

A virulence factor possessed by all *Enterobacteriaceae* (as well as other Gram-negative bacteria) is the lipopolysaccharide, endotoxin, found in the outer membrane of the cell wall. Individuals may experience disseminated intravascular coagulopathy, respiratory tract problems, cellular and tissue injury, fever, and other debilitating problems. Within the hemolymph (bloodlike circulating fluid) of the horseshoe crab, *Limulus polyphemus,* are numerous circulating cells called amebocytes. The lysate of these amebocytes gel in the presence of minute amounts of lipopolysaccharides (endotoxin) from the cell walls of Gram-negative bacteria.

The test for gelation of this material, known as *Limulus* amebocyte lysate (LAL) assay, has been used by the clinical industry to detect Gram-negative bacterial contamination (Baron and Finegold, 1990). High levels of airborne endotoxin have been reported from many environments, including agricultural, industrial, and office workplaces (Rylander and Vesterlund, 1982). Endotoxin aerosol measurement techniques lack comparability between results obtained in different laboratories because of the *Limulus* method *and* differing extraction methods (Rylander and Vesterlund, 1982; Olenchock et al., 1983; Jacobs, 1989; Milton et al., 1990).

8.5.3. Immunoassays

The immunoassay is an analytical technique for measuring a targeted antigen, which is also referred to as an analyte. A critical component of the immunoassay is the antibody, which binds a specific antigen. The binding of antibody and targeted antigen forms the basis for immunoassay, and numerous formats have been devised which permit visual or instrumental measurements of this reaction. Antibodies are commonly employed to detect organisms by binding to antigens, usually proteins or polysaccharides, on the surface or "coats" of organisms. The analysis is usually performed in a complex matrix without the need for extensive sample cleanup. Many immunoassays are now readily available from commercial sources, permitting laboratories to rapidly develop in-house immunochemical analytical capability without lengthy antibody preparation. Some of the more widely used formats are as follows:

8.5.3.1. Radioimmunoassays

Radiolabeled antigen is quantitatively added to antibody along with various concentrations of unlabeled antigen. The unlabeled antigen competes with the radiolabeled antigen for binding to the antibody. Thus, the higher the concentration of unlabeled antigen in the sample, the lower the level of radiolabeled antigen–antibody complexes. The unbound antigens are removed prior to determining the amount of radiolabeled antigen–antibody formed. A standard curve is then constructed showing the effect of the known amount of unlabeled antigen on the amount of radiolabeled antigen-antibody formed. It is now possible to determine the amount of an unknown, unlabeled antigen present in a sample by determining where the value is located on the standard curve (Garvey, 1977). Alternatively, radiolabeled antibody is employed.

8.5.3.2. Fluorescent Immunoassays

Utilization of fluorescent-labeled antibodies to detect bacterial antigens was introduced by Coons et al. (1941, 1942). Various fluorescent immunoassay (FIA) techniques have now evolved. These are referred to as (1) direct FIA, to detect antigen (cell bound) using fluorescent antibody; (2) indirect FIA, to detect antigen (cell bound) using antibody and fluorescent antigamma globulin antibody; and (3) indirect FIA, to detect serum antibody using antigen, serum, and fluorescent antibody. Various fluorescent dyes, such as fluorescein, fluorescein isothiocyanate, and rhodamine isothiocyanate, may be employed. A fluorescent microscope is used to evaluate the samples and to count the number of fluorescent organisms (Garvey et al., 1977). FIA is used to detect viruses and microorganisms.

8.5.3.3. Enzyme Immunoassays

Enzyme immunoassay (EIA) techniques usually involve the binding of an antibody or antigen to an enzyme such as horseradish peroxidase (HRP) or alkaline phosphatase (AP). Enzymatic activity, in the presence of a chromogen, results in a colored end-product. This may be measured by using a spectrophotometer (Monroe, 1984). Many, if not most, commercially available EIAs are enzyme-linked immunosorbent assays (ELISAs). In this competitive-binding EIA, the antibody is coated onto the surfaces of test tubes, or wells of a microtiter plate, antigen-containing samples and enzyme-linked antigen are added, resulting in a color change. The more intense the color, the less antigen or analyte present in the sample. ELISAs can be qualitative or quantitative. A standard curve must be generated if quantitative results are desired. ELISAs are now highly automated and efforts are underway to commercially develop well-standardized kits containing appropriate controls and materials.

8.5.4. Gene Probes

Diagnostic bacteriology, virology, and mycology are rapidly adapting molecular biology techniques in addition to classical identification methods to identify organisms. Thus, diagnostic assays utilizing nucleic acid or DNA probes have now been developed for the detection of numerous pathogenic organisms. Prior to employment of a DNA probe, it must first be demonstrated that the DNA probe is highly specific for the targeted organism. For example, a DNA probe for *Mycobacterium tuberculosis* (Mtb) should not detect other *Mycobacteria* species. However, it should detect all Mtb isolates. Described below are various types of DNA probes and formats used for the detection of organisms.

8.5.4.1. Nick-Translated DNA Probes

An isolated genomic DNA fragment is enzymatically disrupted and some of the DNA bases replaced with highly radioactive DNA bases (Maniatis et al., 1982). The radioactive probe is now tested for its ability to bind (hybridize) the extracted DNA or RNA from the organism of interest. Prior to hybridization, the targeted DNA or RNA is either fixed to a membrane, microscope slide, or resuspended in an aqueous buffer. After hybridization, the unbound DNA probe is then removed and the specimen DNA or RNA analyzed for bound DNA probe. Alternatively, nonradioactive labeling is often employed, but these probes, in general, are less sensitive than using phosphorus-32 (^{32}P) labeled DNA (Goltz et al., 1990).

8.5.4.2. Synthetic Oligomer DNA Probes

A short single-stranded DNA segment, usually 20–60 bases in length, is designed and chemically synthesized. If the precise DNA sequence of the targeted organism is known, a complementary probe, representing a perfect match to the targeted DNA, is designed. However, the precise sequence of the targeted DNA is often not known. Instead, the starting point for probe design is the amino-acid sequence. In this situation, due to codon degeneracy, a single probe exhibiting exact complementarity cannot be designed. Thus, an educated guess, based on understanding the genetic code and codon usage, is used to design the probe. This type of probe usually exhibits a high degree of matching, although seldom is a perfect match achieved. Alternatively, a set or "family" of probes is synthesized. These are designed to cover all possible DNA sequences in the targeted organism. The probes are labeled, usually with ^{32}P, prior to hybridization experiments (Maniatis et al., 1982).

8.5.4.3. Polymerase Chain Reaction

First introduced in 1985, the polymerase chain reaction (PCR) has revolutionized the way DNA analysis is conducted in clinical and research laboratories. Application of the PCR results in the amplification (the in vitro enzymatic synthesis of thousands of copies) of a targeted DNA (Saiki et al., 1985). Two synthetic, single-stranded DNA segments, usually 18–25 bases in length, are bound to the targeted DNA. These serve as primers and permit the rapid enzymatic amplification of complementary DNA. The method is extremely sensitive and specific. Culturing of the targeted organism prior to DNA extraction is often not necessary. This approach has been successfully utilized to detect various organisms including Mtb (Brisson-Noel et al., 1989; Wren et al., 1990; Eisenach et al., 1991).

8.5.4.4. Restriction Fragment Length Polymorphic Analysis

Restriction fragment length polymorphic (RFLP) is widely utilized to distinguish genetic changes within a species. A pure clone of each of the organisms of interest must be generated using standard culturing techniques. The genomic DNA is isolated and cut with a series of restriction enzymes. Each of these enzymes cut double-stranded DNA at a unique, short sequence of DNA bases, generating genomic DNA fragments of various sizes (Maniatis et al., 1982). The DNA fragments are examined by sizing them on agarose gels. Eventually, the region(s) of altered DNA is detected. The fragment appears as a different size when compared to the other isolates. Confirmation may be accomplished by using gene probes as described above.

8.6. Total (Biological and Nonbiological) Aerosol Measurement

Real-time aerosol instruments are used for a variety of tasks in industrial hygiene. Pui and Liu (1988) have summarized the advances in particle measuring and sampling instruments. Aerosol photometers and piezoelectric instruments can be used as survey instruments. Aerosol photometers (and, more recently, condensation nuclei counters) have been used in respirator fit testing (Willeke et al., 1981; NIOSH, 1987; Ernstberger et al., 1988). In conjunction with video recording, aerosol photometers have been used to identify operations or activities causing exposure (Gressel et al., 1987; Gressel et al., 1988).

8.6.1. Concentration Measurement

Aerosol photometers are versatile direct-reading instruments for indicating aerosol concentration. The measurement is based on the amount of light scattered by the aerosol particles. Light scatter is a function of the refractive index, density, and shape of the particles sampled. In environments of high relative humidity, the instrument will detect water vapor as well as the particles (Dimmick, 1961). A beam of light inside the instrument is focused onto a view volume through which the airstream passes. The airstream can be pumped through the sensitive volume or allowed to passively circulate through it. The amount of light scattered is measured by a photosensitive detector. The measured concentration data are relative to the concentration of a calibration aerosol. The optical characteristics of common aerosol photometers are such that they are most sensitive to respirable aerosols. As a first approximation, the aerosol photometers respond roughly to particle volume, so the instrument readings can be corrected for particle density by multiplying by the ratio of the actual particle density to the density of the factory calibration aerosol. Light-scattering photometers have been used to characterize penetration through HEPA filters (Biermann and Bergman, 1988), nuisance dust exposure (Gressel et al., 1987), and welding fume exposure (Glinsmann and Rosenthal, 1985).

Piezoelectric devices, such as TSI Model 3800 (TSI, Inc., St. Paul, MN), are available for the measurement of respirable aerosols (Sem et al., 1977; Swift and Lippmann, 1989). Particles are deposited onto a quartz collection crystal by electrostatic precipitation, after passing through a preclassifier ($< 3.5 \mu m$). Deposited particles increase the mass of the crystal and thereby cause a shift in the oscillation frequency of the crystal. This shift is directly proportional to the mass loading. No correlation for the particle density is required because piezoelectric devices respond directly to mass. Sem et al. (1977) found that for piezoelectric devices calibrated with welding smoke, field and laboratory data showed good agreement ($\pm$ 10%) with parallel filter samples on 10 aerosols including electric arc welding fumes, asbestos mill dust, oil mist, walnut shell abrasive dust, powdered metal dust, and cotton dust. The piezoelectric device consistently measured tobacco smoke 15% low. Experimental data also showed that sensor loading must not exceed certain limits. For most aerosols, the limit is 4 mg min m^{-3} (e.g., 2 min with a concentration of 2 mg m^{-3} or 24 s with a concentration of 10 mg m^{-3} (Sem, et al., 1977).

8.6.2. Particle-Size Measurement

Real-time cascade impactors are available using quartz crystals as the target material (see previous paragraph for discussion on piezoelectric devices). These impactors consist of several inertial impactors of decreasing cut diameter (Carpenter and Brenchley 1972; Fairchild and Wheat 1984).

Single-particle optical counters (OPC) size particles based on the amount of light scattered by individual particles. A beam of light inside the instrument is focused onto a view volume through which the particles theoretically pass one at a time. The amount of light scattered by each particle is measured by a photosensitive detector. The particle-size distribution is then determined from an analysis of the photodetector output by a multichannel (pulse height) analyzer. The size indicated is equivalent to the size of a calibration aerosol, typically polystyrene latex spheres. Estimation of the d_{ae} of an unknown aerosol from the diameter obtained from an optical particle counter (scattered light equivalent) is difficult. Light scatter is dependent on the shape and complex index of refraction of the unknown particles and on instrumental characteristics such as the geometry of the optical system and the photodetector sensitivity. Skillern (1971) used three different light-scattering aerosol photometers for testing high-efficiency filters and clean-room atmospheres, but frequent checking by optical or electron microscopy was necessary to identify the actual particle-size distribution of the aerosol.

Time-of-flight aerosol spectrometers, such as the APS Model 33B and the Aerosizer MACH 2 (TSI, Inc., St. Paul, MN and Amherst Process Industry, Inc., Amherst, MA), share some characteristics of optical particle counters: Particles are counted one at a time, using a light source and a photomultiplier tube. The lower detection limits of both instruments are limited by the amount of light scattered by an individual particle. Although OPCs and time-of-flight aerosol spectrometers are both based on light scattering, the measured quantity is different. The time-of-flight aerosol spectrometer sizes particles by measuring the transit time between two planes of laser light as the particles leave an accelerating flow field. This time is proportional to the aerodynamic size (Remiarz et al., 1983).

The maximum concentration that can be measured by both optical particle counters and time-of-flight spectrometers is limited by coincidence—the simultaneous presence of two or more particles in the viewing volume. The coincidence of two particles in an OPC causes a single larger particle to be counted (Willeke and Liu, 1976); coincidence in time-of-flight aerosol spectrometers results in randomly sized particles (Remiarz et al., 1983); and phantom particles are created when particle concentrations exceed 100 particles cm^{-3} (Heitbrink et al., 1991). These instruments incorporate circuitry or software logic to reduce this effect.

8.7. Bioaerosol Research Equipment

With the advent of large-scale applications of microorganisms in agriculture, forestry, and industry, it is expected these uses will significantly contribute to the natural atmospheric load of microbial bioaerosols creating what may be termed microbial air pollution (MAP). In a typical application, the microorganisms would be aerially dispersed in a target area such as a forest, crop, or industrial spill site, either as a liquid suspension or freeze-dried powder of vegetative, spore, or crystalline formulation. During this process, two types of unwanted drift may occur that contribute to MAP. The first occurs during the initial spray event when very fine droplets evaporate and are advected with the wind, and the second occurs as a result of reentrainment back into the atmosphere of microorganisms deposited on surfaces (i.e., plants) during the original spray event. To determine the potential for these organisms to establish themselves either in the target or nontarget areas, their airborne transport, deposition, and survival must be known. Several experimental systems have been developed or modified in recent years to determine some of these factors particularly for those microorganisms used as microbial pest control agents (MPCAs). A few of these systems include freeze-drying (see Sec. 8.7.1), a computer-controlled dynamic aerosol torroid (or autoDAT; see Sec. 8.7.2), a bioaerosol trajectory room (see Sec. 8.7.3), an environmental bioaerosol research chamber (EBARC; see Sec. 8.7.4), and a microbial bioaerosol meteorological tower (see Sec. 8.7.5) are briefly described.

8.7.1. Freeze-Drying Microorganisms

Perhaps one of the simplest methods for studying the survival of bioaerosols is freeze-drying (FD; Israeli et al., 1992). It is thought that some of the damages that occur in airborne cells also occur in FD and air-dried cells. The cellular damages are thought to be due to dehydration/rehydration and oxidative processes from exposure to atmospheric oxygen. Much of the damage is thought to occur at the cell membrane (Crowe et al., 1989a, 1989b, 1990, 1992).

There are several reasons for the usefulness of this technique: (1) FD is a simple method that may be mastered in a short time, (2) FD reduces the variability inherent in methods requiring aerial dispersal, (3) FD provides good containment conditions for handling genetically engineered microorganisms, and (4) FD can be used to generate cell powders that remain viable for extended periods of time.

An example of the use of FD was shown by Israeli et al. (1992) who demonstrated that a FD genetically engineered strain of *Escherichia coli* lost viability more quickly than the parental strain when exposed to air, light, or combination of both (Fig. 8.6).

8.7.1.1. Procedure: Freeze-Drying Microorganism

Microorganisms to be FD are grown in a shaken liquid culture until early stationary phase, then harvested and washed, and suspended in distilled water with or without protecting agents. Aliquots of the microbial suspension are placed in ampules then frozen by rotating the ampules while emersed in a dry-ice—methanol mixture. The ampules are then attached to the freeze-drier (e.g., Virtis Co. Inc., Gardiner, NY) and dried for 2–3 h under 1–10 μm Hg vacuum. Ampules can then be removed from the FD apparatus and exposed to the test

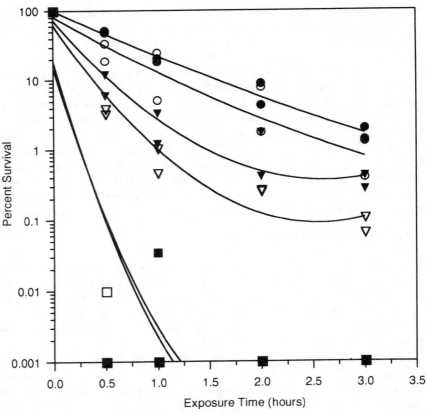

Figure 8.6. Survival of freeze-dried *E. coli* GEM and parental strains after exposure to air and light. [▽ GEM air and dark; ▼ WT air and dark; ● WT-vacuum and light; ○ GEM-vacuum and light; ■ WT-air and light; □ GEM-air and light exposure]

conditions for various lengths of time and cultured on agar plates to determine survival rates.

8.7.2. *Automated Dynamic Aerosol Torroid (AutoDAT)*

Containment of experimental bioaerosols was first done in a settling chamber and later in stirred settling chambers. The latter are described by Dimmick and Akers (1969). Later, aerosols were contained in small circular tubes to study the survival of bioaerosols shortly after their generation (Poon, 1966, 1968) and in large circular tubes called a Dual Aerosol Transport Apparatus (DATA) where effects of changing environmental conditions on the viability in bioaerosols was investigated (Dimmick and Akers, 1969). The most frequently used tool to understand the survival in small-sized microbial bioaerosols droplet/particles is to suspend then in air contained in a slowly (2–3 rpm) rotating airtight drum for extended periods of time (perhaps days). Air samples of the suspension may be periodically removed from the drum to measure the remaining viable microorganisms. The bioaerosol remains suspended due to the rotation of the drum which modifies the trajectory of the particle into a circular orbit spiraling imperceptibly outward toward the periphery of the drum. For an explanation of the DAT theory, see Goldberg et al. (1958, 1971), Dimmick and Wang (1969), and Asgharian and Moss (1992).

Recently, Shaffer, Marthi, and Lighthart (unpublished) modified the DAT to include computer-controlled operations. The computer-controlled modifications of the so-called autoDAT include aerosol generation, influent and effluent airflow, relative humidity control, and sample removal. These modifications significantly reduced experimental variability.

8.7.2.1. *AutoDAT Procedure*

The autoDAT is a 4-ft-diameter by 2-ft-deep stainless-steel drum; one side closed with a welded stainless-steel disc and the other fit with a stainless-steel cover. The cover is made air tight with a rubber gasket between the cover and drum (Fig. 8.7). The drum is mounted on a movable stand and freely rotates around a central axis. Influent and effluent air and aerosol introduction and removal are through a large nonrotating part of the axle on the welded side of the drum. An 8-in.–diameter set of spring-loaded seals seal the rotating and nonrotating parts of the axle. The airflow pattern for the autoDAT system is shown in Fig. 8.8.

In operation, a suspension of a test and a tracer organism (such as *Bacillus subtilis* var. *niger* spores) are suspended in water and nebulized (collison nebulizer, BGI Inc., Waltham, MA) into the drum at the desired test conditions of relative humidity, temperature, and gaseous composition. The tracer organism accounts for the physical decay of the aerosol. At various times thereafter, samples are withdrawn from the drum into a sampler such as an All-Glass

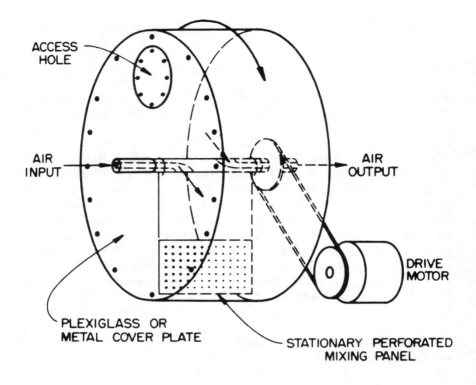

DYNAMIC AEROSOL TORROID

Figure 8.7. Photograph of AutoDAT.

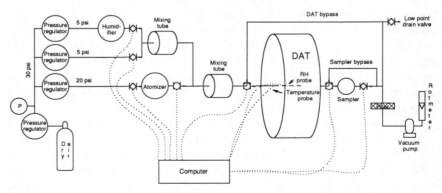

Figure 8.8. Piping diagram of autoDAT.

Impinger (AGI-30; Ace Glass Inc., Vineland, NJ). The impinger fluid is then plated onto culture media. After appropriate incubation, counts of the test and tracer organisms are calculated, and the survival of the test organisms is made using Eq. (8.1):

$$\%S = \left[\left(\frac{\text{Test organism}_{\text{sample size}}}{\text{Test organism}_{\text{initial time}}} \right) \left(\frac{\text{Tracer organism}_{\text{sample time}}}{\text{Tracer organism}_{\text{initial time}}} \right) \right]^{-1} \times 100 \quad (8.1)$$

An example of the autoDATs use is shown in Israeli, Shaffer, and Lighthart (1992) where the airborne survival of the genetically engineered *Escherichia coli* was found to be less than the parental strain (Fig. 8.9.).

8.7.3. Microbial Bioaerosol Trajectory Room

To compare theoretical and observed trajectories of droplets from a polydispersed bioaerosol, a quasi-laminar flow trajectory room was developed (Lighthart et al., 1991). The room (9 × 15 × 2.5 m) constructed of black plastic sheeting is half of an air flow-through greenhouse. The upwind end of the room is closed off with a perforated black plastic sheet (5.1-cm-diameter holes, 30.5 cm apart) which allows the air to flow from the greenhouse through the quasi-laminar flow room. To determine if the room is laminar, visual (smoke releases) and meteorological (sonic anemometer) measurements should be made. If the vertical component of the wind is constant and approaches zero, the room may be considered quasi-laminar. This allows measurements to be evaluated without the expense of a wind tunnel.

8.7.3.1. Procedure for Survival of Airborne Microorganisms— Trajectory Room

An aqueous bacterial spore suspension is sprayed from a nozzle with a known particle-size distribution into the quasi-laminar flow airstream. Downwind aerosol droplets and particles are collected in six-stage replicate air samplers (Grazeby Andersen Inc., Atlanta, GA), placed at various distances and heights downwind of the spray source, depending on the desired particle size. Collections are reported by size class as the percent of the total number of cfu collected.

Lighthart et al. (1991) demonstrated how this system may be used to study the survival of various particle-size classes without the expense of wind tunnels or monodispersed aerosol generators. If a particle > 7 μm in diameter were to be studied, sampling equipment would be set closet to the source (Fig. 8.10B). If particles 1 μm in diameter were to be studied, sampling equipment should be 5–10 m from the source, depending on the desired age, number in droplet, and particle size of the aerosol (Fig. 8.10A).

Based on the model developed by Lighthart et al. (1991), specific size fractions can be studied for survival purposes (Fig. 8.11A, B, C) and the number of bacteria

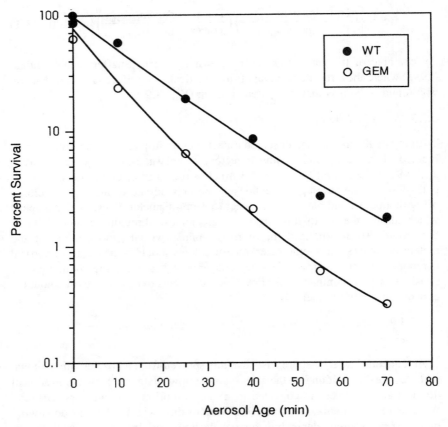

Figure 8.9. Survival of aerosolized *E. coli* GEM and parental strains after exposure to 21°C and 12% relative humidity. WT is wildtype and GEM is genetically engineered microorganism.

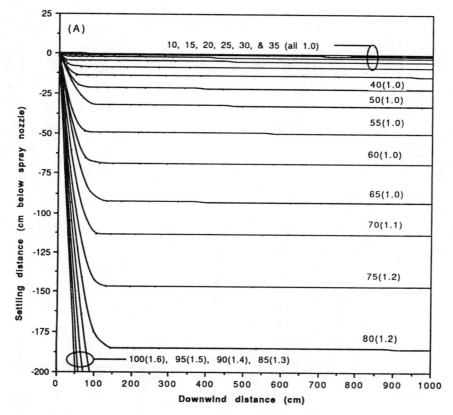

Figure 8.10. Calculated deposition trajectory of various-sized viable, evaporating droplets sprayed (at 20 lb in^{-2}) from a 2-m height in a quasilaminar flow (at 0.163 m s^{-1} airstream at 22°C and 49% relative humidity showing (A) small-droplet and particle trajectories (the numbers in the body of the figure represent particle diameters in micrometers with residue diameters in parentheses) and (B) large-droplet (200–500 μm in diameter) trajectories.

within each droplet may be altered and studied for an additional component to the particle survival function.

8.7.4. *Environmental Bioaerosol Research Chamber*

A solid particle adhering to a surface will be dislodged when the removal force provided by the drag and lift of the airstream on the particle exceeds the force of particle adhesion. This force often occurs during bursts or puffs of wind. To achieve this effect in a controlled area, the Environmental Bioaerosol Research Chamber (EBARC) was developed by Lighthart et al. (in press) (Fig. 8.12).

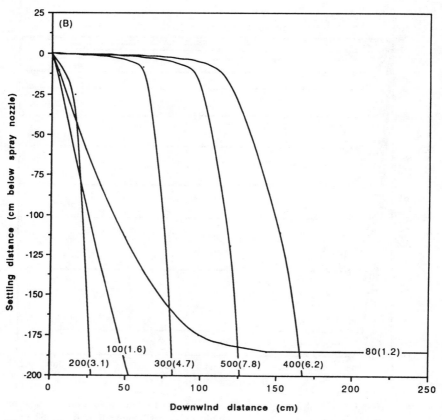

Figure 8.10. (Continued).

The 8-ft-high, 10-ft-diameter cylindrical chamber is constructed with an aluminum-channel frame and covered with several panels of polyvinyl chloride plastic film (Livingston Coatings Corp., Charlotte, NC). Air is passed through the chamber via a manifold tube forming the lower panel and out internally directed 1 inch diameter ports; the influent ports. The airflow is directed to the axis of a rotating floor and exits the building at the top of the chamber. Inoculated plants are placed on the periphery of the rotating floor and pass in front of the influent ports, subjecting them to an airflow of know geometry, consistency, duration, magnitude, and edge acceleration. Viable microbes which have been deposited on the plant surfaces are aerosolized as they pass in front of the ports and are quantitatively monitored in the downwind aerosol.

This chamber is useful in the study of microbial resuspension processes. The

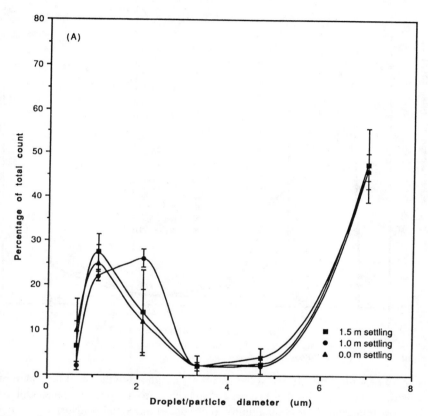

Figure 8.11. Observed percentage distributions of *Bacillus subtilis* var. *niger* median spore particle sizes from six replicate Andersen samplings at three settling heights (0.0, 1.0, and 1.5 m) below the spray nozzle in three air columns 2.5 m (A), 5.0 (B), and 10 m (C) downwind from the spray source. Vertical bars are 95% confidence limits.

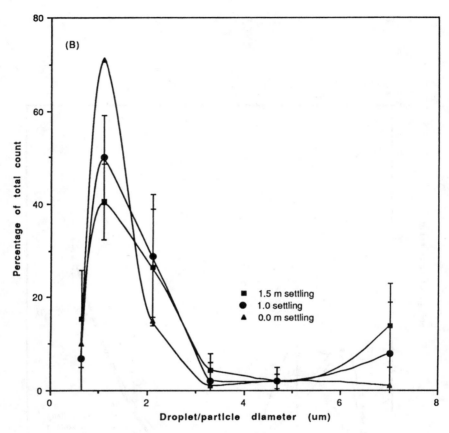

Figure 8.11. (*Continued*).

primary benefit of this design is the capability of adjusting the wind flow pattern which is very crucial to the dispersal of microbes from a plant surface. One of the most important factors thought to contribute to microbial resuspension is the leading edge of a wind puff. It has been theorized to be the critical force necessary for dispersal. The rotation of plants into the influent port wind field is thought to simulate the puff characteristics. Another benefit is that many plants can be simultaneously tested in the chamber. By subjecting many plants to identical puff characteristics, the number of viable microbes being resuspended for sampling is magnified with negligible effects from external factors influencing the resuspension of microbial particles.

8.7.4.1. Procedure for Dispersal of Microorganisms from Plant Surfaces—EBARC

Lighthart et al. (in press) inoculated oat plants with suspensions of *B. subtilis* var. *niger* spores and vegetative *Pseudomonas syringae*. Plants were then placed

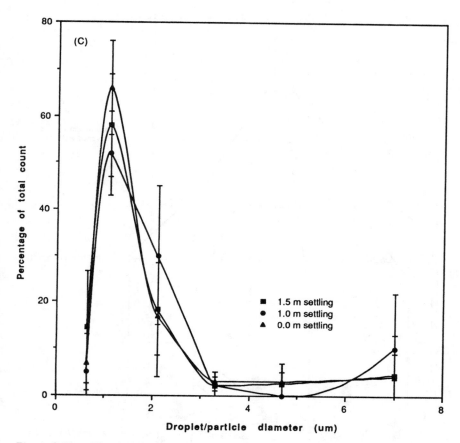

Figure 8.11. (*Continued*).

in the EBARC and subjected to successive wind puffs (either 1, 14, or 28 puffs). This research demonstrated how this chamber is useful for determining the particle-size distribution (Fig. 8.13) for particles emanating from the plant surface and the survival of the organisms being dispersed (Fig. 8.14).

8.7.5. Microbial Bioaerosol Meteorological Tower

It is anticipated that in the future, additions to the natural microbial load will occur from downwind transport of MPCAs including genetically engineered microorganisms used in agriculture, forestry, and other large-scale processes. To assess the present day (or baseline) extent of atmospheric microbial loading, the following system has been developed by Shaffer and Lighthart (1992) to evaluate present-day, large-scale ground-level sources and fluxes of airborne microorganisms (e.g., agricultural cropland, grasslands, and arid lands). Changes

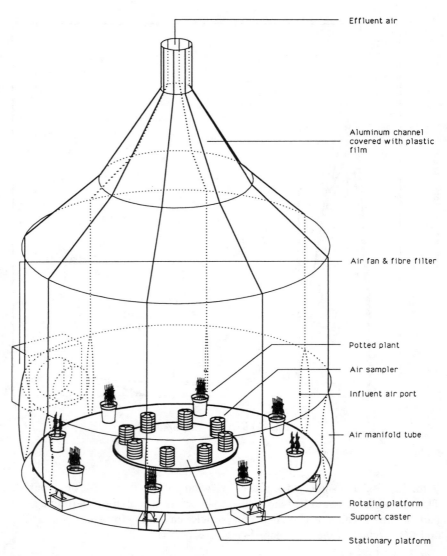

Figure 8.12. Diagram of environmental bioaerosol research chamber (EBARC).

in the microbial flux can also be used to evaluate the effects of environmental conditions.

The flux of microorganisms (e.g., microorganisms $m^{-2} s^{-1}$) emanating from surfaces that contribute to the natural microbial load in the atmospheric boundary layer is very poorly known. Microorganisms have been observed emanating from many experimental and natural sources, but only once has there been an attempt

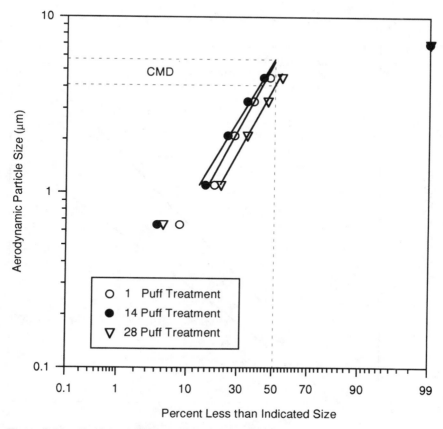

Figure 8.13. Particle-size distribution of *Bacillus subtilis* var. *niger* particles entrained from an oat plant surface following wind puff treatments using an EBARC.

to measure their flux. Lindemann et al. (1982) observed a flux of 57–543 colony forming units (cfu) m^{-2} s^{-1} contributing to a load of 506–2690 cfu m^{-3}, respectively, over several crops.

8.7.5.1. Procedure for Sampling Natural Airborne Microorganisms— Meteorological Tower

Micrometeorological and bacteriological measurements were made by Shaffer and Lighthart (1992) with instruments located at three heights on a 10-m meteorological tower (Fig. 8.15). The tower was situated with an upwind uniform fetch of at least 1000 m (approximately 100 times the sample height). Tower instruments had unimpeded acceptance angles of at least 90°. Vertical wind velocity (at 8 m), temperatures, and bacterial concentrations (at 2, 4, and 8 m)

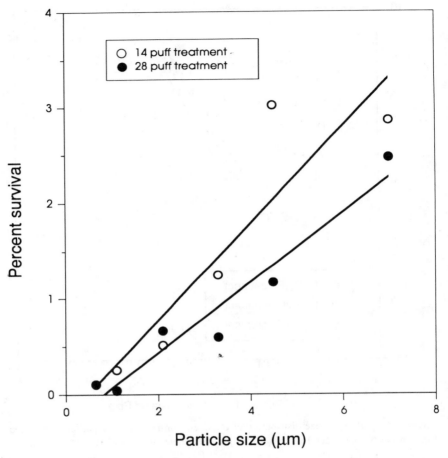

Figure 8.14. Survival of *Pseudomonas syringae* particles entrained from an oat plant surface following wind puff treatments using an EBARC.

were measured with a sonic anemometer, gill radiation shielded probes (Campbell Scientific, Inc., Logan UT), Andersen six-stage microbial samplers (Grazeby Andersen, Inc., Atlanta, GA), and/or a darkened slit sampler (New Brunswick Scientific Co., Edison, NJ), respectively. Sampling occurred between 0600 and 2200 hours. Meteorological observations were made every 2 s and averaged over the 1-h sampling periods. The modified Bowen ratio method (Stull, 1988; quoted from Huebert et al., 1988) was used to relate the bacterial flux from the local vegetation and ground to the local kinematic sensible heat flux [Eq. (8.2)]:

$$Q_B = \frac{\Delta [B] \, \overline{w'T'}}{\rho C_p \, \Delta T},$$ (8.2)

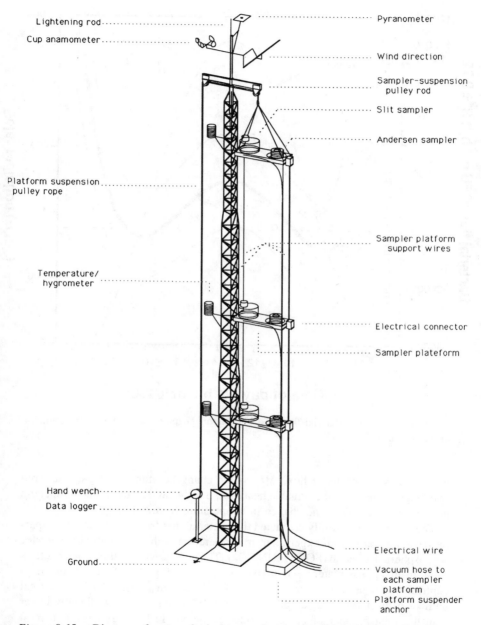

Lightening rod

Cup anamometer

Pyranometer

Wind direction

Sampler-suspension pulley rod

Slit sampler

Andersen sampler

Platform suspension pulley rope

Sampler platform support wires

Temperature/ hygrometer

Electrical connector

Sampler plateform

Hand wench

Data logger

Electrical wire

Vacuum hose to each sampler platform

Platform suspender anchor

Ground

Figure 8.15. Diagram of meteorological tower showing bacteriological sampling platforms and meteorological instruments.

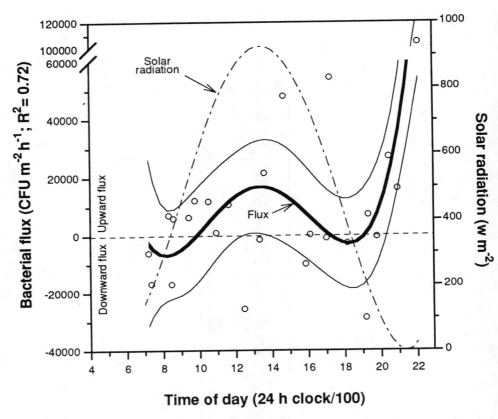

Figure 8.16. Total bacterial flux of a high desert chaparral using the instrumentation tower shown in Fig. 8.15.

where Q_B is the bacterial flux, $\Delta[B]$ is the change in number of bacteria from two heights, $[w'T']$ is the sensible heat flux, ρ is the air density, C_p is the specific heat of air, and, ΔT is the change in temperature from two heights.

The system successfully estimated the bacterial flux from a high desert chaparral terrain as part of an experiment in cooperation with the National Center for Atmospheric Research (FOOTPRINT'92). The hourly average flux of the bacteria went from approximately −6 to 6 cfu m^{-2} s^{-1} (Fig. 8.16) during the day, contributing to a atmospheric load of 20–140 cfu m^{-3} (Fig. 8.17). The total number of bacteria collected between 1100–1600 hours was significantly lower than the total number of bacteria collected before 1100 hours and after 1600 hours (Fig. 8.18). This effect is most likely due to the death of airborne bacteria from high solar radiation and temperature, lower relative humidity, lower bacterial flux emanating from the surface of the vegetation, and dilution in an increasing atmospheric mixed layer because of solar heating of the ground with resultant

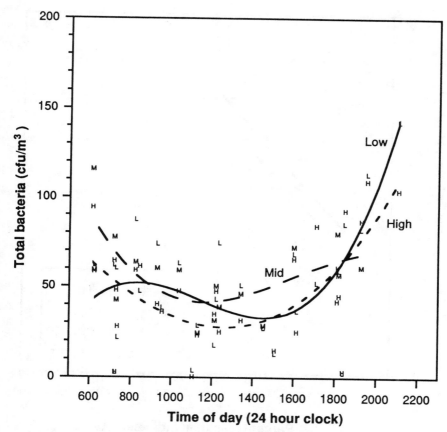

Figure 8.17. Total bacterial concentration of a high desert chaparral observed using the instrumentation tower shown in Fig. 8.15. Low (L), Mid (M), and High (H) at 2, 4, and 8 m, respectively.

convective heating. Greater than 50% of bacteria collected were found in particles ≥ 4.5 μm in diameter (Fig. 8.18), with the particle-size distribution decreasing between 1100 and 1600 hours. This effect is again most likely due to a lower bacterial flux emanating from the surface of the vegetation.

This information may be used in microbial air pollution studies including atmospheric loading of genetically engineered and nonengineered microorganisms, natural and anthropogenic microbial dispersal events, and as input to viable bacterial atmospheric dispersion models [e.g., Lighthart and Kim (1989)].

References

Adams, K. F., and H. A. Hyde. 1965. Pollen grains and fungal spores indoors and out at Cardiff. J. Polynol. 1:67.

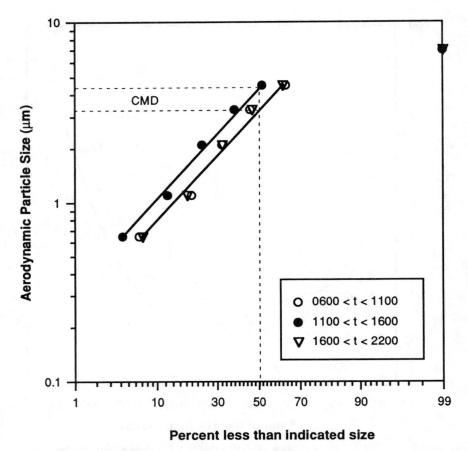

Figure 8.18. Total bacterial particle-size frequency distributions of a high desert chaparral observed using the instrumentation tower shown in Fig. 8.15.

Amner, W., C. Edwards and A. J. McCarthy. 1989. Improved medium for recovery and enumeration of the farmer's lung organism, *Saccharomonospora viridis*. Appl. Environ. Microbiol. 55(10):2669–2774.

Amy, P. S., D. L. Haldemann, D. Ringelberg, D. H. Hall, and C. Russell 1992. Comparison of identification systems for classification of bacteria isolated from water and endolithic habitats within the deep subsurface. Appl. Environ. Microbiol. 58(10):3367–3373.

Andersen, A. A. 1958. New sampler for the collection, sizing, and enumeration of viable airborne particles. J. Bacteriol. 76:471–484.

Andersen Instruments, Inc. 1984. Operating manual for Andersen Samplers, Inc. viable (microbial) particle sizing samplers. Andersen Instruments, Inc., Atlanta, GA.

APHA. 1989. Standard methods for the examination of water and waste water, 17th ed. American Public Health Association, Washington, D.C.

Asgharian, B., and O. R. Moss. 1992. Particle suspension in a rotating drum chamber when the influence of gravity and rotation are both significant. Aerosol. Sci. Tech. 17(4):263–277.

Banaszak, E. F., W. H. Thiede, and J. N. Fink. 1970. Hypersensitivity pneumonitis due to contamination of an air conditioner. New Engl. J. Med. 283(6):271–276.

Baines, W. D., and E. G. Peterson. 1951. Investigation of flow through screens. Trans. ASME 7, 467–480.

Baron, E. J. and S. M. Finegold. 1990. Bailey & Scott's diagnostic microbiology, 8th ed. The C. V. Mosby Company, St. Louis, MO.

Biermann, A. H., and W. Bergman. 1988. Filter penetration measurements using a condensation nuclei counter and an aerosol photometer. J. Aerosol Sci. 19(4):471–483.

BIOTEST-Serum-Institut GmbH. n.d. The Biotest RCS centrifugal air sampler. BIO-TEST-Serum-Institut GmbH, Frankfurt.

Block, S. S. ed. 1991. Disinfection, sterilization, and preservation, 4th ed. Lea and Febiger, Philadelphia, PA.

Brisson-Noel A., D. Lecossier, X. Nassif, B. Gicquel, V. Levy-Frebault, and A. J. Hance. 1989. Rapid diagnosis of tuberculosis by amplification of mycobacterial DNA in clinical samples. Lancet 2(8671):1069–1071.

Buckland, F. E., and R. Tyrrell. 1964. Loss of infectivity on drying various viruses. Nature 195:1063–1064.

Burge, H. A., J. C. Feeley, K. Kreiss, D. Milton, P. R. Morey, J. A. Otten, K. Peterson, J. J. Tulis, and R. Tyndall. 1989. Guidelines for the assessment of bioaerosols in the indoor environment. American Conference of Governmental Industrial Hygienists, Cincinnati: OH.

Burge, H. P., J. R. Boise, J. A. Rutherford, and W. R. Solomon. 1977. Comparative recoveries of airborne fungus spores by viable and non-viable modes of volumetric collection. Mycopatholgia 61(1):27–33.

Cannon, W. C., E. F. Blanton, and K.E. McDonald. 1983. The flow-past chamber: an improved nose-only exposure system for rodents. Amer. Ind. Hyg. Assoc. 44(12):923–928.

Carpenter, T. E., and D. L. Brenchley. 1972. A piezoelectric cascade impactor for aerosol monitoring. Amer. Ind. Hyg. Assoc. J. 33(8):503–510.

CDC/NIH. 1993. *Biosafety in microbiological and biomedical laboratories, 3rd ed.* U.S. Department of Health and Human Services, Public Health Service, Centers for Disease Control, and National Institutes of Health, Washington, D.C. HHS (CDC) Publication No. 93-8395.

Coons, A. H., H. J. Creech, and R. N. Jones. 1941. Immunological properties of an antigen containing a fluorescent group. Proc. Soc. Exp. Biol. Med. 47:200–205.

Coons, A. H., H. J. Creech, R. N. Jones, and E. Berliner. 1942. The demonstration of pneumococcal antigen in tissues by use of fluorescent antibody. J. Immunol. 45:159–165.

Cown, W. B., T. W. Kethley, and E. L. Fincher. 1957. The critical-orifice liquid impinger as a sampler for bacterial aerosols. Appl. Microbiol. 5:119–124.

Cox, C. S. 1968. The aerosol survival of *Escherichia coli* B in nitrogen, argon and helium atmospheres and the influence of relative humidity. J. Gen. Microbiol. 50(1):139–147.

Cox, C. S. 1970. Aerosol survival of *Escherichia coli* B disseminated from the dry state. Appl. Microbiol. 19(4):604–607.

Cox, C. S. 1971. Aerosol survival of *Pasteurella tularensis* disseminated from the wet and dry states. Appl. Microbiol. 21(3):482–486.

Cox, C. S. 1987. The aerobiological pathway of microorganisms. John Wiley & Sons Ltd., Chichester.

Crowe, J. H., J. F. Carpenter, L. M. Crowe, and T. J. Anchordoguy. 1990. Are freezing and dehydration similar stress vectors? A comparison of modes of interaction of stabilizing solutes with biomolecules. Cryobiology 27:219–231.

Crowe, J. H., L. M. Crowe, and F. A. Hoekstra. 1989a. Phase transitions and permeability changes in dry membranes during rehydration. J. Bioenerg. Biomemb. 21(1):77–91.

Crowe, J. H., F. A. Hoekstra, and L. M. Crowe. 1989b. Membrane phase transitions are responsible for imbibitional damage in dry pollen. Proc. Nat. Acad. Sci. USA 86:520–523.

Crowe, J. H., F. A. Hoekstra, and L. M. Crowe. 1992. Anhydrobiosis. Annu. Rev. Physiol. 54:579–599.

Dennis, P. J. L. 1990. An unnecessary risk: Legionnaires' disease. In P. R. Morey, J. C. Feeley, and J. A. Otten (eds.), Biological contaminants in indoor environments. American Society for Testing Materials, Philadelphia, PA: pp. 84–98.

DIFCO Laboratories. 1984. DIFCO manual: Dehydrated culture media and reagents for microbiology. DIFCO Laboratories, Inc., Detroit, MI.

Dimmick, R. L. 1961. A light-scatter probe for aerosol studies. Amer. Ind. Hyg. Assoc. J. 22(1):80–82.

Dimmick, R. L., and A. B. Akers. 1969. An introduction to experimental aerobiology. Wiley–Interscience, New York.

Dimmick, R. L., and L. Wang. 1969. Rotating drum. Pp. 164–176. In An Introduction to experimental aerobiology. Wiley–Interscience, New York.

Donham, K. J., L. J. Scallon, W. Popendorf, M. W. Treuhaft, and R. C. Roberts. 1986. Characterization of dusts collected from swine confinement buildings. Amer. Ind. Hyg. Assoc. J. 47(7):404–410.

Druett, H. A. 1969. A mobile form of the Henderson apparatus. J. Hyg. Camb. 67:437–448.

Eduard, W., J. Lacey, K. Karlsson, U. Palmgren, G. Strom, and G. Blomquist. 1990. Evaluation of methods for enumerating microorganisms in filter samples from highly contaminated occupational environments. Amer. Ind. Hyg. Assoc. J. 51(8):427–436.

Eisenach, K. D., M. D. Cave, J. H. Bates, and J. T. Crawford. 1991. Polymerase chain reaction amplification of a repetitive DNA sequence specific for *Mycobacterium tuberculosis*. J. Infect. Dis. 161:977–981.

Ernstberger, H. G., R. B. Gall, and C. W. Turok. 1988. Experiments supporting the use of ambient aerosols for quantitative respirator fit testing. Amer. Ind. Hyg. Assoc. J. 49(12):613–619.

Faber, H. K., R. J. Silverberg, and L. Dong. 1944. Poliomyelitis in the cynomolgus monkey. J. Exp. Med. 80:39–57.

Fairchild, C. I., and L. D. Wheat. 1984. Calibration and evaluation of a real-time cascade impactor. Amer. Ind. Hyg. Assoc. J. 45(4):205–211.

Fields, N., G. Oxborrow, J. Puleo, and C. Herring. 1974. Evaluation of membrane filter field monitors for microbiological air sampling. Appl. Microbiol. 27(3):517–520.

Garvey, J. S., N. E. Cremer, and D. H. Sussdorf. 1977. Methods in immunology. 3rd ed. W. A. Benjamin, Inc., Reading, MA, pp. 301–312.

Gherna, R., P. Pienta, and R. Cote (eds.). 1992. Catalogue of bacteria and phages, 18th ed. American *Type Culture* Collection. Rockville, MD. [Catalogs of other protists and cell lines available upon request.]

Glinsmann, P. W., and F. S. Rosenthal. 1985. Evaluation of an aerosol photometer for monitoring welding fume levels in a shipyard. Amer. Ind. Hyg. Assoc. J. 46(7):391–395.

Goldberg, L. J. 1971. Naval Biomedical Research Laboratory, programmed environment, aerosol facility. Appl. Microbiol. 21(2):244–252.

Goldberg, L. J., H.M.S. Watkins, E. E. Boerke, and M. A. Chatigny. 1958. The use of a rotating drum for the study of aerosol over extended periods of time. Am. J. Hyg. 68:85–93.

Goltz, S. P., J. J. Donegan, H.-L. Yang, M. Pollice, J. A. Todd, M. M. Molina, J. Victor, and N. Kelker. 1990. The use of nonradioactive DNA probes for rapid diagnosis of sexually transmitted bacterial infections. In A. J. L. Macario and E. C. de Macario (eds.), Gene probes for bacteria. San Diego, CA: Academic Press, Inc. pp. 1–44.

Gressel, M. G., W. A. Heitbrink, and J. D. McGlothlin. 1988. Advantages of real-time data acquisition for exposure assessment. Appl. Ind. Hyg. 3(11):316.

Gressel, M. G., W. A. Heitbrink, J. D. McGlothlin, and T. J. Fischbach. 1987. Real-time, integrated and ergonomic analysis of dust exposure during manual materials handling. Appl. Ind. Hyg. 2(3):108–113.

Gruel, R. L., C. R. Reid, and R. T. Alleman. 1987. The optimum rate of drum rotation for aerosol aging. J. Aerosol Sci. 18:17–22.

Harper, G. J. 1963. The influence of environment on the survival of airborne virus particles in the laboratory. Arch. Gesamte. Virusforsh. 13:64–71.

Heitbrink, W. A., P. A. Baron, and K. Willeke. 1991. Coincidence in time-of-flight aerosol spectrometers: Phantom particle creation. Aerosol Sci. Tech. 14(1):112–126.

Henderson, P. W. 1952. An apparatus for the study of airborne infection. J. Hyg. Camb. 50:52–68.

Henningson, E. W., I. Fangmark, E. Larsson, and L. E. Wikstrom. 1988. Collection efficiency of liquid samplers for microbiological aerosols. J. Aerosol Sci. 19(7):911–914.

Hinds, W. C. 1982. Aerosol technology. John Wiley & Sons, New York, pp. 104–126, 379.

Hocking, A. D., and J. L. Pitt. 1980. Dichloran-glycerol medium for enumeration of xerophilic fungi from low-moisture foods. Appl. Environ. Microbiol. 39(3):488–492.

Hood, A. M. 1971. An indoor system for the study of biological aerosols in open air conditions. J. Hyg. Camb. 69(4):607–617.

Huebert, B. J., W. T. Luke, A. T. Delany, and R. A. Brost. 1988. Measurements of concentrations and dry surface fluxes of atmospheric nitrates in the presence of ammonia. J. Geophys. Res. 93(D6):7127–7136.

Hunter, C. A., C. Grant, B. Flannigan, and A. F. Bravery. 1988. Mould in buildings: The air spora of domestic dwellings. Int Biod 24(2):81–101.

Ibach, M. J., H. W. Larsh, and M. L. Furcolow. 1954. Isolation of *Histoplasma capsulatum* from the Air. Science 119:17.

Israeli, E., B. T. Shaffer, J. A. Hoyt, B. Lighthart, and L. M. Ganio. 1993. Survival differences among freeze-dried genetically engineered and wild type bacteria. Appl. Environ. Microbiol. 59(2):594–598.

Israeli, E., B. T. Shaffer, and B. Lighthart. 1992. Survival and enumeration of aerosolized and freeze-dried genetically engineered *E. coli* under controlled environmental conditions. Mol. Ecol. (in review).

Jacobs, R. R. 1989. Airborne endotoxins: An association with occupational lung disease. Appl. Ind. Hyg. 4(2):50–55.

Jensen, P. A., W. F. Todd, G. N. Davis, and P. V. Scarpino. 1992. Evaluation of eight bioaerosol samplers challenged with aerosols of free bacteria. Amer. Ind. Hyg. Assoc. J. 53(10):660–667.

Johnston, A. M., J. H. Vincent, and A. D. Jones. 1985. Measurements of electric charge for workplace aerosols. Annu. Occup. Hyg. 29(2):271–284.

Jong, S. C., and M. J. Edwards (eds.). 1991. Catalogue of filamentous fungi. 18th ed. American *Type Culture* Collection, Rockville, MD. [Catalogs of other protists and cell lines available upon request.]

Karlsson, K. and P. Malmberg. 1989. Characterization of exposure to molds and actinomycetes in agricultural dusts by scanning electron microscopy, fluorescence microscopy and the culture method. Scand. J. Work Environ. Health 15(5):353–359.

Koneman, E. W., S. D. Allen, A. R. Dowell, W. M. Janda, H. M. Sommers, and W. C. Winn. 1988. Color atlas and textbook of diagnostic microbiology, 3rd ed. J. B. Lippincott Company, Philadelphia, PA.

Lacey, J., and B. Crook. 1988. Fungal and actinomycete spores as pollutants of the workplace and occupational allergens. Annu. Occup. Hyg. 15(4):515–533.

Lach, V. 1985. Performance of the surface air system air samplers. J. Hosp. Infect. 6(1):102–107.

Larson, E. W., H. W. Young, and J. S. Walker. 1976. Aerosol evaluations of the DeVilbiss no. 40 and Vaponefrin nebulizers. Appl. Microbiol. 31:150–151.

Lembke, L. L., R. N. Kniseley, R. C. V. Nostrand, and M. D. Hale. 1981. Precision of the All-Glass Impinger and the Andersen Microbial Impactor for air sampling in solid-waste handling facilities. Appl. Environ. Microbiol. 42(2):222–225.

Lennette, E. H., A. Balows, W. J. Hausler, Jr., and H. J. Shadomy (eds.). 1985. Manual of clinical microbiology. 4th ed. American Society for Microbiology, Washington DC.

Leong, K. H. 1986. On the continuous operation of the vibrating orifice aerosol generator. J. Aerosol Sci. 5:855–858.

Leopold, S. S. 1988. "Positive hole" statistical adjustment for a two-stage, 200-hole-per-stage Andersen air sampler. Amer. Ind. Hyg. Assoc. J. 49(2):A88–A90.

Lighthart, B., and J. Kim. 1989. Simulation of airborne microbial droplet transport. Appl. Environ. Microbiol. 55(9):2349–2355.

Lighthart, B., B. T. Schaffer, B. Marthi, and L. Ganio in press, 1993. Artificial wind puff liberation of microbial bioaerosols deposited on plants. Aerobiologia.

Lighthart, B., B. T. Shaffer, B. Marthi, and L. Ganio. 1991. Trajectory of aerosol droplets from a sprayed bacterial suspension. Appl. Environ. Microbiol. 57(4):1006–1012.

Lindemann, J., H. A. Constantinidou, W. R. Barchet, and C. D. Upper. 1982. Plants as sources of airborne bacteria, including ice nucleation-active bacteria. Appl. Environ. Microbiol. 44(5):1059–1063.

Lippmann, M. 1989. Sampling aerosols by filtration. Pp. 305–336. In S. V. Hering, (ed.), Air sampling instruments for evaluation of atmospheric contaminants, 7th ed. American Conference of Governmental Industrial Hygienists, Cincinnati, OH.

Liu, B. Y. H., R. N. Bergland, and J. K. Agarwal. 1974. Experimental studies of optical particle counters. Atmos. Environ. 8:717–732.

Lundholm, I. M. 1982. Comparison of methods for quantitative determination of airborne bacteria and evaluation of total viable counts. Appl. Environ. Microbiol. 44(1):179–183.

Macher, J. M. 1989. Positive-hole correction of multiple-jet impactors for collecting viable microorganisms. Amer. Ind. Hyg. Assoc. J. 50(11):561–568.

Macher, J. M., and M. W. First. 1983. Reuter centrifugal air sampler: Measurement of effective airflow rate and collection efficiency. Appl. Environ. Microbiol. 45(6):1960–1962.

Macher, J. M., and H. C. Hansson. 1987. Personal size-separating impactor for sampling microbiological aerosols. Amer. Ind. Hyg. Assoc. J. 48(7):652–655.

Maniatis, T., E. F. Fritsch, and J. Sambrook. 1982. Molecular cloning: A laboratory manual. Cold Spring Harbor Laboratory Press, Cold Spring Harbor, NY.

Marthi, B. and B. Lighthart. 1990. Effects of betaine on enumeration of airborne bacteria. Appl. Environ. Microbiol. 56(5):1286–1289.

May, K. R. 1973. The collison nebulizer: description, performance and application. Aerosol Sci. 4:235–243.

McFeters, G. A., S. C. Cameron, and M. W. LeChellier. 1982. Influence of diluents, media and membrane filters on detection of injured waterborne coliform bacteria. Appl. Environ. Microbiol. 43(1):97–103.

Miller, J. M., and D. L. Rhoden. 1991. Preliminary evaluation of Biolog, a carbon source utilization method for bacterial identification. J. Clin. Microbiol. 29(6):1143–1147.

Milton, D. K., R. J. Gere, H. A. Feldman, and I. A. Greaves. 1990. Aerosol sampling and application of a new *Limulus* method. Amer. Ind. Hyg. Assoc. J. 51(6):331–337.

Mitchell, J. P. 1984. The production of aerosols from aqueous solutions using the spinning top generator. J. Aerosol Sci. 1:35–45.

Monroe, D. 1984. Enzyme immunoassay. Anal. Chem. 56:920–931.

Morey, P., J. Otten, H. Burge, M. Chatigny, J. Feeley, F. M. LaForce, and K. Peterson. 1986. Airborne viable microorganisms in office environments: Sampling protocol and analytical procedures. Appl. Ind. Hyg. 1:R19–R23.

Morring, K. L., W. G. Sorenson, and M. D. Attfield. 1983. Sampling for airborne fungi: A statistical comparison of media. Amer. Ind. Hyg. Assoc. J. 44(9):662–664.

Muilenberg, M. L., T. M. Sweet, and H. A. Burge. (1992). A comparison of methods of estimating bacterial levels in machining oil mists (Abstract #20). Unpublished paper presented at the Aerobiology 1992 Symposium of the Pan-American Aerobiology Association, Scarborough, Ontario, Canada, June 11, 1992.

Nevalainen, A. 1989. Bacteria aerosols in indoor air. National Public Health Institute, Helsinki, Finland.

National Institute for Occupational Safety and Health (NIOSH) 1987. NIOSH guide to industrial respiratory protection. U.S. Department of Health and Human Services, Public Health Service, Centers for Disease Control, National Institute for Occupational Safety and Health, Cincinnati, OH. DHHS (NIOSH) Publication No. 87-116.

Olenchock, S. A., J. C. Mull, and W. G. Jones. 1983. Endotoxins in cotton: Washing effects and size distribution. Amer. J. Ind. Med. 4(4):515–521.

Otten, J. A., P. R. Morey, and H. A. Burge. 1986. Airborne microorganisms in office environments: Sampling protocol and analytical procedures. In Proceedings of the 1986 EPA/APCA Symposium on Measurement of Toxic Air Pollutants, EPA Report No. 600/9-86-013, pp. 36–44.

Palmgren, U., G. Ström, G. Blomquist, and P. Malmberg. 1986. Collection of airborne microorganisms on nuclepore filters, estimation and analysis—CAMNEA method. J. Appl. Bacteriol. 61(5):401–406.

Phillips, W. 1990. Telephone conversation on June 21, 1990, between W. Phillips, Andersen Instruments, Inc., and P. Jensen, Division of Physical Sciences and Engineering, National Institute for Occupational Safety and Health, Centers for Disease Control, Public Health Service, U.S. Department of Health and Human Services.

Pitt, J. L., A. D. Hocking, and D. R. Glenn. 1983. An improved medium for the detection of *Aspergillus flavus* and *A. parasiticus*. J. Appl. Bacteriol. 54(1):109–114.

Poon, C. P. C. 1966. Studies on the instantaneous death of airborne *Escherichia coli*. Amer. J. Epidemiol. 84(1):1–9.

Poon, C. P. C. 1968. Viability of long-storaged airborne bacterial aerosols. J. Sanitary Eng. Div., Proc. Amer. Soc. Civil Eng. SA6:1137–1146.

Pramer, and M. Rogul (eds.), Engineered organisms in the environment: scientific issues. American Society for Microbiology, Washington, DC.

Pui, D. Y. H., and B. Y. H. Liu. 1988. Advances in instrumentation for atmospheric aerosol measurement. Physica Scripta 37:252–269.

Ramiarz, R. J., J. K. Agarwal, and E. M. Johnson. 1982. Improved polystyrene latex and vibrating orifice monodispersed aerosol generators. TSI Quart. 7 (3).

Remiarz, R. J., J. K. Agarwal, F. R. Quant, and G. J. Sem. 1983. Real-time aerodynamic particle size analyzer. In V. A. Marple, and B. Y. H. Liu (eds). Aerosols in the mining and industrial work environments. Vol. 3. Ann Arbor Science Publishers, Ann Arbor, MI, pp. 879–895.

Rosebury, T. 1947. Experimental air-borne infection, Williams and Wilkins, Baltimore.

Rylander, R., and J. Vesterlund. 1982. Airborne endotoxins in various occupational environments. Proj. Clin. Biol. Res. 93:399–409.

Saiki, R. K., S. Scharf, F. Faloona, K. B. Mullis, G. T. Horn, H. A. Erlich, and N. Arnheim. 1985. Enzymatic amplification of beta-globin genomic sequences and restriction site analysis for diagnosis of sickle cell anemia. Science 230(4732):1350–1354.

Salem, H. 1987. Principals of inhalation toxicology. Pp. 1–29. In H. Salem (ed.), Inhalation toxicology—research methods, applications, and evaluation. Marcel Decker, Inc., New York.

Salvin, S. B. 1949. Cysteine and related compounds in the growth of the yeast-like phase of *Histoplasma capsulatum*. J. Infect. Dis. 84:275–283.

Sasser, M. 1990a. Identification of bacteria through fatty acid analysis. In Z. Klement, K. Rudolph, and D. C. Sands (eds.), Methods in phytobacteriology. Akademia Kiado, Budapest.

Sasser, M. 1990b. Identification of bacteria by gas chromatography of cellular fatty acids. Microbial Identification, Inc., Newark, NJ (MIDI), Technical Note #101.

Sem, G. J., K. Tsurubayashi, and K. Homma. 1977. Performance of the piezoelectric microbalance respirable aerosol sensor. Amer. Ind. Hyg. Assoc. J. 38(11):580–588.

Shaffer, B. T., and B. Lighthart. 1992. Baseline sources and fluxes of airborne microorganisms from plant surfaces. Proc. 11th Annual Meeting, Amer. Assoc. Aerosol Res., San Francisco, CA, Oct. 12–16, p. 183.

Skillern, C. P. 1971. Problems using Mie scattering photometers for in-place HEPA filter tests and aerosol studies. Amer. Ind. Hyg. Assoc. J. 32:96–103.

Smid, T., E. Schokkin, J. S. M. Boleji, and D. Heederik. 1989. Enumeration of viable fungi in occupational environments: A comparison of samplers and media. Amer. Ind. Hyg. Assoc. J 50(5):235–239.

Solomon, W. R., H. P. Burge, and J. R. Boise. 1980. Exclusion of particulate allergens by window air conditioners. J. Allergy Clin. Immunol. 65(4):305–308.

Strachan, D. P., B. Flannigan, E. M. McCabe, and F. McGarry. 1990. Quantification of airborne moulds in the homes of children with and without wheeze. Thorax 45(5):382–387.

Strugger, S. 1948. Fluorescence microscope examination of bacteria in soil. Can. J. Res. Series C. 26:188–193.

Stull, R. B. 1988. An introduction to boundary layer meteorology. Kluwer Academic Publishers, Boston, p. 666.

Swift, D. L., and M. Lippmann 1989. Electrostatic and thermal precipitators. In S. V. Hering, (ed.) Air sampling instruments for evaluation of atmospheric contaminants, 7th ed. American Conference of Governmental Industrial Hygienists, Cincinnati, OH. pp. 387–404.

Thompson, S. V., M. N. Schroth, W. J. Moller, and W. O. Reil. 1976. Efficacy of bactericides and saprophytic bacteria in reducing colonization and infection of pear flowers by *Erwinia amylovora*. Phytopathology 66:1457–1459.

Tortora, G. J., B. R. Funke, and C. L. Case. 1989. Microbiology—An introduction. 3d ed. The Benjamin/Cummings Publishing Company, Redwood City, CA, pp. 162, 215–216.

Verhoeff, A. P., J. H. van Wijnen, J. S. M. Boleij, B. Brunekreef, E. S. van Reenen-Hoekstra, and R. A. Samson. 1990. Enumeration and identification of airborne viable mould propagules in houses. A field comparison of selected techniques. Allergy 45(4):275–284.

Walter, M. V., B. Marthi, V. P. Fieland, and L. M. Ganio. 1990. Effects of aerosolization on subsequent bacterial survival. Appl. Environ. Microbiol. 56(11):3468–3472.

White, L. A., J. D. Hadley, J. E. Davids, and R. I. Naylor. 1975. Improved large-volume sampler for the collection of bacterial cells from aerosol. Appl. Microbiol. 29(3):335–339.

Willeke, K., H. E. Ayer, and J. D. Blanchard. 1981. New methods for quantitative respirator fit testing with aerosols. Amer. Ind. Hyg. Assoc. J. 42(2):121–125.

Willeke, K., and B. Y. H. Liu. 1976. Single particle optical counter: principle and application. In B. Y. H. Liu (ed.), Fine particles: Aerosol generation, measurement, sampling, and analysis. Academic Press, New York, pp. 697–729.

Wolf, H., P. Skaliy, L. Hall, M. Harris, H. Decker, L. Buchanan, and L. Dahlgren. 1959. Sampling microbiological aerosols, Public Health Monograph No. 60, U.S. Department of Health, Education, and Welfare, Public Health Service.

Wren, B., C. Claytin, and S. Tabaqchali. 1990. Rapid identification of toxigenic *Clostridium difficile* by polymerase chain reaction. Lancet 335(8686):423.

Young, H. M., J. W. Dominik, J. S. Walker, and E. W. Larson. 1977. Continuous aerosol therapy system using a modified collison nebulizer. J. Clin. Microbiol. 5(2):131–136.

Zimmerman, R., and L. A. Meier-Reil. 1974. A new method for fluorescence staining of bacterial populations on membrane filters. Kieler Meeresforchung 30:24–27.

9

Dispersion Models of Microbial Bioaerosols

Bruce Lighthart

9.0. Introduction

Many bioaerosol models have been prepared ranging from compartment models (Forrester, 1961; Atkins, 1969), describing the downwind concentrations and flux (i.e., D/P transfer rate where D/P is a droplet/particle; D/P / M^{-2} s^{-1}) of bioaerosols from a source that contributes to the loading of the bulk atmosphere (Fig. 9.1) through comprehensive, theoretical, and multiple regression models characterizing the factors that affect the survival of airborne microbes (Larson, 1973) [Eq. (9.1)].

$$R = a + bx_1 + cx_2 + dx_3 + ex_4 + fx_5 + gx_6 + h_1x_1x_2 + \cdots + h_2x_1x_3x_4x_5x_6 + E, \quad (9.1)$$

where R is the microbial response, x_1 is the moisture effect (e.g., RH), x_2 is the temperature effect, x_3 is the chemical toxicity (e.g., OAF, oxygen, CO, SO_2, NO_x), x_4 is the radiation toxicity (e.g., solar radiation), x_5 is the particle size (e.g., packed cells in a particle), x_6 is the microbial resistance effect (e.g., difference between species and physiology), and E is the inherent variability (i.e., statistical error).

Note that the response (R) may mean more than just survival of the airborne microbe. For example, it may mean the loss of the ability to infect but not the loss of viability as shown in Fig. 9.1. The multiple x terms account for the interactions that might occur between factors. The factor of time is implicit in each term in Eq. (9.1).

In the past, microbial bioaerosol models describing single-factor effects on bioaerosol survival over time were exponential fits of the data. Equation (9.2) is the general form of such a model:

$$S_t = S_0 e^{-kt}, \quad (9.2)$$

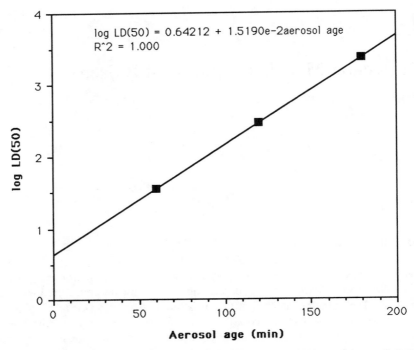

Figure 9.1. Graph showing effect of *Franciella tularensis* bioaerosol age on the lethality (as LD_{50}) to monkeys [Adapted from Larson (1973).]

where S_t is the percent of surviving microorganisms at time t (i.e., aerosol age), S_0 is the percent of surviving microorganisms at time $t = 0$, and k is the Exponential death rate constant [see Eq. (9.3)] usually expressed as percent decay of the bioaerosol per minute.

$$\text{Death rate constant} = k = (1/t)[\log_{10}(N_0/N_t)], \qquad (9.3)$$

where t is the time, N_0 is the number of cfu[1] at time $t = 0$ bioaerosol age, and N_t is the number of cfu at some time t of bioaerosol age.

To define a more mechanistic explanation of bioaerosol survival, Cox (1987) describes a chemical model to fit bioaerosol survival data.

Recently, a microbial death term (i.e., the inverse of microbial survival) has been included in models that describe the dispersion of pollutant gases and particles from a source in a downwind plume. The death term was first included

---

[1]cfu = colony-forming unit. Note: This means that one or more microorganism in a D/P could form a colony on culture media which is the usual counting technique.

in models based on statistical distribution of bioaerosol D/P in a plume (Lighthart and Frisch, 1976; Peterson and Lighthart, 1977; Lighthart and Mohr, 1987), the so-called Gaussian plume model, and later in particle models describing the random distribution of many individual D/Ps in a plume (Lighthart and Kim, 1989; Lighthart et al., 1991), termed the random-walk models. The random particle distribution models simulate D/P trajectories up to ~ 10 m (Lighthart et al., 1991) and less than 1 km (Lighthart and Mohr, 1987), whereas Gaussian models are thought to simulate dispersion processes from 100 m to 10 km.

9.1. Gaussian Plume Dispersion Model

The Gaussian plume model is based on the idea that the mean concentration of bioaerosol D/Ps are normally distributed about the downwind plume axis from a point source (Fig. 9.2). Intuitively, one would think that the central axis of

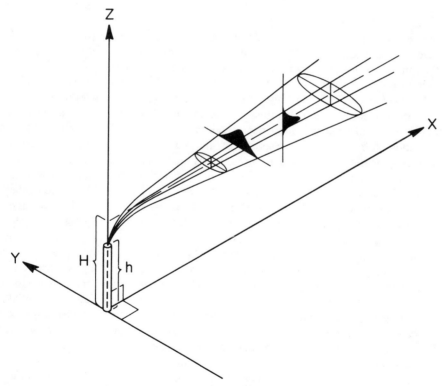

Figure 9.2. Gaussian plume diagram showing statistical distributions (black vertical and horizontal distribution curves) of source material downwind from a source with height *h*.

the plume would contain the highest concentration of the source D/Ps. It would also seem intuitive that turbulent diffusion would disperse the D/Ps horizontally and vertically, resulting in a concentration gradient diminishing downwind and away from plume axis. Because less turbulent energy is needed to counteract the force of gravity in the horizontal plane, dispersal of the D/Ps is greater in the horizontal plane. Justification for the geometry in this model is given by Pasquill (1974).

The form of the Gaussian plume model given in Peterson and Lighthart (1977) is

$$
\chi = \underbrace{\left(\frac{bQ}{2\pi u \sigma_y \sigma_z}\right)}_{A} \underbrace{\left[\exp\left(\frac{y_2}{2\sigma_y^2}\right)\right]}_{B} \underbrace{\left\{\sum_i q_i \exp\left[\frac{-\left(H - V_{s_i}\frac{x}{u}\right)^2}{2\sigma_z^2}\right]\right\}}_{C} \underbrace{\exp\left(-\lambda\frac{x}{u}\right)}_{D},
\tag{9.4}
$$

where χ is the mean number of microbes per unit volume of air (m^{-3}) at ground level and is a function of downwind and lateral distance from the source, b is the number of microbes per unit volume of source $(\text{m}^{-3}$ at the source, Q is the volume emission rate of the source $(\text{m}^3 \text{ s}^{-1})$, u is the mean wind speed (generally a 1-h average) (m s^{-1}) and must be greater than zero for this model to be valid, σ_y is the standard deviation of the horizontal plume spread (m) and is a function of downwind distance and atmospheric stability (see Table 9.2), σ_z is the standard deviation of the vertical plume spread (m) and is a function of downwind distance and atmospheric stability (see Table 9.2), y is the lateral distance from the mean position of the plume axis (m), q_i is the mass-weighted proportion of particles in the ith size category (dimensionless), H is the height above the ground of the plume axis after the initial plume rise (m), V_{si} is the settling speed of average microbe-containing droplets in the ith size category (m s^{-1}) (The settling rate may be approximated by $V_{s_i} = 0.025d^{1.47}$, where d is the drop diameter in microns), and λ is the death rate of the microbes (s^{-1}).

Term A represents the concentration D/Ps on the plume axis at a given distance, x, from the source with no microbial death. Term B accounts for lateral dispersion of the D/Ps in the plume. Term C accounts for vertical dispersion and differential droplet settling rates of the D/Ps. It also modifies the formula so that the computed concentration D/Ps is the ground-level concentration. Term D accounts for the loss of viable microbes in the bioaerosol due to death.

The Guassian plume model is valid for estimating mean concentrations only when meteorological and source conditions are stable for periods longer than the distance of the receptor from the source divided by the mean wind speed. The σ_y and σ_z values are based on 1-h averages. Therefore, the wind speed, u, to enter into the equation is a mean wind, averaged over the appropriate time period.

If real wind is used in the equation, then the wind speed used in the model must be representative of the wind at the plume centerline, not the local surface wind.

The expressions for the plume-spread factors, σ_y and σ_z, suggested in Table 9.1 were determined empirically (Smith, 1968) based on an analysis of the available data on plume spread. These expressions are not appropriate for all situations and should not be taken as universally valid. The variation in plume-spread rates is determined primarily by the turbulence of the air in the vicinity of the plume. The level of the turbulence depends on the atmospheric stability, whether the air is being heated or cooled by the ground, the strength of the wind, and the roughness of the terrain. The effect of the rough terrain, although quite difficult to assess quantitatively, tends to cause increased turbulence and, thus, a faster spreading of the plume. The vertical spread of the plume varies much more with the level of turbulence than the lateral spread. The expressions for the plume-spread factors are divided into four categories: 1 through 4, where 1 is the most turbulent condition and 4 is for the least turbulent case.

The maximum height, *H*, of the axis of the plume after leaving the source, in this case a stack, is determined mainly by the characteristics of the stack, the

Table 9.1. *List of bacteria and references used in this analysis, and their comparison fit (=R^2) to the regression model in Eq. (9.4).*

Bacterium	N[a]	R	Reference
Corynebacterium xeros	11	0.94	Graham et al. (1979)
Erwinia carotovora var. *carotovora*	4	NA[b]	Lighthart (1973)
Escherichia coli	79	0.97	Ferry et al. (1958); Poon (1966, 1968)
Escherichia coli MRE 162	1	NA	Graham, et al. (1979)
Flavobacterium sp.	25	0.91	Ferry et al. (1958)
Klebsiella pneumoniae Type A	12	0.33	Goldberg et al. (1958)
Micrococcus candidus	2	NA	Webb (1959)
Mycobacterium phlei	2	NA	Graham et al. (1979)
Mycoplasma gallisepticum	6	NA	Wright et al. (1968)
Mycoplasma laidianii	6	NA	Wright et al. (1968)
Mycoplasma pneumoniae	4	0.88	Wright and Bailey (1969); Wright et al. (1969)
Pasteurella pestis	3	NA	Goldberg et al. (1958)
Pasteurella tularensis	12	0.71	Cox and Goldberg (1972); Ehrlich and Miller (1973)
Pseudomonas pseudomallei	2	NA	Goldberg et al. (1958)
Sarcina lutea	9	0.35	Lighthart (1973)
Serratia marcescens	1		Webb (1959, 1961a, 1961b)
Serratia marcescens ATCC 274	30	0.87	Lighthart (1973); Lightart et al. (1971)
Serratia marcescens 8UK (ATCC 14041)	1		Ferry et al. (1958)
Staphylococcus albus	4	0.98	Goldberg, et al. (1958); Webb (1960)

[a]Number of death rate constant values used in the analysis.

[b]NA = not available.

temperature of the effluent gases, the wind speed, and the vertical temperature structure of the bulk atmosphere. Because all of these variable are not usually known, the problem of estimating the plume height will not be dealt with here. For a discussion of plume rise, refer to Briggs (1969).

Estimations made using this model are not valid for distances less than 100 m or more than 10 km from the source because the plume-spread factors are not well determined outside that range. The model is also invalid for calm and very light wind situations.

The death rate, λ, used in the model may vary depending on many factors [e.g., those in Eq. (9.1)] and has usually been derived experimentally using pure cultures. This is shown in Eq. (9.5) where a multiple regression was fit to the death rates of 16 species of bacteria (Table 9.2; Fig. 9.3). These death rates were determined as a function of two meteorological variables [i.e., temperature and RH (Lighthart, 1989)]. Because so few quantitative measurements have been made on how light affects airborne bacterial survival, this factor was not included in the regression analysis, although it is known to be very important.

$$\log_{10} \text{(death rate constant)} = -0.245 - 0.892 \times \log_{10} \text{(aerosol age (h))}$$
$$+ 0.0055 \times \text{temperature (°C)} - 0.15$$
$$\times \text{Gram reaction} \times \text{RH(\%)}, \qquad (9.5)$$

where the value for the Gram reaction is zero for Gram-negative bacteria and one for Gram-positive bacteria.

Both simpler and more complex formulations for the death rate constant are given in Lighthart (1989).

The Guassian plume model was used by Lighthart and Mohr (1987) to estimate the dispersion pattern of a "composite virus," developed by combining the charac-

Table 9.2. *Plume spread factors for various atmospheric conditions.*

Category	Plume Spread Factors	State of Atmosphere	Possible Causes, Symptoms, and Effects
1	$\sigma_y = 0.40x^{0.91}$ $\sigma_z = 0.40x^{0.91}$	Highly turbulent, extremely unstable	Strong solar heating or cold air moving over warm stable ground often, but not necessarily, resulting in cumulus clouds and occasionally thunderstorms.
2	$\sigma_y = 0.36x^{0.86}$ $\sigma_z = 0.36x^{0.86}$	Moderately turbulent	Ordinary sunny day in summertime (middle latitudes).
3	$\sigma_y = 0.32x^{0.78}$ $\sigma_z = 0.32x^{0.78}$	Wind caused turbulence	Windy, overcast if during the day.
4	$\sigma_y = 0.31x^{0.71}$ $\sigma_z = 0.31x^{0.71}$	Stable; turbulence suppressed.	Nighttime and light winds, or daytime in winter (middle latitudes) with low-level temperature inversion.

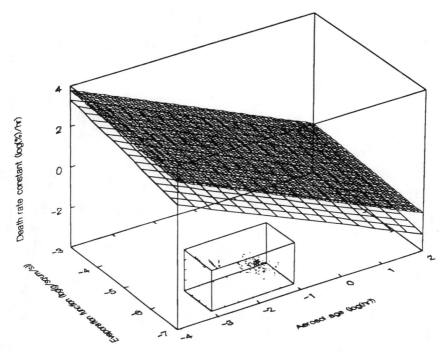

Figure 9.3. Response surfaces for airborne death-rate equation using data (198 variates) shown in the insert scatter plot of Gram-positive (+, lower surface) and Gram-negative (−, upper surface) bacteria.

teristics of two viruses showing how wind speed could have a major modulating effect on near-source viable concentrations. For example, at high wind speeds such as those occurring during the day, or over short time intervals, near-source locations experience high viable concentrations because the microorganisms in the plume have not had time to be inactivated or die. As the travel time increases, because of slow wind speed or longer distances, die-off modulation by sunshine, RH, temperature, and so forth becomes increasingly important (Fig. 9.4).

9.2. Bioaerosol Droplet/Particle Dispersion Model

9.2.1. Laminar Airflow Regime

A bioaerosol may be dispersed in either a laminar air flow regime as described in this section or a turbulent flow regime as described in Sec. 9.2.2. In the laminar flow regime, D/Ps are passively carried by the air at a constant velocity and direction. Changes due to biological factors that affect the D/Ps survival still occur over time. The following example discusses the simulated D/P trajectory for

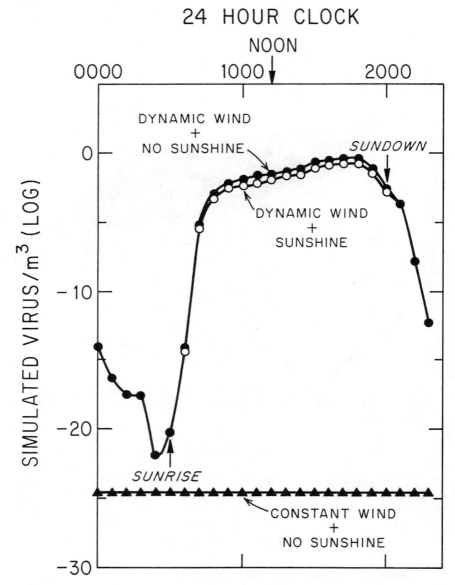

Figure 9.4. Atmospheric dispersion model estimation of the concentration of a simulated virus at ground level, 20 m downwind from an elevated source (0.1 m) under either dynamic wind speed (1 m s^{-1}) and no sunshine (0 W−h m^{-2} h^{-1}; ●), dynamic wind speed and sunshine (○), or constant wind speed and no sunshine (▲). Inert or nonviable particles plot slightly above points for dynamic wind speed and no sunshine. Environmental variables used to drive the model include 10-year hourly July means for wind speed, RH, temperature, and solar radiation in Eugene, Or. Microbial survival factors include RH, temperature, solar radiation, and time. Ordinary summer-day plume spread factors were used in the model [after Mohr (1984) and Berendt et al. (1972)].

a short distance after it is emitted from a spray nozzle into a laminar airflow regime. Figure 9.5 describes an algorithm of the droplet trajectory simulation program. In the program, an aqueous bacterial spore suspension is sprayed from an elevated nozzle into a laminar flow airstream. The initial environmental input (or driving) variables to the program include the ambient air temperature (t_∞),

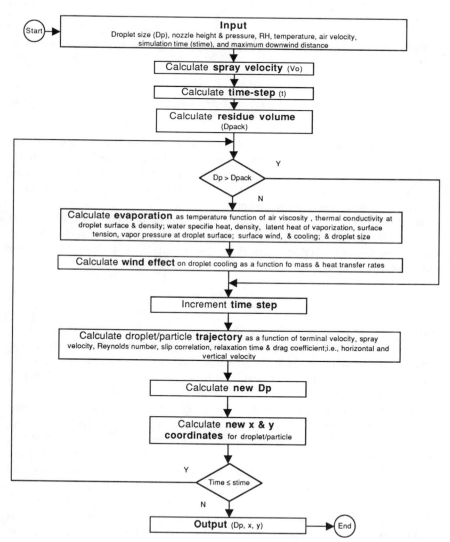

Figure 9.5. Flow diagram of the bacterial aerosol spray droplet trajectory simulation program.

RH, and wind velocity (V_f) over a series of time steps. The spray variable includes spray pressure from which the initial spray droplet velocity (V_0) leaving the nozzle is calculated [Eq. (9.6)] and the initial droplet and final evaporate or packed bacterial particle diameter [Eq. 9.7)].

$$V_0 = \sqrt{\frac{2P}{\rho_p}} , \qquad (9.5)$$

where P is the pressure in the spray nozzle and ρ_p is the density of water at ambient temperature.

$$D_{pack} = 3\sqrt{\frac{v}{0.71}} , \qquad (9.6)$$

where D_{pack} is the residual particle diameter as calculated as the packed diameter [defined by Dallavalle (1948) for rhombohedral-type packing] and v is the volume of the mean number of bacteria (n) that would be suspended in an initial spray droplet with an initial diameter D_p [i.e., $v = n[4/3)\pi(D_p/2)^3$]. The bacteria in the suspension are assumed to be spheres.

Horizontal or vertical droplet velocities (V_t) at each simulated time step (t) may be calculated using Eq. (9.7) and summing the distance traveled at each step in the horizontal (x) or vertical (y) direction (i.e., velocity multiplied by time equals distance).

$$V_t = V_f - (V_f - V_0)E^{-t/\tau}, \qquad (9.7)$$

where V_f is the airstream velocity (for the horizontal-axis calculations) or the terminal settling velocity (for the vertical-axis calculations), V_0 is the initial spray velocity for the horizontal-axis calculations and zero for the vertical-axis calculations, t is the time step, and, τ is the relaxation time [see Sec. 2.1.1.4, Eq. (2.6) of Chapter 2.

The rate of change in droplet size as a result of evaporation at each time step was calculated by using Eq. 2.25 in Sec. 2.2 where the change in droplet diameter (dD_p) with respect to the simulation time step (dt) is the product of the first two terms on the right of the equation describing evaporative effects, including the Kelvin effect on droplet size, and the last term, which describes the Fuchs effect on the mass transfer rate of water vapor away from the droplet.

For each simulated time step, the change in droplet surface temperature, δT, the critical site for droplet evaporation, is given in Eq. (2.26) in Sec. 2.2. The surface cooling of the droplet at each time step, δt, is subtracted from the present droplet temperature, T_{d-1} and the new temperature, T_d, used for the next simulated time step (i.e., $T_d = T_{d-1} - \delta T$). This equation expresses the heat loss by droplet evaporation in the left-hand term and heat gain by the conduction from the air to the droplet in the right-hand term.

The number of spores contained in a droplet is modeled as a random quantity that depends on the diameter of the droplet emitted by the sprayer. The initial droplet diameter is also a random variable that can be described by the probability distribution supplied by the manufacturer of the spray nozzle. Refer to Lighthart et al. (1991) for further details on the distribution of microorganisms containing simulated droplets.

The results of simulation trajectories are shown in Fig. 9.6. These trajectories have three main features: (i) Droplets ≥ 80 μm hit the ground before they evaporate (at the observed RH and temperature). (ii) Droplets ≤ 80 μm in diameter evaporate to residue size before they reach the ground. The smaller the initial droplet, the quicker it evaporates and the less distance it falls. Thus, because of the differential evaporation and settling, a vertical particle-size fractionation occurs downwind from the spray source. (iii) Droplets up to 35 μm evaporate very rapidly (in less than 1 s) and therefore, travel short distances ($<$ 6 cm) and form a dense plume downwind at the spray nozzle height.

Observation of a polydispersed bioaerosol of bacterial spores in air columns (at 0.0, 1.0, and 1.5 m below and 2.5, 5.0 and 10.0 m downwind from the spray nozzle) in a quasi-laminar airflow regime are shown in Fig. 9.7. These distributions have several features in common:

(i) All the air columns have a prominent small-particle (or droplet residue) population approximately 1 μm in diameter at the nozzle height;

(ii) the entire proximal air column contained a large (diameter, $\geq$ 7 μm) droplet or particle-size fraction;

(iii) the column contained a prominent component comprising particles 2–3 μm in diameter;

(iv) this size fraction extended from the proximal to the intermediate and lowest levels in the two distal air columns, and particle sizes tended toward the smallest aeroplanktonic sizes at the highest elevation and further downwind in the air columns;

(v) samples from the distal air columns had no statistical differentiable large droplet particle sizes, except for a remnant near the ground at the 5-m sample distance;

(vi) there were very few to no observed spore particles in the size range from 3 to 5 μm in diameter in any column.

The simulated trajectories of viable droplets from a polydispersed aerosol show that deposition patterns may be dependent on droplet size as they are affected by environmental conditions. Droplets that have not evaporated to particle sizes prior to hitting a surface (i.e., the ground) will fall to the ground near the aerosol

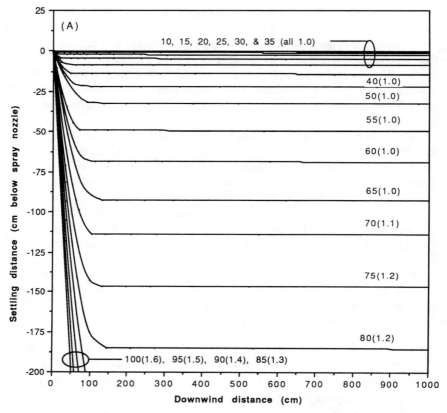

Figure 9.6. Calculated deposition trajectory of various-sized viable, evaporation droplets sprayed (at 20 lb in^{-2}) from a 2-m height in a quasilaminar flow (at 0.163 m s^{-1}) airstream at 21.8°C and 48.3% relative humidity showing (A) small droplet and particle trajectories (the numbers in the body of the figure represent particle diameters in micrometers with residue diameters in parentheses and (B) large droplet (200–500 μm in diameter) trajectories.

source, whereas those small enough to evaporate to particle sizes before hitting a surface can remain aloft for extended periods of time. For example, particles of unit density and less than 10 μm in diameter have a settling rate less than 0.31 cm s^{-1} (Hinds, 1982). This is shown in the spectrum of predicted (Fig. 9.6A) droplet/particle sizes at ground level, 2.5, 5, and 10 m downwind from the source when compared to size spectra at the corresponding downwind observation sites (Fig. 9.6). They are very similar. Thus, near the spray source, a broad spectrum of droplet sizes are present with two peaks; one for small and one for large droplet/particles. Distal from the source, there is only one peak for small droplets. Thus, only small particles remain downwind from the source and as

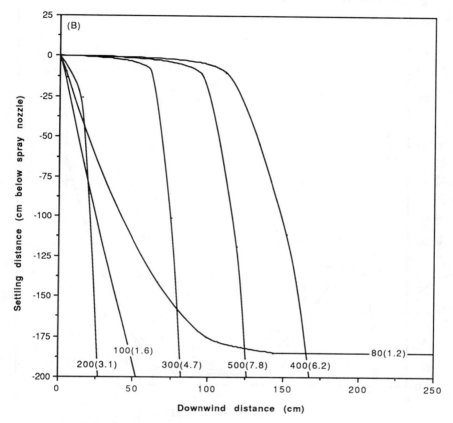

Figure 9.6. (Continued).

indicated by the calculated trajectories, are the residue of droplets that have evaporated prior to any deposition processes. This demonstrates that, although it is often done, one cannot compare the chronology of any one size droplet fraction unless it has attained a constant evaporated size, that is, the small D/Ps near the source are probably not the same population of small particles appearing downwind at different heights sometime later. One must take samplers along the trajectory line if one is to evaluate the survival in a uniform droplet sized generated in a polydispersed aerosol.

There is one area of nonsimilarity between the simulated and observed distribution patterns. The simulated trajectories are projections of droplets containing a single packed cell residue, and do not take into account the effects of larger residue diameters expected from droplets that contain a Poisson distribution of spores and consequent residue diameters. It is the larger residue diameters that

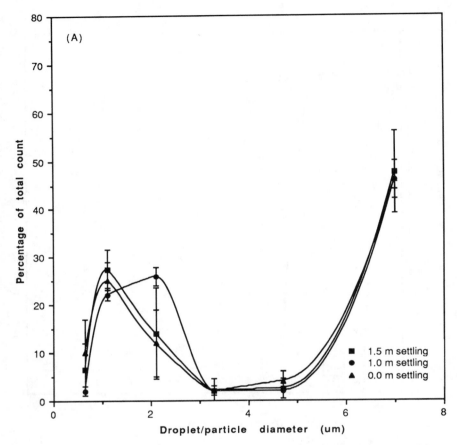

Figure 9.7. Observed percentage distributions of *Bacillus subtilis* var. *niger* median spore particle sizes from size-replicate Andersen samplings at three settling heights (0.0, 1.0, and 1.5 m) below the spray nozzle in three air columns 2.5 m (A), 5.0 m (B), and 10 m (C) downwind from the spray source. (Note that the simulated 10–80-μm droplets in Fig. 9.6 have a packed residue volume of 1.6 μm or less and, therefore, probably account for the increasing size of the 1–2-μm size fraction downwind from the source.) Vertical bars are 95% confidence limits.

are thought to be measured in the 2–3-μm-diameter particles found in the observed distributions (Fig. 9.7A, B, C).

Logically, the survival of microbes in evaporated microbial aerosols would depend on several factors, one of which would be the number of microbes in the original droplet that might pack together in the residue particle. Past reports suggest that the larger the droplet/particle, the longer the survival (Bausum et al., 1976; Shaw, 1978) of microbes in the residue. Presently, because there is

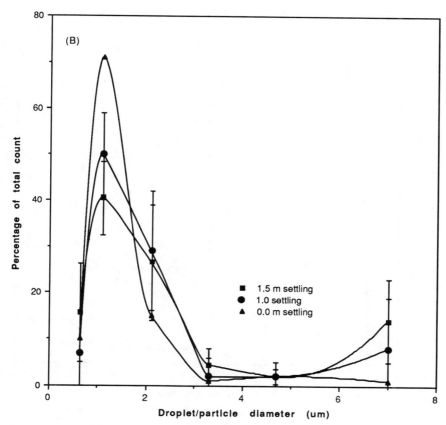

Figure 9.7. (*Continued*).

no practical way available to generate large numbers of droplets of known size from suspensions of viable bacteria, the survival of aerosolized bacteria as a function of residue size has not been investigated. With the characterization of the droplet trajectories of a polydispersed viable aerosol as described herein, it is now possible to evaluate the chronological survival of microbes in droplets of different sizes by calculating the droplet/particle-size trajectory spectrum, and then sampling at those loci corresponding to the calculated trajectory for a particular droplet size. The trajectory depends on the initial droplet size and environmental conditions of laminar airflow rate, RH, and temperature. The notion of aerosol residue trajectory fractionation should provide a useful tool to investigate and understand viable aerosol survival and deposition patterns.

In conclusion, it is speculated that as the droplets dry, those that dry first that is, the smallest droplets die first, and the larger droplets with many microbes

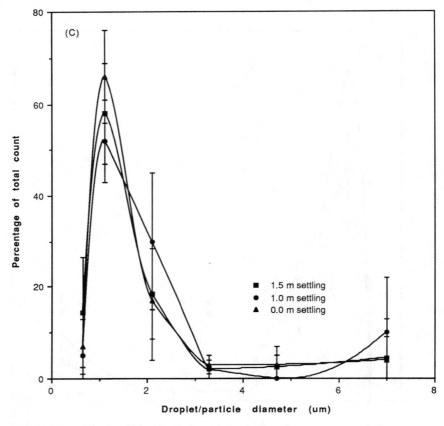

Figure 9.7. (*Continued*).

form aggregates which, as might be expected, dry from the outside in. Thus, it would take longer for all cells in the larger droplets to die, with the more internal cells drying and dying last. The preceding assumes that desiccation is the only death mechanism.

9.2.2. Turbulent Airflow Regime

In the turbulent flow regime, such as the extramural atmosphere, a D/P has a trajectory that can be thought of as the summation of wind direction and wind speed alterations at each of many small time steps. The droplets are assumed to be passively carried by the mean wind. Many such droplets would form a cloud.

For a short time interval (δt), the positional change of a droplet in one dimension (x) from the initial position ($D[0]$) to the final position ($D[t]$) may be calculated from Eq. (9.7):

$$D_{i,x}(t) = D_{i,x}(t-\delta t) + \mu\delta t, \tag{9.7}$$

where the total speed, μ, is the sum of a mean (μ) and a turbulent component, μ':

$$\mu = \mu + \mu'. \tag{9.9}$$

The turbulent component is the summation of a correlated component, the first term in Eq. (9.10), and a random component:

$$\mu'(t) = \mu'(t-\delta t)R(\delta t) + \mu'', \tag{9.10}$$

where the random component μ'' is assumed to have a Gaussian distribution with a zero mean and variance of $\sigma^2_{\mu''}$, Eq. (9.11):

$$\sigma^2_{\mu''} = \sigma^2_{\mu'} [1-R^2(\delta t)], \tag{9.11}$$

where $R(\delta t)$ is the correlation coefficient of wind speed for the time lag δt. The exact value of $R(\delta t)$ is obtained from meteorological data. It is also assumed that the correlation coefficient for different wind components (u and v, for example) is zero for all δt. The droplet motion dynamics are detailed by Hanna et al. (1982) and van Dop et al. (1985). Calculations (such as in Chapter 2) may be easily incorporated into this model at each time step to account for evaporation, including evaporative cooling during evaporation, the Kelvin effect, the Fuchs effect, microorganisms affecting droplet surface properties such as changes in evaporation rate when they become trapped in the droplets surface film, solute effect, and droplet size trajectory changes; all in changing meteorological conditions of RH and temperature. The method is used in Lighthart and Kim (1989) to estimate the downwind drift of genetically engineered, frost-formation-reducing bacteria sprayed onto strawberries (Lindow et al., 1988). Plate I in the front of this book illustrates the dispersion pattern that may be generated from this kind of simulation model.

References

Atkins, G. L. 1969. Multicompartment models for biological systems. Methuen & Co. Ltd., London, p. 153.

Bausum, H. T., S. A. Schaub, M. J. Small, J. A. Highfill, and C. A. Sorber. 1976. Bacterial aerosols resulting from spray irrigation with wastewater. Technical Report 7602. U.S. Army Medical Bioengineering Research and Development Laboratory, Fort Detrick, MD.

Berendt, R. F., E. L. Dorsey, and H. J. Hearn. 1972. Viricidal properties of light and

SO_2. I. Effect of aerosolized Venezuelan equine encephalomyelitis virus. Proc. Soc. Exp. Biol. Med. 130:1–5.

Briggs, W. K. W. 1969. "Plume rise." U.S. A.E.C., Technical Information Division-25075, Dept. of Communications, Springfield, VA.

Cox, C. S. 1987. The aerobiological pathway of microorganisms. John Wiley & Sons, New York, NY.

Cox, C. S., and L. J. Goldberg. 1972. Aerosol survival of *Pasteurella tularensis* and the influence of relative humidity. Appl. Microbiol. 23(1):1–3.

Dallavalle, J. M. 1948. Micromeritics. The technology of fine particles. Pitman Publishing Corp., New York.

Ehrlich, R., and S. Miller. 1973. Survival of airborne *Pasteurella tularensis* at different atmospheric temperatures. Appl. Microbiol. 25(3):369–372.

Ferry, R. M., W. F. Brown, and E. B. Damon. 1958. Studies on the loss of viability of stored bacteria aerosols. II. Death rates of several non-pathogenic organisms in relation to biological and structural characteristics. J. Hyg. 56:125–150.

Forrester, J. W. 1961. Industrial dynamics. M.I.T. Press, Cambridge, MA.

Goldberg, L. J., H. M. S. Watkins, E. E. Boerke, and M. A. Chatigny. 1958. The use of a rotating drum for the study of aerosols over extended periods of time. Amer. J. Hyg. 68:85–93.

Graham, D. C., C. E. Quinn, I. A. Sells, and M. D. Harrison. 1979. Survival of strains of Soft Rot Coliform bacteria on microthreads exposed in the laboratory and in the open air. J. Appl. Bacteriol. 46:367–376.

Hanna, S. R., G. A. Briggs, and R. P. Hosker, Jr. (eds.). 1982. Handbook on atmospheric diffusion. P102. DOE/TIC-11223. Technical Information Center, Department of Energy, Washington, DC.

Hinds, W. C. 1982. Aerosol Technology. John Wiley & Sons, New York, p. 424.

Larson, E. W. 1973. Environmental variables and microbial survival. Pp. 81–86. In J. F. Ph. Hers and K. C. Winkler (eds.), Airborne transmission and airborne infection. Oosthoek Publishing Co., Utrecht, The Netherlands.

Lighthart, B. 1973. Survival of airborne bacteria in a high urban concentration of carbon monoxide. Appl. Environ. Microbiol. 25(1):86–91.

Lighthart, B. 1989. A statistical model of laboratory death rate measurements for airborne bacteria. Aerobiology 5:138–144.

Lighthart, B., and A. S. Frisch, 1976. Estimation of viable airborne microbes downwind from a point source. Appl. Environ. Microbiol. 31(5):700–704.

Lighthart, B., V. E. Hiatt, and A. T. Rossano, Jr. 1971. The survival of airborne *Serratia marscensens* in urban concentration of sulfur dioxide. Air Pollut. Control Assoc. 21(10):639–642.

Lighthart, B., and J. Kim. 1989. Simulation of airborne microbial droplet transport. Appl. Environ. Microbiol. 55(9):2349–2355.

Lighthart, B., and A. J. Mohr. 1987. Estimating downwind concentrations of viable airborne microorganisms in dynamic atmospheric conditions. Appl. Environ. Microbiol. 53(7):1580–1583.

Lighthart, B., B. T. Shaffer, B. Marthi, and L. Ganio. 1991. Trajectory of aerosol droplets from a sprayed bacterial suspension. Appl. Environ. Microbiol. 57(4):1006–1012.

Lindow, S. E., G. R. Knudsen, R. J. Seidler, M. V. Walter, V. W. Lambou, P. S. Amy, D. Schmedding, V. Prince, and S. Hern. 1988. Aerial dispersal and epiphytic survival of *Pseudomonas syringae* during a pretest for the release of genetically engineered strains into the environment. Appl. Environ. Microbiol. 54:1557–1563.

Mohr, A. J. 1984. Doctoral dissertation. Utah State University. Logan, UT.

Pasquill, F. 1974. Atmospheric diffusion, 2nd. ed. John Wiley & Sons, New York.

Peterson, E. W., and B. Lighthart. 1977. Estimation of downwind viable airborne microbes from a wet cooling tower—Including settling. Microbial Ecol. 4:67–79.

Poon, C.P.C. 1966. Studies on the instantaneous death of airborne *Escherichia coli*. Amer. J. Epidemiol. 84:1–9.

Poon, C.P.C. 1968. Viability of long-storaged airborne bacterial aerosols. J. Sanitary Eng. Div., Proc. Amer. Soc. Civil Eng. SA6:1137–1146.

Shaw, D. T. 1978. Fundamentals of aerosol science. John Wiley & Sons, New York.

Smith, M. 1968. Recommended guide for the prediction of the dispersion of airborne effluents. American Society of Mechanical Engineering, New York.

van Dop, H., F. T. M. Nieuwstadt, and J. C. R. Hunt. 1985. Random walk models for particle displacements in inhomogeneous unsteady turbulent flows. Phys. Fluids 28:1639–1653.

Webb, S. J. 1959. Factors affecting the viability of air-borne bacteria. I. Bacteria aerosolized from distilled water. Can. J. Microbiol. 5:649–669.

Webb, S. J. 1960. Factors affecting the viability of air-borne bacteria. III. The role of bonded water and protein structure in the death of air-borne cells. Can. J. Microbiol. 6:89–105.

Webb, S. J. 1961a. Factors affecting its E viability of air-borne bacteria. IV. The inactivation & reactivation of airbonding *Serratia marcuscans* by ultraviolet & visable light. Can. J. Microbiol. 7:607–619.

Webb, S. J. 1961b. Factor affector the viability of air-bonding bacteria V. The effect of dessication on some metabolic systems of *Escherichia coli*. Can. J. Microbiol. 7:621–631.

Wright, D. N., and G. D. Bailey. 1969. Effect of relative humidity on the stability of Mycoplasma pneumoniae exposed to simulated solar ultraviolet and to visible radiation. Can. J. Microbiol. 15:1449–1452.

Wright, D. N., G. D. Bailey, and L. J. Goldberg. 1969. Effect of temperature on survival of airborne *Mycoplasma pneumoniae*. J. Bacteriol. 99(2):491–495.

Wright, D. N., G. D. Bailey, and M. T. Hatch. 1968. Survival of airborne Mycoplasma as affected by relative humidity. J. Bacteriol. 95(1):251–252.

10

Health Aspects of Bioaerosols

Harry Salem and Donald E. Gardner

Introduction

Bioaerosols have been defined as colloidal suspensions in air of liquid droplets or solid particles, containing, or having attached to them, one or more living organisms. These organisms include viruses, bacteria, fungi, protozoa, or algae. Bioaerosols may range in size from a single microorganism to large droplets containing many microorganisms. Microorganisms may also be attached to pollen grains, plant debris, skin flakes, and/or soil particles. Liquid droplets may change in size upon evaporation or condensation, which may or may not result in the loss of viability of the organism.

Bioaerosols are known to have a major impact on vegetation, livestock, ecology, the environment, and on the health of humans. The major human health aspect of bioaerosols is their use in the study of etiologic agents, control, prevention, and treatment in the spread of infection. Current concerns of airborne contagion include the environment, especially indoor air pollution in homes, schools, buildings, and hospitals. The incidence of airborne disease has been correlated with the extent of crowded conditions and poverty levels in our cities.

Most reviews of inhalation toxicology indicate that we live in a sea of chemicals. These chemicals can gain access to the body by means of inhalation and may affect the respiratory system directly, or by using this route as a portal of entry and then adversely affect some other part of the body (Salem, 1987). It is estimated that there are already almost 100,000 chemicals in commerce, with approximately 25,000 new entities synthesized yearly. Bacteria, viruses, and fungi have also been identified in ambient air. They can be transmitted to cause infection, but their survival in air is more limited than chemicals. In air, these microorganisms cannot grow or multiply because of the absence of nutrients, but they can be transmitted for long distances. Different microorganisms are found at various heights from the ground, and amounts and varieties of microbial populations in air are dependent on location and environmental conditions such as humidity and temperature, as well as the density and activity of humans (Al-Dagal and Fung, 1990). Viable microorganisms are known to occur up to altitudes of about 20 miles, and fungal spores have been found in air flights over the North Pole. The potential adverse effects of chemicals and biologicals in the environment on domestic animals, wildlife, fish, and plants must also be considered so that a balanced ecosystem is maintained. It must also be recognized that in spite of the sea of chemicals and biologicals we live in, life expectancy has increased over the last few decades, and the quality of life has improved.

This chapter will review the status of aerobiology in terms of respiratory infections with emphasis on airborne contagion, defense mechanisms, and the maintenance of homeostasis. In addition, where possible, a parallelism will be drawn between biologicals and chemicals.

Molecular biology and biotechnology have made genetic engineering and designer drugs a reality as well as blurring the demarcation between chemical and biological agents; the connecting link being agents of biological origin (i.e., poisonous products of microorganisms, animals, and plants) such as toxins. During World War II, various countries developed extensive biological arsenals and maintained them for years afterward. Much has been written on aerobiology and numerous conferences were held until 1969 when the United States under President Richard Nixon abandoned its research into biological weapons. Unfortunately, evidence points to just the opposite for the former Soviet Union. Since the 1970s, and until recently, they appear to have been pursuing an active program of research into biological agents for military use. The deaths caused by anthrax spread from a Soviet weapons laboratory in Sverdlovsk in 1979 is an example of this. The advances of genetic technology may have provided the Soviets with newer biological agents for which there are no vaccines or therapeutic regimens.

In the 1977 Senate Hearings, the Department of Defense acknowledged that in the 1950s, it had engaged in massive open-air tests in which it had released vast quantities of mildly pathogenic bacteria (*Serratia marcescens* and *Bacillus subtilis*) over San Francisco Bay and in the New York subway system. These tests appeared to have triggered disease outbreaks that caused at least one death (Yamamoto, 1989). Such studies were conducted to assess the survivability and dispersion patterns of the organisms. Over 250 years ago, smallpox and possibly typhus were reported to have been used as biological weapons during the French, British, and Indian wars in North America (Poupard et al., 1989). As recently as 1981, the Department of State declared that clouds of yellow rain (trichothecene toxins from fungi) were released over Laos, Kampuchen, and Afghanistan, causing illness and death.

Some of the current concerns include the sources of bioaerosols, their particle-size distribution and dose responses, and airborne contagion.

10.1. Sources of Bioaerosols

Bioaerosols are aerosols containing living organisms such as viruses, bacteria, fungi, protozoa, or algae, as well as products of their metabolism or their decomposition such as toxins. Bioaerosols, in addition to causing infectious diseases, can act as sensitizing agents in susceptible hosts, similar to chemical aerosols. In addition, agents of biological origin such as toxins can also cause acute and chronic effects as do chemical aerosols.

Obligate parasitic microorganisms such as viruses, bacteria, and fungi require a living host for growth and reproduction. Although some of these organisms may survive on environmental surfaces for various durations of time, they only very rarely cause disease when they are reentrained into the air (Burge, 1990), except for organisms that have protective devices, such as spores.

Microorganisms, although frequently found in the air, do not multiply there. Outdoor air rarely contains pathogens due, in part, to the bactericidal effects of dessication and ozone and ultraviolet irradiation. Indoor air, however, may contain pathogenic organisms that are shed from the skin, hands, and respiratory tract of humans as well as from their clothing. Talking, coughing, and sneezing can produce respiratory droplets containing bacteria, and viruses can spread infection only to susceptible individuals in close proximity to the source. (Gallis, 1976).

The particulate form of matter containing bacterial or viral pathogens, or their products, determines and limits the atmospheric spread of infectious diseases. The size or settling velocity determines the duration of time that pathogens are in the air. This limits the probability of transfer between persons in the environment. Hatch and Gross (1964) have shown that for a 63% reduction in the concentration of particles by gravity settlement in a confined space 10 ft high, requires less than 10 min for particles greater than 13 μm and several hours for single bacterial particles of 2–3 μm. The removal of particles from the air is also dependent on ventilation. It is apparent that the risk of direct respiratory exchange of infectious particles is from the small particles dispersed from the respiratory tracts of infected individuals and that the hazard from the inhalation of particles resuspended from dust deposits on the floor is limited by their relatively large size.

Microorganisms are dispersed into the air as liquid droplets varying in size from greater than 100 μm to less than 10 μm. Salem and Aviado (1970) have reported that of the respiratory activities of sneezing, coughing, and talking, sneezing produces the highest proportion of the small droplets. Nearly all of the small droplets containing airborne bacteria have been reported to originate from the front of the mouth. Only a few originate from the nose (Duguid, 1945). The spread has been observed by high-speed stroboscopic light photography. The smaller droplets remain suspended in air, evaporate quickly, and leave droplet nuclei a few micrometers in diameter which may or may not contain microorganisms. The droplet nuclei settle very slowly and may remain suspended in air almost indefinitely, especially in a room with people where the air currents keep them suspended. This creates an environment with a high concentration of potentially infective particles. Larger particles are expelled up to a distance of 2 m and at a velocity of at least 152 ft/s (Gallis, 1976; Jennison, 1942). These large particles fall rapidly to the floor, where they dry and may attach to dust particles which could be resuspended by sweeping, dusting, movement such as walking, and wind.

As far back as 1934, Wells demonstrated that large particles evaporate rapidly and these residual particles which he called droplet nuclei were only a few micrometers in size. He also showed that even in an atmosphere of 90% relative humidity, droplets with diameters in the order of 80 μm evaporate before settling to the floor from a height of 6 ft. These particles which may contain virulent pathogens are, thus, capable of remaining in atmospheric suspension over long periods of time during which they can be inhaled by susceptible individuals sharing the same confined atmosphere.

10.2. Particle Size and Dose-Response

The particle size of the bioaerosol determines the duration of its availability in the atmosphere, the distance it can travel, and the site of its ultimate deposition in the respiratory tract of the next host. Hatch (1961) has shown that for droplet nuclei (2–3 μm), pulmonary deposition is higher than that for the upper respiratory tract, whereas for dust-borne bacteria and droplets, deposition is essentially limited to the nasopharyngeal area (see Fig. 10.1). Wells et al. (1948) suggested that to establish disease, the infectious particles must be deposited as virulent organisms at the critical site and that only droplet nuclei can contribute significantly to the atmospheric spread of diseases initiated by deposition of infectious particles in the lungs. This was demonstrated in their quantitative study of tuberculosis infection in rabbits. Using suspensions of tubercle bacilli as single organisms of 2–3 μm and 13 μm, they found that the number of tubercles developing in the rabbits' lungs approximated the number of inhaled organisms of 2–3 μm, whereas only 6% of the 13-μm bacilli reached the lungs to produce tubercles.

These studies not only confirmed their prediction that larger particles would not reach the depths of the lungs but also demonstrated the selectivity of tissue susceptibility. Tubercle bacilli implanted on the mucosa of the upper respiratory tract proved to be innocuous. Effective contact requires implantation on a particu-

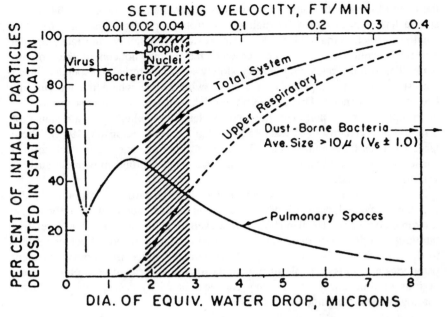

Figure 10.1. Total and regional deposition of inhaled particles in relation to aerodynamic particle size, showing relative positions of viral and bacterial particles, droplet nuclei, and dust-borne bacteria on the size-deposition curve [From Hatch (1961).]

lar part of the respiratory tract, or a part from which susceptible tissue can be reached. This is in contrast to chemical aerosols where deposition, also dependent on particle size, can cause damage and lethality at any region of the respiratory tract. Therefore, it was recommended that for toxicity testing in rodents, particle sizes between 1 and 4 μm be used (Salem et al., 1992).

Studies by Druett et al. (1953) also found striking differences in the atmospheric concentration of anthrax spores of different sizes required to produce disease in 50% of the exposed animals. There was a 17-fold increase in the concentration of particles larger than 12 μm to produce disease than with anthrax spores smaller than 5 μm. The results were independent of the number of spores in a particle.

Goodlow and Leonard (1961) confirmed this finding with aerosol particles of *Pasturella tularensis* of different sizes. To produce 50% mortality in guinea pigs and rhesus monkeys required 3 and 17 cells of 1 μm particle size, 6500 and 240 cells of 7 μm particle size, 20,000 and 540 cells of 12 μm particle size; and 170,000 and 3000 cells of 22 μm particle size, respectively.

Brucellosis, a zoonotic disease, can be transmitted from its animal reservoir to humans by the inhalation of infectious aerosols in addition to skin and conjunctival contact as well as ingestion. Aerosol transmission of brucellosis has been widely accepted as a potential biohazard in laboratories and its potential spread at abattoirs has been reviewed by Kaufmann et al. (1980).

In a cross-sectional epidemiological study associating air quality with swine health, Donham (1991) was able to demonstrate that bacterial concentration (respirable and total microorganisms) was highly correlated to pneumonia. Other air contaminants (dust, ammonia, and carbon dioxide) were also found to be correlated with swine health problems. On the basis of dose–response correlation to swine health or human health problems, the recommended maximal safe microbial concentration of total microbes was 1×10^5 colony forming units per meter square (cfu).

10.3. Airborne Contagion

Wells (1955) originally described airborne contagion as the chain reaction indoors resulting from person-to-person transfer of droplet nuclei implicated in respiratory tract infection. At a Conference on Airborne Contagion sponsored in 1979 by the New York Academy of Sciences, the concept of airborne contagion was expanded to include aerosols from inanimate sources such as air conditioners, dental drills, fungal spores from soil, and insulating materials, as well as long-range outdoor transfer of microorganisms. Air is a vehicle for dissemination of microorganisms, whether dynamic contagion is involved with a geometric increase in cases or static infection arising from a single source and whether humans, animals or plants are implicated as hosts (Kundsin, 1980).

The trachea, bronchi, lungs, and sinuses are usually sterile. The nasopharynx is the natural habitat of the common pathogenic microorganisms that cause infections in the nose, throat, bronchi, and lungs. Humans are the primary hosts for these microorganisms. The reservoir from which infections are spread are patients and healthy carriers. Some people become nasal carriers for streptococci and staphylococci and discharge these organisms in large amounts into the air from their noses (Gallis, 1976). In any case, the microorganisms or chemicals must reach a susceptible and/or immunologically compromised individual to cause infection and/or adverse effects. Interaction of infectious microorganisms with other air pollutants can increase susceptibility of the individual or the severity of the effect (Gardner, 1988; Graham et al., 1987; Parker et al., 1989; Spengler and Sexton, 1983).

Many microorganisms can cause human disease when they are airborne in sufficient numbers. Many environmental and physiological factors play a significant role in the pathogenesis of infectious diseases. These factors are discussed in detail in later sections of this chapter. Bacteria and fungi can produce spores that are hardy and may persist in the environment for many years. Some can cause infection long after the initial contact, and others can cause hypersensitivity disease. Spores produced by fungi are easily transported through the air, and their concentrations in the air are subject to seasonal, diurnal, and geographic variations. The outdoors are abundant with fungal spores which freely penetrate indoors by many routes, including open windows and mechanical air intakes. In ventilated buildings, the indoor air concentrations are directly correlated with the outdoor concentrations. Additionally, many thousands of spores are present per gram of surface dust in most enclosed spaces (Spendlove and Fannin, 1983).

It has been shown by Kaminski et al. (1974) that many microorganisms are capable of producing volatile organic compounds (VOCs) which can be irritants or systemic toxins. Low levels of these compounds have been associated with moldy or mildewy odors. Burge (1987) has reported that microorganisms also produce a variety of higher-molecular-weight toxins.

Mannis et al. (1986) have reported that protozoa are unicellular, often motile organisms grown in indoor reservoirs and can cause severe infections as well as producing antigenic or toxic metabolites. These may contribute to hypersensitivity pneumonitis and humidifier fever (Edwards et al., 1976).

The emphasis on energy conservation will require the recirculation of air, whether heated or conditioned. Thus, less make up outdoor air from the outside will be required.

In addition, making buildings more air-tight will result in higher exposures to more recirculated air with an increased potential for exposure to more airborne microorganisms. Although there have been threshold limit values (TLVs) for chemicals published since 1971, there are no threshold levels for viable particles as of the latest TLVs published by the American Conference of Governmental Industrial Hygienists (ACGIH) for 1991–1992. Perhaps limits will be established for them so that they will be controlled along with temperature and humidity.

Examples of microorganisms causing respiratory infection or sensitization when inhaled are presented in Table 10.1.

10.4. Biotechnology and Genetic Engineering

Molecular biology research on the manipulation of the gene inaugurated genetic engineering. New strains of microbes and plants are possible through biotechnology and genetic engineering. In addition, vaccines can be produced from microbial and viral genes by these techniques. Biotechnology utilizes living organisms and their products for the industrial production of biologically active mammalian proteins such as insulin and the human growth factor. This technology also permits research on the functional organization of genetic information. Genetic engineering allows analysis of the gene and DNA sites that control genetic expression. The potential of this technology is limited only by the creativity and imagination of the human intellect.

The discipline of biotechnology emerged as a result of breakthroughs in recombinant DNA and hybridomas. Recombinant DNA technology is gene splicing or genetic engineering which involves inserting genetic material from one organism into another. Utilizing this technology it is hoped that many of the genetic misinstructions which lead to genetic disorders including hemophilia, spina bifida, Down's syndrome, cystic fibrosis, and diabetes can be erradicated.

Biotechnology has already led to the development of new pharmaceuticals, crop plants, livestock varieties, and other products. Genetic engineering uses recombinant DNA techniques to rearrange genes by removing, adding, or transfering them from one organism or location to another. Restriction enzymes are employed to cut out a gene sequence from one organism and insert it into the plasmid of another. The genes are usually implanted in microorganisms such as single-cell bacteria, yeast, or Chinese hamster ovary cells. The recipient organism then carries out the instructions of the inserted gene. Human insulin, for example, is manufactured by inserting the human gene into the bacteria *Escherichia coli* which multiply in fermentation tanks, producing the polypeptide hormone. Other pharmaceutical products which have been or are being developed through biotechnology are presented in Table 10.2 (Baum, 1987a).

In 1992, the U.S. Food and Drug Administration proposed to regulate products of genetic engineering, and not the process by which they are created. The proposed guidelines for regulating new varieties of foods suggest that the agencies charged with preserving public health do not need special rules to oversee genetically engineered organisms. Foods developed through genetic engineering including fruits, vegetables, and grains will be regulated within the framework of the Federal Food, Drug and Cosmetic Act. The level of oversight will be based on the characteristics of the food and its intended use rather than the method by which it was produced. It is expected that many of the gene-altered foods being introduced will not require premarket FDA approval, unless the process increases the concentration of naturally occurring toxicants in the plant, introduces an allergen not normally found in the plant, or alters the nutritional composition of the plant (Gershon, 1992).

Important therapeutic products to treat cancer and other diseases are being developed using monoclonal antibodies. Immunotherapy for cancer is based on the assumption that tumor cells possess immunogenic characteristics different from those of normal cells. Monoclonal antibodies have been developed that bind preferentially to tumor cells. Such antibodies after binding to tumor cells can activate host defense mechanisms against those cells. Immunotoxins on the other hand, utilize the antibody as a vehicle for a toxic agent that kills the tumor cells directly. Antibody conjugates have been made with chemotherapeutic agents such as methotrexate or with ribosomal inhibitor proteins such as ricin to prepare immunotoxins. Monoclonal antibodies are also being designed to bind to and neutralize endotoxins. Endotoxins are released by a broad spectrum of bacteria to cause septic shock which is characterized by a precipitous drop in blood pressure, blood clotting abnormalities, failure of major organ systems, and death.

Agricultural biotechnology products are also being developed with the potential

Table 10.1. Microorganisms Causing Respiratory Infection or Sensitization When Inhaled

Disease	Causative Organism	Primary Reservoir	Reference
Bacterial Disease			
Pneumonia	*Streptococci pneumoniae*	Humans	Ketchum (1988)
Pneumonia, nosocomial infection	*Klebsiella pneumoniae*	Humans	Ketchum (1988)
Pneumonia	*Haemophilius influenzae*	Humans	Ketchum (1988)
Walking pneumonia	*Mycoplasma pneumoniae*	Humans	Ketchum (1988)
Q fever	*Coxiella burnetii*	Animals (sheep)	Ketchum (1988)
Ornithosis, psittacosis, parrot fever	*Chlamydia psittaci*	Birds (domestic and wild)	Ketchum (1988)
Brucellosis[a]	*Brucella melitensis*	Animals	Ketchum (1988); Kaufmann et al. (1990)
Legionnaires' disease, Pontiac fever	*Legionella pneumophila, Legionella anisa*	Water (indoor and outdoor)	Fraser (1980); Mallison, (1980); Breiman et al. (1990); Fenstersheib et al. (1990); O'Mahony et al. (1990); Lee and West (1991)
Tuberculosis	*Mycobacterium tuberculosis*	Humans	Ketchum (1988)
Hypersensitivity pneumonitis	*Thermoactinomyces*	Heated water, soil, compost, moldy hay	Burge (1990)
Diphtheria	*Corynebacterium diphtheriae*	Humans	Ketchum (1988)
Pertussis, whooping cough	*Bordetella pertussis*	Humans	Ketchum (1988)
Inhalation anthrax	*Bacillus anthracis*	Animals	Brachman (1980); Ketchum (1988); Titball et al. (1991)
Bubonic plague[a], pneumonic plague	*Yersinia pestis*	Animals	Ketchum (1988)
Tularemia	*Francisella tularensis*	Animals	Ketchum (1988)
Viral Disease			
Influenza	Influenza A, B, and C viruses	Humans	Ketchum (1988)
Croup	Parainfluenza viruses	Humans	Ketchum (1988)
Bronchiolitis, pneumonia	Respiratory syncytial virus (RSV)	Humans	Ketchum (1988)
Mumps[a]	Mumps virus	Humans	Ketchum (1988)
Measles	Rubella virus	Humans	Ketchum (1988)
Common cold	Rhinoviruses, coronaviruses parainfluenza viruses	Humans	Gwaltney (1980); Ketchum (1988)

continued

Table 10.1. (Continued).

Disease	Causative Organism	Primary Reservoir	Reference
Viral Disease			
Chicken pox[a] (varicella) Zoster[a] (shingles)	Varicella–Zoster virus	Humans	Lange (1976); Ketchum (1988)
Smallpox[a]	Variola virus	Humans	Joklik (1976)
Fungal Disease			
Asthma, rhinitis	*Alternaria*	Outdoor air, dead plants	Burge (1990)
Asthma, rhinitis	*Cladosporium*	Outdoor air, dead plants	Burge (1990)
Asthma, rhinits	*Penicillium*	Indoor/outdoor damp organic material	Burge (1990)
Pulmonary aspergillosis	*Aspergillus*	Soil, compost	Herman (1980)
Coccidioidomycosis	*Coccidioides immitis*	Soil of arid regions	Ketchum (1988)
Histoplasmosis	*Histoplasma capsulatum*	Animals, soil, keratonaceous material (feathers) in soil	Conant (1976)
Protozoan Disease			
Hypersensitivity pneumonitis	Protozoa	Water reservoirs	Burge (1990)
Algal Disease			
Asthma, rhinitis	Alga	Lighted water reservoirs	Burge (1990)

[a] Diseases transmitted via respiratory tract, but signs of infection are seen elsewhere in the body.

to revolutionize the practice of agriculture. These products range from animal vaccines to microbial pesticides to herbicide resistant plants as well as transgenic "pharming" (Baum, 1987b; Glanz, 1992; Powledge, 1992). The technology for producing vaccines and therapy for diseases of farm animals and pets and for producing proteins such as bovine somatotropin (BST) is similar to the technology used in developing corresponding products for humans. Porcine somatotropin is also being developed. Consensus interferon was designed to have optimum characteristics for use in preventing bovine respiratory disease in cattle. A vaccine against feline leukemia, a leading cause of death in cats, was developed based on agricultural biotechnology. The major impediment for the interactions among microorganisms and the plants they colonize or infect has been the fear of releasing such genetically engineered microbes into the environment. Although our knowledge of plant biochemistry is considerably more limited than mammalian biochemistry, progress has been made in model systems such as petunia,

Table 10.2. Genetically-Engineered pharmaceutical products

Biotechnology Product	Type	Use
Atrial natriuretic factor (ANF)	Peptide hormone	Reduce blood pressure
Erythropoietin (EPO)	Peptide hormone	Stimulates red blood cell production to treat anemia associated with dialysis
Factor VIII	Protein in blood clot cascade reaction (Protease)	Hemophelia
Tissue-type Plasminogen Activator (TPA)	Protein enzyme (Protease)	Dissolves clots
Superoxide dismutase (SOD)	Protein enzyme (Protease)	Scavenges Superoxide radicals With TPA to treat heart attacks and transplant patients
Epidermal growth factor (EGF)	Protein enzyme	Epidermal cell proliferation in burn and wound healing Cataract surgery Opthalmic conditions
Fibroblast growth factor (FGR)	Protein	Angiogenesis or growth of blood vessels Treat burns and wounds Assists vascularizations of skin grafts
Human growth hormone (HGH)	Protein	Pituitary dwarfism
Interleukin-2 (IL-2)	Lymphokine	Immune system Cancer
Interleukin-3 (IL-3)	Lymphokine	Blood cell growth factor Bone marrow transplants
Tumor necrosis factor (TNF)	Lymphokine	Cancer Used with IL-2 and Interferon
Granulocyte colony-stimulating factor (G-CSF)	Protein	Leukemia Cancer
Granulocyte–monocyte colony-stimulating factor (GM-CSF)	Protein	Immune deficiency
Macrophage colony-stimulating factor (M-CSF)	Protein	Infectious diseases

tobacco, and tomato species. The element unique to agricultural biotechnology is that regulatory approval is required for field tests of genetically engineered microbes and plants. Plants, to be agriculturally useful, must grow in a field and microbial pesticides have to be applied to those plants. Tobacco plants have been engineered to be resistant to crown gall disease and to manufacture medicine and proteins that would be too expensive to make by other means. Tobacoo leaves are infected with a genetically altered virus which prompts the leaves to make proteins on request. Genes are inserted to tell the virus how to make hemoglobin or other desired protein. An experimental AIDS drug, Compound Q, has been extracted from engineered ground-up tobacco leaves. Hybrid corn is being developed to be herbicide-resistant, and *Pseudomonas fluorescens* was engineered to protect plants from frost damage. Products are also being created that possess specific desired traits in a fruit or seed. Bovine somatotropin (BST), a recombinant DNA version of growth hormone produced in the pituitary glands of cows, when administered to mature cows increases their milk production from 10 to 25% with only 6% more feed. It has no effect on humans or other primates. The BST gene is expressed in the bacteria *Escherichia coli*, which is how BST is produced. The bacteria *Pseudomonas fluorescens* has been engineered as an insecticide to protect the roots of corn plants from a pest called black cutworm.

Ways to utilize toxins produced by the bacteria *Bacillus thuringiensis* (Bt) are being studied. The toxins interfere with insect digestive processes but are nontoxic to mammals, birds, fish, and other wildlife. The Bt toxins are rapidly biodegraded. *P. fluorescens* colonize the surfaces of a wide range of plants. A gene that encodes a B6 toxin has been transferred into a strain system that colonizes the roots of cornplants. An elegant marker system has been developed that allows genetically modified microorganisms in the environment to be followed. The technique can detect as few as one recombinant microbe in a gram of field soil. The system involves inserting *E. coli* genes (lac Z, lac Y) into *P. fluorescens* which gives it the ability to grow on lactose, a property no other *P. fluorescens* has. Progress is being made on inserting Bt toxin genes in plant tissues so that when insects or caterpillars eat the plant leaves or roots, they ingest the toxin and die. Bt toxin is expressed in tomato plants which become resistant to attacks by pests such as tomato hookworms and fruitworms. Bt toxins are toxic to caterpillars, beetles, and weevils. Some strains of *P. fluorescens* inhibit the growth of soil-borne fungi that either kill crop plants or destroy sufficient root mass to limit the plants growth significantly. These bacteria produce a hydrolytic enzyme chitinase which breaks down chitin, a common fungal cell-wall material. The genes that encode chitinase can be transferred to strains of *P. fluorescens* that colonize a specific crop's root system. Ice-minus *P. fluorescens* and *P. syringae* protect crops from frost damage. Tobacco and tomato plants are being engineered to be herbicide resistant. The gene that encodes the polygalacturonase enzyme which causes fruit to soften has been isolated from tomatoes. Reinserting this gene into tomatoes in a reverse or "antisense" orientation blocks most of the enzymes production and thus prolongs the life of the tomatoes.

Rapeseed, a plant related to mustard, is a very efficient oil producer. Efforts in Canada have resulted in a strain Canola which produces an edible oil. Rapeseed has also been engineered to produce high-value specialty oils that the plant does not normally produce.

Genetic engineers have also created "pharm" animals that produce therapeutically important proteins in bulk. Genetic manipulation has shown the feasibility of producing therapeutic proteins in the milk and blood of genetically altered cows, sheep, goats, pigs, and mice. TPA, Tissue Plasminogen Activator, an anti-blood-clotting agent used to treat heart attack victims is manufactured in bioreactor vats of recombinant bacteria and costs thousands of dollars a gram. Current research has produced transgenic goats which can produce as much as 3 g/L of TPA in their milk. A single optimally producing goat could make as much therapeutic protein as a 1000-L bioreactor so that TPA is available to more patients at much lower cost. A hydrophobic membrane protein was produced in the milk of transgenic mice. This protein which is a potential therapy for cystic fibrosis is difficult to make by other techniques. Lactoferrin, an antibacterial protein which is also difficult to make, can be produced in significant quantities in the milk from transgenic bull calf female offspring. Bovine systems could be a most plentiful source for lactoferrin and other proteins. Lactoferrin can be used to treat immune-suppressed patients and as a component of infant formula. Transgenic pigs have been developed that produce human hemoglobin in their blood for use in transfusions. Transgenic sheep were developed whose eggs were implanted with the gene for alpha-1-trypsin, a glycoprotein used to treat life-threatening emphysema. These transgenic animals appear to be normal except for their drug-producing capabilities. Transgenics should be considered as an extension of traditional breeding practices, that is, a very sophisticated way of breeding for some desired trait. After the first transgenic animal is made, it is bred to produce more of the same. Products developed utilizing transgenic animal technology are efficacious and safe. Xenograph research is in progress to grow quasi-human organs in pigs for human transplant.

Many transgenic mice have been created not for producing pharmaceuticals, but to be genetically susceptible to certain human ailments so that drug treatment can be tested on them.

The revolution in vaccine development, fueled by progress in genetic engineering and immunology, offers hope in combating pathological disorders and even possibly AIDS. Previously, vaccines were developed empirically, mutating or killing disease-causing organisms and then testing to see if the preparation made from them were effective. At least two human vaccines were created using the new genetic engineering approach, both for hepatitis B, and both on the market. The tools being used in the new approaches are also improving the understanding of the biology of infection and the immune response. Today, the new way to develop vaccines is with genetic-engineering tools. Genes from infectious pathogens that contain instructions for specific antigens can be isolated and inserted into a harmless bacteria or virus, which then produce the antigen. To the immune system, the harmless bacteria or virus appears to be the real pathogen, and the immune cells learn how to attack the pathogen without exposure to it. Before designing vaccines this way, investigators must understand the biology of each pathogen. Viruses and bacteria have many ways of invading the human body and its cells and, therefore, have evolved various mechanisms to evade the host defenses. The immune system displays an extraordinary array of cells during infection. These include macrophages which engulf and break down the invaders. B-lymphocytes which make antibodies that bind and kill bacteria and viruses before they infect cells. Cytotoxic T-lymphocytes kill infected cells in an attempt to prevent the spread of the pathogens, whereas helper-T-lymphocytes send signals to other immune system cells (Aldovini and Young, 1992).

Critics of biotechnologically derived food and drugs ignore the fact that selective breeding has produced desirable products in the cattle and horse industries, whereas agriculture has depended both on natural selection and selective breeding. With the use of biotechnology, desired genes can be inserted into bacteria, plants and animals to produce the desired end product quicker and more efficiently. Many viruses and bacteria have mutated and produced many drug-resistant strains. This emerging threat is caused by these microbes as they develop new pathways, new proteins, and new strategies for survival. Bacteria, even of different species, can exchange genetic material, including the gene for drug resistance, and that resistance can be transfered to other bacteria as well.

10.4.1. Inhalation of Bioaerosols

Both the site of deposition and the total dose of an airborne infectious agent delivered to the respiratory tract can be significantly affected by the uniqueness of the pulmonary anatomy; the route of breathing (nose versus oral-nasal); the depth and rate of airflow; and the physical properties which governs particle transport and deposition. Having a knowledge of these various factors can significantly improve our ability to (1) understand the pathogenesis of the infectious disease process; (2) extrapolate laboratory animal studies to humans; (3) predict and assess health risk associated with airborne pathogens; and (4) provide guidance in identifying a most effective mode of treatment for respiratory infections.

10.4.1.1. Anatomical and Physiological Factors in Respiratory Deposition of Infectious Organisms

The respiratory system can be conveniently divided into three major compartments or regions (Fig. 10.2). Each of these regions have unique anatomical features that can influence the fate, deposition, transport, and clearance pattern of entry of airborne microorganisms. It is not appropriate to go into great detail regarding the anatomic structure of the respiratory system in this chapter, but there are several excellent reviews devoted to this topic (Gardner et al., 1988; Crystal et al., 1991; McClellan et al., 1989; Miller et al., 1989). It is evident that the anatomical structure and the physical dimensions of the respiratory system are important factors that must be considered in understanding the deposition and fate of inhaled organisms. This brief review will provide a basic understanding of the anatomical and physiological factors associated with the inhalation of bioaerosols.

The upper respiratory portion (nasal pharyngeal region) consists of the area extending from the nares down to the epiglottis and larynx at the entrance to the

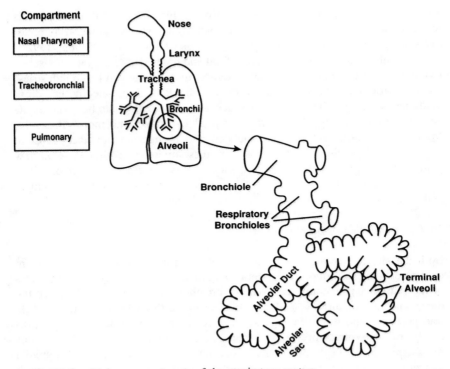

Figure 10.2. Major compartments of the respiratory system.

trachea. The mouth is included in this region and is important during mouth breathing.

This region has a complicated morphology consisting of the turbinates, epiglottis, glottis, pharynx, and larynx. The complicated shape of the nasal passages is not solely related to the role of the nose as an olfactory organ, but it also plays a major physiological function in the modification of the inspired air prior to its access to the lungs. Upon inspiration, the air enters the nose through the anterior nares passing into the airway passages where it is ultimately carried to the pharynx. These airway passages are lined with small hairs and ciliated mucous membranes which is rich with goblet and mucous-secreting cells. The high surface-to-volume ratio facilitates humidification and warming of the incoming air. In this region, the largest inhaled particles are removed by impaction and filtration. This uniqueness of the anatomical structure of this region plays a significant role in removing potentially harmful substances present in inhaled air from being deposited deeper in the respiratory system.

The tracheobronchial (TB) region consists of the conducting airways beginning at the trachea and extending down to the terminal bronchioles. The TB region functions to deliver inspired air to the deeper portions of the lung. The dimensions and number of branching airways varies from species to species, making definitive extrapolation of animal studies to man difficult. The trachea is an elastic tube supported by 16–20 cartilagineous rings that circle about three-fourths of its circumference and is the first and largest of a series of branching airways. The left and right lung are entered by two major bronchi that branches off the trachea into five separate lobes. In man, there are five lobes. The left lung consists of an upper and lower lobe, whereas the right lung has an upper, middle, and lower lobe. This may differ in laboratory animals, for example, although the rat also has five lobes, they have only a single lobe on the left and four on the right. The conducting airways in each lobe consists of up to 18–20 dichotomous branches from the bronchi to the terminal bronchiole. Beyond the terminal bronchi, the airways become very thin-walled and are referred to as respiratory bronchiole which have numerous small air sacs (alveoli) protruding from its walls.

Although the pulmonary region primarily functions in gas exchange, it also functions in pulmonary clearance and immunological defense. This region begins with the partially alveolated respiratory bronchioles.

The epithelium of the respiratory bronchioles is nonciliated. Each respiratory bronchiole divides into alveolar ducts. These ducts are actually bronchioles where almost their entire wall has been completely alveolated. These ducts lead ultimately into the alveolar sacs. The total number of ducts and sacs has been estimated to be 7×10^6 and 8.4×10^6, respectively (Gardner et al., 1988). The alveoli, which are evaginations of the alveolar sacs, are thin-walled, surrounded by a meshwork of blood capillaries. In the alveolus, the atmosphere and the blood are brought into an intimate contact where an equilibration between O_2 and CO_2 can take place. The surface of the alveoli is made up of primarily two types of cells. The Type-1 cells are very thin and covers the greatest surface area. Type-2 cells are larger, contain numerous microvilli, and produce and secrete a surface-lining fluid (surfactant) that functions in reducing surface tension which reduces the tendency for alveoli to collapse. The total respiratory surface of the lung has been estimated to be 50 m^2 during expiration and as much as 100 m^2 during the deepest inspiration.

10.4.2. Dosimetry Factors for Inhaled Microorganisms

The goals of experimental inhalation studies of bioaerosols in animal models are to be able to estimate infectivity in man and to be able to define the mechanisms of the pathogenesis of the airborne disease. Animal studies can provide a scientifically sound approach to providing cause–effect data under well-controlled and defined conditions with virtually unlimited exposure condition.

Aerosolization of viable organisms can be a serious threat to the microbes and, once launched into the air, the viability of the test organism and the actual number of airborne organisms reaching the respiratory tract may be greatly reduced. In bioaerosol studies, it is important to distinguish between the concentration generated and the dose. Posology is the study of dose and dosage. Although the term dose and dosage are usually used interchangeably, dose is total amount of a test article administered, whereas dosage is a relative amount. A dose of 10 mg, for example, means that this is the amount of material administered whether it is to a mouse, rat, dog, or human. Dosage, on the other hand, is the amount of material administered relative to body weight, body surface, and/or time ($mg\ kg^{-1}$, $mg\ m^{-2}$). Inhalation dosage can be expressed as concentration or concentration and duration (i.e., LC_{50} as $mg\ L^{-1}$, $mg\ m^{-3}$, ppm, ppb, or LCt_{50} as mg-$min\ m^{-3}$. Concentration or dosage in inhalation studies is the amount of material per unit volume, that is, the amount of a substance in the air or in the test medium. Dose refers to the total amount of the microorganism that is actually inhaled, delivered, and deposited to the tissue.

Developing a quantitative understanding of the relationship among exposure concentration, dose delivered to the tissue, and a specific pulmonary response is a basic fundamental goal of inhalation scientists. To achieve these goals requires information on deposition and the ultimate fate of the microorganism once deposited in the tissue. The term deposition refers specifically to the amount of inhaled airborne agent in the inhaled air that are deposited in regions of the respiratory tract. The fate of the organism can be expressed in terms of clearance, which refers to the subsequent translocation of and removal of deposited substances from the respiratory tract, and retention, which refers to the temporal pattern of residual substance that stay in the respiratory tract and are not cleared.

The infective dose for microbial aerosols has been shown in animals to be greatly influenced by particle size which, in turn, controls the site of deposition. Particle size is also critical where one part of the respiratory system is more susceptible to the inhaled organism than another. Airborne microbes may include particles varying in size from a single virus unit which could be as small as 0.1 μm in diameter to the largest fungal spores and pollens reaching sizes of 50–100 μm.

There are a number of generic factors that can significantly influence the deposition of particles. Ventilation is important because (1) the physics of airflow is important to deposition mechanisms and (2) the rate and depth of breathing influence the volume of air and, hence, the mass of infectious agent entering the respiratory tract and the total surface area over which deposition can occur. Another important element is the route of breathing (oral, nasal, or oronasal). This influences filtering efficiency of inhaled materials and, thus, impacts the dose delivered to the lower respiratory tract.

Although most adults are nasal breathers at rest, they may resort to chronic or periodic mouth breathing under certain conditions such as exercise, nasal obstruction, or in the presence of chemical irritants. As respiratory demands increase, the proportion of air entering via the mouth also increases. Such action can significantly alter the pattern of deposition and, thus, the response to the inhaled substance. For instance, nearly 100% of particles having an aerodynamic size of about 10 μm or larger are deposited in the nasopharyngeal region during nasal breathing. This compares to only about 65% deposition of such particles under conditions of oronasal breathing. There is also increased penetration of larger particles deeper into the respiratory tract with oronasal breathing. Although nasal breathing offers an effective means of protecting sensitive lower respiratory tract tissues from airborne particles, it should be remembered that rodents cannot breathe through their mouth, a factor that must be taken into consideration when extrapolating such animal data to man. Also, there exists a great difference in the complexity of the nasal passages resulting in differences in nasal airflow patterns between man and the laboratory animal which, in turn, may account for species-specific lesion distribution following inhalation exposure to certain substances. Although some of these morphological factors may be useful in protecting the deeper regions of the respiratory tract, it also makes the nose more vulnerable to the effects of particles, soluble gases, and vapors.

Chemical agents can alter physiological responses during exercise by causing pulmonary function changes (e.g., increased airway resistance through constriction) which tend to decrease the volume of air penetrating to the alveoli and can result in a shift to rapid, shallow breathing. Exercise has been shown to have a pronounced effect on pulmonary tissue uptake of gaseous pollutants but to have little effect on TB tissue. The deposition, clearance, and retention of inhaled particles has been extensively reviewed (National Research Council, 1991; Bates et al., 1989; Gardner et al., 1988).

For particles, the overriding factors influencing regional respiratory tract deposition are those based on aerodynamic properties, which, in turn, depend on a variety of physical properties. Particles of the same physical size do not necessarily behave the same aerodynamically. For example, a denser particle will tend to fall faster than a less dense particle of the same size. Therefore, the size of airborne particles is expressed in terms of their "aerodynamic diameter" which is the diameter equivalent to that of a theoretical spherical particle with a density of 1 that has the same terminal settling velocity (i.e., behaves aerodynamically in the same way) as the particle in question. Using such an adjustment permits a more valid comparison among particles of different physical sizes. These differences between actual and aerodynamic sizes are important in predicting respiratory deposition of inhaled particles.

After an infectious agent has entered the airways, it must first be deposited in a significant dose onto susceptible tissue to produce an effect. Figure 10.3 illustrates the five mechanisms by which particle deposition can occur: impaction, sedimentation, Brownian diffusion, interception, and electrostatic precipitation. Electrostatic attraction of particles to the walls of respiratory airways is a minor mechanism and is not important for the inhalation deposition of most environmental contaminants. Interception is more important for fiber deposition.

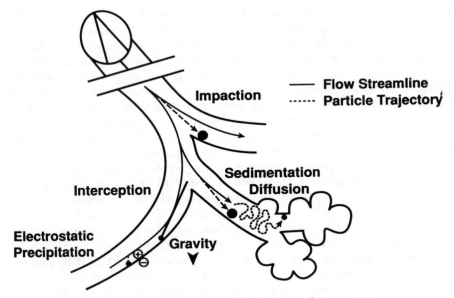

Figure 10.3. Mechanisms of particle deposition in the respiratory system.

Larger particles are removed from the inhaled air by the mechanism of impaction at the various bifurcations. Impaction onto an airway surface occurs when a particle's momentum prevents it from changing course in an area where there is a rapid change in the direction of the airflow. It is the main mechanism for the deposition of particles having an aerodynamic size of ≥ 2.0 μm. The largest sized particles are deposited in the anterior regions of the nose and are then most effectively cleared by mechanical means, including sneezing, coughing, and nose blowing. Deposited microbes may also be cleared from the upper airways by an effective mucociliary system which can transport the deposited particle from the airways to the throat, where it may be swallowed or expectorated.

The probability of impaction increases with increased air velocity, rate of breathing, particle size, and density. Sedimentation is deposition due to gravity and is important for particles with an aerodynamic size of ≥ 0.5 μm in medium and small airways where the air velocity is relatively low. In this case, the particles will fall out of the airstream at a constant rate when the gravitational forces on the airborne particle is balanced by the total force due to air buoyancy and air resistance. The smallest particles (≥ 0.5 μm) may be deposited onto the airway walls due to their bombardment by surrounding air molecules. Brownian diffusion is a major mechanism for deposition in airways where the airflow is low, such as found in the alveoli and bronchiole. Figure 10.1 depicts the regional deposition of inhaled aerosols as a function of aerodynamic diameter.

10.4.3. Infection

The process of infection has been defined as an interaction of a host, a microorganism, and the environment. In the natural environment, healthy beings exist in an equilibrium with numerous potentially pathogenic microbes. Infection has been considered as a normal state of life and the process of disease merely as a disturbance of this equilibrium between the host and the parasite. In recent years, it has been shown that the environment may also play a major role in this process (Gardner, 1988; Graham et al., 1987; Parker et al., 1989; Coffin and Gardner, 1972).

Once deposited in the lung, the invading microorganism must be able to grow and multiply to produce disease. This is accomplished by increasing in number in a local lesion or by spreading systemically throughout the host. The success of the microbe depends on (1) the specific properties of the invading organism to produce disease and (2) the manner in which the infected host responses to the microbial invasion.

The pathogenicity or virulence of a infectious organism depends on its ability to grow and multiply in susceptible tissue. This requires an inherent biochemical ability to grow and to obtain the necessary nutrients within the respiratory tract and an ability to combat the pulmonary defense mechanisms that is present in the respiratory tract, which functions in clearing or killing the invading microbes. These include various humoral and cellular defense mechanisms of the host.

When clinically manifested injury to the host becomes evident, an overt infectious disease has occurred. At the other extreme, when the organism is deposited in the host and proliferates, at least enough to maintain its original numbers, but there is no evidence of inciting any adverse lost reaction, then colonization takes place. Between colonization and overt infections lies covert or subclinical infection.

In general, pathogenic microorganisms express their disease producing properties through two kinds of mechanisms: (1) the invasion and destruction of the host tissue and (2) the production of toxins. Table 10.3 lists examples of toxigenic pathogens. This can occur at the site of invasion or distally if metastatic infection occurs via bacteremia spread. Although many organisms cause disease by both invasion and toxin production, other are pathogenic primarily due to their invasive properties, for example, *pneumococci aureus*. Extracellular pathogens, (e.g., *Streptococcus pyogenes, Staphylococcus aureus*, and *Diplococcus pneumoniae*) are capable of producing disease only as long as they remain outside of defense (phagocytic) cells, whereas intracellular pathogens (e.g., *Brucella abortus, Salmonella typhosa*) can survive and multiply within these phagocytic cells and may actually destroy them.

The degree of virulence of an organism may also be due to certain enzymes and other metabolic substances that are produced by the organisms. These biological

Table 10.3. Principal bacterial toxins.

Toxigenic Species	Toxin	Disease
Clostridium botulinum	Type-specific neurotoxins	Botulism (neurotoxin, paralytic)
Clostridium perfringens	α-Toxins and others	Gas gangrene
Clostridium tetani	Tetanospasmin; tetanolysin	Tetanus
Corynebacterium diphtherine	Diphtheria toxin	Diphtheria (modifies enzymes)
Staphylococcus aureus	α, β, γ-toxins; leukocidin Enterotoxin	Staphylococcal pyogenic infections
	Erythrogenic toxin	Staphylococcal food poisoning Staphylococcal scarlet fever
Streptococcus pyogenes	Streptolysins O and S Erythrogenic toxin	Streptococcal pyogenic infections Streptococcal scarlet fever
Shigella dysenteriae	Neurotoxin	Bacillary dysentery
Salmonella Typhi	Endotoxins	Gastroenteritis; enteric fever
Vibrio cholerae	Endotoxins	Cholera

Table 10.4. Factors influencing virulence.

Factor	Action
Hyaluronidase (spreading factor)	Facilitates diffusion of pathogens and toxin materials through host tissues. Increases permeability of tissue spaces.
Coagulase	Causes resistance to phagocytosis.
Hemolysins	Destroy red blood cells and other tissue cells.
Lecithinase "alpha toxin"	Causes lysis of red blood cells and other tissue cells.
Collagenase	Dissolves collagen.
Leucocidin	Kills leucocytes.
Streptokinase (streptococcal fibrinolysin), streptodornase	Dissolves human fibrin.
Exotoxins	Cause degeneration of host cells. Block essential metabolites, i.e., enzymes. Render substrates unsusceptible to bacterial enzymatic digestion. Prevent enzyme synthesis by the cell.
Endotoxins	Liberated on cell lysis, less potent than exotoxins. Causes diarrhea, shock, and circulatory disturbances.
Capsules	Enable some bacterial to resist phagocytosis.

products can play a significant role in determining the outcome of an infection. Table 10.4 lists some of these microbial factors that influences virulence).

10.5. Factors Influencing Airborne Contagion

Many factors influence the inhalation, deposition, and retention of inhaled materials whether microorganisms or chemicals is in the respiratory tract. These factors may also influence the response of the lungs and other organs as well as dose–

response relationships to various inhalation exposures of chemicals, microorganisms, or their products.

The number of microbes required to cause a disease in a given host is called the infective dose. The infective dose varies not only with the strain of the microorganism, but also with the host. A number of factors play a role in making the host susceptible to an infectious organism. Exposure to cold, heat, certain stress like exercise, poor nutrition, age, secondary infections, and the interaction of the infectious agent with numerous noxious substances found in the host's environment (pollution) can modify and complicate the etiology and pathogenesis of the disease (Gardner, 1988; Coffin and Gardner, 1972; Graham et al., 1987).

In general, the number of organisms required to cause disease is usually proportional to their virulence. When an extremely virulent organism enters the body through its natural portal of entry, very few organisms are required to establish an infection. Less virulent strains may require considerably greater numbers to establish infection. Some microorganisms have a special affinity for the respiratory tract and establishes infection in the bronchi and lungs; tuberculosis and diphtheria usually enter by this portal. To produce lobar pneumonia, the pneumococcus, regardless of its virulence, must enter the lungs via the respiratory passages. The infectivity dose also varies with the host species.

10.5.1. Methods of Study

There are three general approaches for studying the pathogenesis of infectious disease: these include epidemiological, human clinical, and animal studies. The optimal database for assessing health risk has information from all three approaches because this minimizes the inherent limitations of each. Epidemiological studies are conducted to determine the correlation between exposure and disease. Unfortunately, epidemiological studies alone have great difficulties in proving causative relationships because of the many confounding variables. Such studies can show an association between exposure and effects and identify qualitatively the potential for human risk.

Although controlled human clinical studies offer the best opportunity to directly relate the cause of an infectious agent to human effects, they too have deficiencies. Because the safety of the volunteer is of paramount importance, the test substance can have no long-term residual effects, restricting such studies to exposure to organism with limited pathogenicity or completely nonpathogenicity, and complete recovery must be ensured. Only rather typical medical diagnostic procedures can be performed to follow the course of the disease.

The strengths of well-designed animal studies are that they can provide complete evaluation of the infectious aerosol. The researcher has the choice of a wide range of concentrations, exposure regimens, biological agents, biological parameters, and test species. Such studies are most useful in studying the pathogenesis of the disease. Because many physiological mechanisms are common to animals and humans, scientists hypothesize that if an agent is infective in several animal species, it is likely to cause a similar effect in humans. However, care must be taken in attempting to quantitatively extrapolate the effective airborne infectious concentration in animals to the human and that certain defense mechanisms may differ in various species.

10.5.2. Experimental Infections of the Lung

Laboratory-induced pulmonary infections have been useful in understanding the (1) pathogenicity of microorganisms, (2) environmental factors that alter susceptibility to respiratory disease, and (3) mechanisms of host resistance to infection. By using the appropriate animal model, one can measure subtle defects of the total pulmonary defense system by testing the lungs ability to efficiently defend itself against invading, potentially pathogenic organisms. The three experimental techniques that have been most widely used in the laboratory to expose a test animal to an airborne infectious organism includes (1) aerosolization, (2) intratracheal instillation, and (3) intranasal inoculation.

Aerosolization of the microorganism most closely mimics or simulates a natural exposure. This method provides the most uniform exposure of the animal and is applicable to a wide variety of pathogenic organisms. To conduct such exposures requires expensive equipment and a team of scientists with expertise in aerosol sciences, engineering, and biology as well as specialized containment to ensure the health and safety of the investigators (Phalen, 1984; Gardner, 1988).

Organisms may also be instilled intratracheally in an anaesthetized animal. Although these studies are less costly, one cannot guarantee a uniform distribution throughout the lung as with inhalation exposure (Hatch et al., 1981; Phalen, 1984). Also, the need to anesthetize the test animal results in an unnatural condition of exposure.

Intranasal inoculation of microorganisms has been used for years. With this method, one must depend on the aspiration of the inoculum to spread the infectious organism from the upper respiratory tract to the conducting airways and deep lung. Both this technique and intratracheal instillation can result in considerably more animal-to-animal variation than with the aerosol challenge.

Ideally, whichever experimental model is chosen, certain criteria needs to be considered. Any experimental model should (1) yield reproducible data; (2) permit correlation among species of test animals; (3) ensure the sensitivity of the test animal to the microorganism, and (4) facilitate the ability for direct comparison of data with that from mechanistic studies. The test organism must be able to (1) multiply within the infected susceptible host tissue, (2) be quantifiable at the administered dose, (3) trace and follow during the infectious and recovery phase; (4) not greatly influence mortality rate with small variations in dose or virulence and (5) withstand the vigors of aerosolization if the route of administration is by inhalation.

Such in vivo exposure studies can be used to provide vital information on survival statistics, growth, replication, and fate of the infectious organism in the host tissue and mechanistic studies using sensitive biochemical and immunological measurements

10.6. Summary

The health aspects of bioaerosols have been reviewed in terms of sources, particle size, and dose–response. Airborne contagion and factors influencing this are also discussed. The anatomical and physiological factors in respiratory deposition and dosimetry of infectious organisms are described as are the processes of infectious methods of study and experimental infections of the lung. The advent of biotechnology and genetic engineering will expand the horizons of bioaerosols. Hopefully, these processes will be employed for the benefit of mankind.

References

Al-Dagal, M., and D. Y. Fung. 1990. Aeromicrobiology—A review. Crit. Rev. Food Sci. Nutr. 29:333–340.

Aldovini, A., and R. A. Young. 1992. The new vaccines. Technol. Rev. 95:24–31.

Bates, D. V., D. L. Dungworth, P. N. Lee, R. O. McClellan, and F. J. C. Roe. 1989. Assessment of inhalation hazards. Springer-Verlag, New York.

Baum, R. M. 1987a. Biotech industry moving pharmaceutical products to market. Chemical and Engineering News 65(29):11–32.

Baum, R. M. 1987b. Agricultural biotechnology advances toward commercialization. Chemical and Engineering News 65(32):9–14.

Brachman, P. S. 1980. Inhalation anthrax. Pp. 83–93. In R. B. Kundsin, (ed.), Airborne contagion. Annals of New York Academy of Sciences, New York. 353.

Breiman, R. F., W. Cozen, B. S. Fields, T. D. Mastro, S. J. Carr, J. S. Spika, and L. Mascola. 1990. Role of air sampling in investigation of an outbreak of Legionnaires' disease associated with exposure to aerosols from an evaporative condenser. J. Infect. Dis. 161:1257–1261.

Burge, H. A. 1987. Toxigenic potential of indoor microbial aerosols. In S. S. Sandhu, D. M. Demarini, M. J. Mass, M. M. Moore, and J. L. Mumford, (eds.), Short-term bioassays in the analysis of complex environmental mixtures. Plenum, New York.

Burge, H. 1990. Bioaerosols: Prevalence and health effects in the indoor environment. J. Allergy Clin. Immunol. 86:687–701.

Coffin, D. L., and D. E. Gardner. 1972. Interaction of biological agents and chemical air pollutants. Ann. Occup. Hyg. 15:219–235.

Conant, N. F. 1976. Fungous diseases involving internal organs, Pp. 1056–1057. In W. K. Joklik and H. P. Willett (eds.), Zinsser microbiology. Appleton-Century-Crofts, New York.

Crystal, R. G., J. B. West, P. J. Barnes, N. S. Cherniack, and E. R. Weibel (eds.). 1991. The lung: Scientific foundations. Raven Press, New York.

Donham, K. J. 1991. Association of environmental air contaminants with disease and productivity in swine. Amer. J. Vet. Res. 52:1723–1730.

Druett, H. A., D. W. Henderson, L. Packman, and S. Peacock. 1953. Studies on respiratory infection. 1. The influence of particle size on respiratory infection with anthrax spores. J. Hyg. 51:359.

Duguid, J. P. 1945. The number and sites of origin of droplets expelled during expiratory activities. Edinb. Med. J. 52:385–401.

Edwards, J. H., A. J. Griffiths, and J. Mullins. 1976. Protozoa as sources of antigens in humidifier fever. Nature 264:438.

Fenstersheib, M. D., M. Miller, C. Diggins, S. Liska, L. Detwiler, S. B. Werner, D. Lindquist, W. L. Thacker, and R. F. Benson. 1990. Outbreak of Pontiac fever due to Legionella Anisa. Lancet 336:35–37.

Fraser, D. W. 1980. Legionellosis: Evidence of airborne transmission. Pp. 61–66. In R.B. Kundsin, (ed.), Airborne contagion. Annals of New York Academy of Sciences, New York. 353.

Gallis, H. A. 1976. Microbial ecology and normal flora of the human body. Pp. 404–412. In W. K. Joklik and H. P. Willett (eds.), Zinsser microbiology. Appleton-Century-Crofts, New York.

Gardner, D. E., J. D. Crapo, and E. J. Massaro. (eds.) 1988. Toxicology of the lung. Target Organ Toxicology series. Raven Press, New York.

Gardner, D. E. 1988. The use of experimental infections to monitor improvements in pulmonary defense. J. Appl. Toxicol. 6:385–388.

Goodlow, R. J., and F.A. Leonard. 1961. Viability and infectivity of microorganisms in experimental airborne infection. Bacteriol. Rev. 25:182.

Gershon, D. 1992. Genetically engineered foods get green light. Nature 357:352.

Glanz, J. 1992. Herman: The pharmaceutical industry's next star? R&D Mag. 34:36–42.

Graham, J. A., D. E. Gardner, E. J. Blommer, D. E. House, M. G. Menache, and F. J. Miller. 1987. Influence of exposure patterns of nitrogen dioxide and modifications by ozone on susceptibility to bacterial disease in mice. J. Toxicol. Environ. Health 21:113–125.

Gwaltney, J. M., Jr. 1980. Epidemiology of the common cold. Pp. 54–60. In R. B. Kundsin, (ed.), Airborne contagion. Annals of New York Academy of Sciences, New York.

Hatch, G. E., R. Slade, E. Boykin, F. G. Miller, and D. E. Gardner. 1981. Correlation of effects of inhaled versus intratracheally injected metals on susceptibility to infection. Amer. Rev. Respir. Dis. 124:167–173.

Hatch, T. F. 1961. Distribution and deposition of inhaled particles in respiratory tract. Bacteriol. Rev. 25:237–240.

Hatch, T. F., and P. Gross. 1964. Pulmonary deposition and retention of inhaled aerosols. Academic Press, New York. Pp. 137–145.

Herman, L. G. 1980. Aspergillus in patient care areas. Pp. 140–146. In R. B. Kundsin, (ed.), Airborne contagion. Annals of New York Academy of Sciences, New York.

Jennison, M.W. 1942. Atomizing of mouth and nose Secretions into the air as revealed by high-speed photography. Pp. 106–128. In F. R. Moulton, (ed.), Aerobiology. American Association for the Advancement of the Sciences, Washington, D.C.

Joklik, W. K. 1976. Poxviruses. Pp. 937–944. In W. K. Joklik and H. P. Willett (ed.), Zinsser microbiology. Appleton-Century-Crofts, New York.

Kaminski. E., S. Stawicki, and E. Wasowicz. 1974. Volatile flavor compounds produced by molds of aspergillus, penicillium, and fungi Imperfecti. Appl. Microbiol. 27:1001.

Kaufmann, A. F., M. D. Fox, J.M. Boyce, D. C. Anderson, M. E. Potter, W. J. Martone, and C. M. Patton. 1980. Airborne spread of brucellosis. Pp. 105–114. In R. B. Kundsin, (ed.), Airborne contagion. Annals of New York Academy of Science, New York.

Ketchum, P. M. 1988. Microbiology: Concepts and applications. John Wiley & Sons, New York.

Kundsin, R. B. 1980. Opening remarks. Pp. 1–2. In R. B. Kundsin, (ed.), Airborne contagion. Annals of New York Academy of Sciences, New York.

Lange, D. J. 1976. Herpesviruses. Pp. 947–949. In W. K. Joklik and H. P. Willett (eds.), Zinsser microbiology. Appleton-Century-Crofts, New York.

Lee, J. V., and A. A. West. 1991. Survival and growth of Legionella species in the environment. J. Appl. Bacteriol. (Symp. Suppl.) 70:121S–129S.

Mallison, G. F. 1980. Legionellosis: Environmental aspects. Pp. 67–70. In R. B. Kundsin, (ed.), Airborne contagion. Annals of New York Academy of Science, New York.

Mannis, M. J., R. Tamaru, A. M. Roth, M. Burns, and C. Thirkill. 1986. Acanthamoeba Sclerokeratitis. Arch. Ophthalmol. 104:1313.

McClellan, R. O. and F. F. Henderson. (eds.). 1989. Concepts in inhalation toxicology. Hemisphere Publishing Co., New York.

Miller, F. J., and D. B. Menzel. (eds.). 1989. Extrapolation modeling of inhaled toxicants. Hemisphere Publishing Co., New York.

National Research Council. 1991. Human exposure assessment for airborne pollutants: Advances and opportunities. National Academy Press, Washington, D.C.

O'Mahony, M. C., R. E. Stanwell-Smith, H. E. Tillett, D. Harper, G. P. Hutchison, I. D. Farrell, D. N. Hutchinson, J. V. Lee, P. J. Dennis, H. V. Duggal, J. A. Scully, and C. Denne. 1990. The Stafford outbreak of Legionnaires' disease. Epidemiol. Infect. 104:361–380.

Parker, R. F., J. K. Davis, and G. H. Cassell. 1989. Short term exposure to nitrogen dioxide enhances susceptibility to respiratory mycoplasmosis and decreases in intrapulmonary killing of mycoplasma pulmonia. Amer. Rev. Resp. Dis. 140:502–512.

Phalen, R. F. 1984. Inhalation studies: Foundation and techniques. CRC Press, Boca Raton, FL.

Poupard, J. A., L. A. Miller, and L. Granshaw. 1989. The use of smallpox as a biological weapon in the French and Indian War of 1763. ASM News 55:122–124.

Powledge, T. M. 1992. Gene pharming. Technol. Rev. 95:61–66.

Salem, H., and D. M. Aviado. 1970. Physiology of the cough reflex. Pp. 233–270. In International encyclopedia of pharmacology and therapeutics. Section 27, Vol.1. Antitussive agents. H. Salem and D.M. Aviado (eds.).

Salem, H. 1987. Principles of inhalation toxicology. Pp. 1–33. In H. Salem (ed.), Inhalation toxicology. Marcel Dekker, Inc. New York.

Salem, H., G. L. Kennedy, J. B. Morris, M. V. Roloff, C. E. Ulrich, R. Valentine, and R.K. Wolff. 1992. Recommendations for the conduct of acute inhalation limit tests. Fund. Appl. Toxicol. 18:321–327.

Spendlove, J. C., and K. F. Fannin. 1983. Source, significance, and control of indoor microbial aerosols: Human health aspects. Public Health Rep. 98:229.

Sprengler, J. D., and K. Sexton. Indoor air pollution: A public health perspective. Science 221:9–17.

Titball, R. W., P. C. B. Turnbull, and R. A. Hutson. 1991. The monitoring and detection of Bacillus-anthracis in the environment. J. Appl. Bacteriol. (Symp. Suppl. 70:9S–18S.

Wells, W. F. 1934. On air-borne infection: II. Droplets and droplet nuclei. Amer. Hyg. 20:611.

Wells, W. F. 1955. Air-borne contagion and air hygiene. Harvard University Press, Cambridge, MA.

Wells, W. F., H. L. Ratcliffe, and C. Crumb. 1948. On the mechanism of droplet-nucleus infection. II. Quantitative experimental air-borne turberculosis in rabbits. Amer. J. Hyg. 47:11.

Yamamoto, K. R. 1989. Retargeting research on biological weapons. Technol. Rev. 92:23–24.

11

Regulatory Issues for Bioaerosols

Philip Sayre, John Burckle, Gregory Macek, and Gerald LaVeck

Introduction

In broad terms, biotechnology is the harnessing of biological processes for the production of commercially useful substances. The basic concepts have been practiced for thousands of years, yielding many products used in commerce today. But the applications of biotechnology based on genetic engineering support a new industry with tremendous potential to develop even more useful products. An improved capability to address potential concerns posed by these products is needed. Commercial biotechnology applications can be divided into two loosely defined categories: (1) contained product manufacturing processes and (2) uncontained and semicontained processes.

Contained product manufacturing processes are analogous to in-plant chemical processing and use the microbe or its enzymes to convert a feedstock (substrate, to a useful product by fermentation. A broad range of products are currently produced by fermentation processes including organic acids, alcohols, methane, amino acids, vitamins, various pharmaceuticals, enzymes, polymers, surfactants, hydrocarbons, and single-cell protein.

There are four major process operations in contained product manufacturing processes: substrate preparation, inoculum preparation, full-scale fermentation, and product isolation and purification. Substrate preparation may involve pretreatment of the feedstock material to convert it to a form suitable for microbial conversion, sterilization, nutrient addition, and pH adjustment. This step does not involve the microorganism. Inoculum preparation involves culturing of a small volume of cells that will be used to begin the full-scale fermentation in the fermentation reactor. The fermentation reactor is the heart of the process. The microorganism is first introduced into the process in this step, in which the substrate and other nutrients are contacted with the microorganism under conditions which encourage the production of the desired product. The final process involves recovery of the product through a variety of separation technologies. These processes typically consist of solids separation to isolate the broth from the solid culture, cell disruption if the product is formed within the cell (as opposed to excreted through the cell wall into the broth), and a variety of product concentration and purification steps such as distillation, steam stripping, and chromatography.

Whereas contained product manufacturing processes involve use of a microrganism in a closed, somewhat controlled environment, uncontained and semicontained processes involve the direct release of the microorganism to the environment. These processes use the microoganism to recover a product by nonfermentation processes such as adsorption or to effect a specific physical or chemical change in the environment. Examples of such applications include use of the organism to extract minerals from ore or from metal-bearing wastes, to enhance oil recovery, or to clean up chemical spills and hazardous waste sites.

Currently, there are a number of federal and international organizations that address concerns related to microbes and microbial products produced in fermentation and laboratory facilities. A small number of these concerns focus directly on the release of aerosolized microorganisms. The purpose of this chapter is to review these regulations and guidance documents, along with guidance documents for intentional release of microorganisms to the environment. The experience gained in the U.S. EPA's Office of Pollution Prevention and Toxics (OPPT) is examined for both aerosolized microbes released for environmental use and those released from fermentation facilities. Models and other estimation techniques that apply to intentional environmental releases and to releases related to fermentation waste streams are discussed. Respiratory testing protocols relevant to fermentation plant worker effects and avian effects are also noted.

11.1. Current Regulations and Guidelines

There are regulations which are intended for commercial products and those which are intended to apply to laboratory research. In the United States, the *Coordinated Framework for Regulation of Biotechnology* (Office of Science Technology and Policy, 1986) explains how biotechnology research and products are reviewed in accordance with the use of each product (Milewski, 1990). Many agricultural uses of microbes, plants, and animals are regulated by the Department of Agriculture (USDA); foods, drugs, cosmetics, and biologics are regulated by the Food and Drug Administration (FDA); microbial pesticides are regulated by the EPA Office of Pesticide Programs (OPP) in accordance with the Federal Insecticide, Fungicide, and Rodenticide Act (FIFRA). Uses of microbes for commercial purposes (at both the research and development, and commercial marketing stages) not regulated by the OPP, USDA, or FDA are reviewed by OPPT under the Toxic Substances Control Act or TSCA. The Coordinated Framework addresses (1) commercial biotechnology products at the research and marketing stages and (2) federally funded biotechnology research. A table summarizing the federal oversight of biotechnology research and products is provided on page 23305 of the 1986 Coordinated Framework document (EPA, 1986). United States guidance applicable for U.S. research laboratories has been provided by such agencies as NIH (NIH, 1986).

A number of international organizations, as well as U.S. agencies, have provided guidance related to containment and inactivation of microbes and microbial products. This guidance applies to fermentors and field tests.

Certain recombinant microorganisms have been identified as posing low risk to humans and the environment when used in fermentation facilities due to factors such as a history of safe use. These microbes have very few containment restrictions placed on them (such as restrictions reducing aerosol emissions) and have been referred to as ones which can be commercially cultured under "good industrial large-scale practice" (GILSP) procedures. Other microorganisms are classified as pathogenic and have strict controls placed on their release. In between these two extremes is a third larger group of microbes which have not been used in either commercial release to the environment or in commercial fermentation and are not, therefore, recognized as having a history of safe use.

11.1.1. OECD GILSP Criteria

The Organization for Economic Cooperation and Development (OECD) has developed a set of general principles and criteria for the safe handling of recombinant microorganisms and cell cultures in large-scale industrial production (OECD, 1991)— these constitute GILSP. A number of OECD countries adopted the GILSP principles put forth in 1987 (OECD, 1987) as part of national guidelines because these principles and guidelines assist in identification of low-risk organisms. GILSP organisms fall into two categories:

1. Recombinant organisms that are constructed entirely from a single prokaryotic or eukaryotic host and/or consisting of DNA segments from species that are natural gene exchangers (provided that the hosts are nonpathogenic and without adverse environmental impacts).

2. Recombinant organisms whose hosts are nonpathogenic, well-defined taxonomically, bear no adventitious agents, and have a history of safe use or limited survival without adverse environmental impact. The vector/insert of the recombinant should be well characterized and free from known harmful sequences, limited in size to the DNA required to perform the intended function, poorly mobilizable, and should not transfer any resistance markers to microbes not known to acquire them naturally.

The resultant recombinant organisms should be nonpathogenic and as safe as the host organism (or have limited survival), without adverse environmental consequences. These microorganisms are considered low risk, but zero risk is not realistic even for these microbes.

Recombinant organisms that do not meet all the criteria for (1) or (2) above are not GILSP organisms. However, they may be found, on case-by-case review, to be low risk. In such cases, these organisms may be handled using GILSP. Examples of hosts that are currently used in GILSP applications are *Saccharomyces cerevisiae, Escherichia coli* K-12, *Bacillus subtilis*, and *Aspergillus oryzae*.

The OECD has identified fundamental principles of good occupational and environmental safety for processes using GILSP organisms. These principles include the following, which are relevant to containment in general (including aerosols): (1) keep workplace and environmental exposure to microbes and microbial products to a level commensurate to the characteristics of the organism, product, and process; (2) exercise engineering control measures at the sources and supplement engineering controls with necessary worker-protective clothing/equipment; and (3) test as appropriate for the presence of viable microorganisms in the workplace and surrounding environment.

11.1.2. NIH GLSP and BL-LS Descriptions

The NIH established two sets of containment criteria for fermentation applications, dependent on the volume of the cultures under fermentation. One set of criteria which applied to cultures of less than 10 liters can be found in Appendix G of the NIH "Guidelines for Research Involving Recombinant DNA Molecules" (NIH, 1986). Appendix G listed—from least to most stringent—BL1 through BL4 containment levels (the "BL" abbreviation stands for "biological containment level"). The four BL1 through BL4 standards were based on containment criteria for pathogens developed by the CDC/NIH (CDC & NIH, 1984), and on criteria developed for oncogenic viruses by the National Cancer Institute (NCI, 1974). Briefly, BL1 and BL2 criteria specify that all fermentation procedures be performed carefully to minimize the creation of bioaerosols. BL3 criteria repeat this same language, and add that molded surgical masks or respirators should be worn in rooms containing experimental animals, and that vacuum lines are protected with high efficiency particulate air (HEPA) filters and liquid disinfectant traps. BL4 criteria go beyond the lower biosafety levels by adding to the BL3 criteria such specifications as the placement of HEPA filters on sewer and other ventilation lines (including all exhaust air from the facility), the use of Class III biological safety cabinets for all procedures involving microorganisms, and the isolation of central vacuum systems from any use beyond the facility.

A second set of containment criteria for large-scale fermentation applications was developed by the NIH Recombinant DNA Advisory Committee (NIH, 1986). Three different containment categories were established. They are—listed from least to most stringent-BL1-LS, BL2-LS, and BL3-LS (where BL denotes biological containment level and "LS" denotes large scale or any culture of 10 liters or more). In 1991 (NIH, 1991), NIH amended the guidelines to include a GLSP containment level which is less stringent than the BL1-LS standard. The NIH guidelines apply to researchers/institutions who receive Federal government funding. However, the NIH guidelines do not extend to commercial submissions, although commercial requestors do ask the NIH to review their applications.

The NIH adopted a less rigorous level of containment termed "GLSP" (Good Large-Scale Practices) which is recommended for large-scale research or production involving recombinant strains. Recombinant strains to which GLSP containment would apply include strains based on the four GILSP taxa listed above. The NIH containment standards for the GLSP standard were originally proposed by the Bioprocessing Committee of the Industrial Biotechnology Association and Pharmaceutical Manufacturers Association, and were based on OECD principles.

A partial listing of the NIH physical containment standards for large-scale fermentations that either apply directly, or are related to, aerosol emissions from fermentors are noted in Table 11.1. The criteria in Table 11.1 address only the biological hazard associated with organisms containing recombinant DNA. Other hazards such as toxic properties of products and downstream processing are not addressed.

The specifics of the three standards from Table 11.1 that apply directly to aerosol emissions should be examined. For item 1, control of aerosol releases from fermentors, the text of the NIH guidelines do not provide details beyond that in the table. For item 2, treatment of exhaust gases, BL1-LS standards do not allow opening of a sealed fermentor unless contents are sterilized by a technique demonstrated to be effective on the recipient cell used to construct the recombinant under production. No restrictions are placed on the exhaust gases themselves. Under BL2-LS and BL3-LS containment procedures exhaust gases are treated by HEPA filters or an equivalent method (such as incineration) to prevent releases of aerosolized microbes. For item 3, ventilation systems for controlled access areas, only BL3-LS facilities have requirements. These facilities must have a ventilation system that controls air movement such that the air moves from areas of lower contamination potential to areas of higher potential. If the ventilation system provides positive pressure supply air, the system should prevent the reversal of airflow or should be equipped with an alarm that would actuate if the flow is reversed. The exhaust air from the controlled area may be discharged to the outside without filtration or other treatment, provided it can be released in an area removed from occupied buildings or air intakes. It is important to note that the containment standards listed are cumulative in that a

Table 11.1. *GLSP and BL-LS Containment Standards which apply, or are relevant to, control of aerosol emissions. Parenthetic references are to sections of Appendix K of the NIH Guidelines for Recombinant DNA Molecules (1986).*

Aerosol Containment Standard	GLSP	BL1-LS	BL2-LS	BL3-LS
1. Control of aerosols by engineering or procedural controls to prevent or minimize release of organisms during: a. sampling from a system, b. addition of materials to a system, c. transfer of cultivated cells, d. removal of material, products, and effluents from a system	Minimize by proc. controls (K-II-F)	Minimize by proc. controls (K-III-D)	Prevent by eng. controls (K-IV-C)	Prevent by eng. controls (K-V-C)
2. Treatment of exhaust gases from a closed system to minimize or prevent release of viable organisms	NR[a]	Minimize releases (K-III-E)	Prevent releases (K-IV-D)	Prevent releases (K-V-D)
3. Controlled access area to have appropriate ventilation system	NR	NR	NR	Required (K-V-M-S)
4. Viable organisms should be handled in a system that physically separates the process from the external environment	NR	Required (K-III-B)	Required (K-IV-A)	Required (K-V-A)
5. Culture fluids not removed from a system until organisms inactivated	NR	Required (K-III-C)	Required (K-IV-B)	Required (K-V-B)
6. Inactivation of waste solutions and materials with respect to their biohazard potential	Required (K-II-E)	Required (K-III-C)	Required (K-IV-B)	Required (K-V-B)
7. Closed system that has contained viable organisms not to be opened until sterilized by a validated procedure	NR	Unspecif.[a] (K-III-F)	Unspecif. (K-IV-E)	Unspecif. (K-V-E)

Description				
8. Closed system to be maintained at as low a pressure as possible to maintain integrity of containment features	NR	NR	NR	Required (K-V-F)
9. Rotating seals and other penetrations into closed system designed to prevent or minimize leakage	NR	NR	Prevent (K-IV-F)	Prevent (K-V-G)
10. Controlled area designed to preclude release of culture fluid in the vent of closed-system failure	NR	NR	NR	Unspecif. (K-V-M-7)
11. Provide protective clothing appropriate to the risk, to be worn during work.	Required (K-II-C)	Required (G-II-A-1h)	Required (G-II-B-2f)	Required (G-II-C-2i)

NR = Not required.

[a]Unspecif. = Not specified in NIH documentation.

given containment level (such as BL3-LS) consists of those containment procedures noted for that level (BL3-LS) as well as those procedures listed for lower
levels (BL2-LS and BL1-LS). Detailed descriptions of how large-scale fermentation facilities have met the NIH standards are available (Maigetter et al.,
1990).

11.1.3. Current and Proposed TSCA Regulations

In its 1986 Policy Statement for TSCA-covered microbial products (51 FR23313),
microorganisms were defined to be "chemical substances" and, therefore, subject
to TSCA regulations. The Toxic Substances Control Act (TSCA) applies to
microbial uses which do not fall under the purview of other federal agencies (as
noted above). TSCA applies, for example, to microorganisms used for fermentation production of detergent surfactants, increased nitrogen fixation in crop le-
gumes, waste treatment (including bioremediation of hazardous wastes), metal
mining, and biomass conversion. In addition to the type of use, two additional
factors define the types of microorganism applications reviewed under TSCA:
whether the microorganism is intergeneric (composed of DNA from at least two
different genera) and whether the microorganism is for commercial use. Details
regarding TSCA application to biotechnology products can be found in the work
of Sayre (1990). Under the current policy, there are no specific restrictions that
apply to microbial aerosols. However, restrictions may be placed on proposed
fermentation applications, small-scale field tests, and large-scale environmental
releases through Section 5 of TSCA.

A new Rule implementing TSCA for microbial products has been drafted
(EPA, 1991), although it has not been used for review of microbial products
under TSCA (the EPA still operates under the 1986 Policy statement). Under
the 1991 draft Rule, certain microorganisms such as *Bacillus subtilis*, *Escherichia
coli*, and *Aspergillus niger* are proposed as recipient microorganisms for novel
DNA. Such microorganisms would be exempt from review under Section (5h)
(4) of TSCA, providing certain criteria are met for the identity of the recipient,
the introduced DNA, and the physical containment/inactivation/worker protection
practices. The criteria used in the TSCA draft Rule are extensions of the OECD
GILSP criteria. The fermentation applications which fall under the proposed
TSCA exemption are required to have certain procedures which would limit
exposure of workers and the environment to microorganisms and microbial products. These procedures include (1) inactivation of microorganisms in liquid and
solid wastes to achieve a 6 log reduction and (2) inactivation of microorganisms
in fermentor gaseous exhaust by 2 logs. The solid/liquid waste log reduction
would result in an inactivation efficiency of at least 99.9999% because bacteria
normally reach 10^{10}–10^{11} cfu ml^{-1} and fungal populations normally reach 10^6–
10^7 microorganisms ml^{-1}. The remaining viable microorganisms originating

from fermentor aerosols, after a 2 log reduction, would be 5×10^2 microorganisms ft^{-3} (Batelle, 1988). In addition, exposure to workers during the processing of microbial biomass is limited by (1) general worker hygiene procedures and (2) entry, which is restricted to essential personnel.

11.1.4. U.S. Federal Subtilisin Regulations

One of the few proposed U.S. federal restrictions on workplace air contaminants that focuses on aerosolized bacterial products exposure limits is the 60-min short-term exposure limit (STEL) of 0.06 $\mu g/m^3$ for occupational exposure to subtilisins (OSHA, 1992). Subtilisins are proteolytic bacterial enzymes produced by *Bacillus* spp. that are used in laundry detergents and, to a lesser degree, in contact lens cleaners, film processing, and the food industry. Adverse effects include immune-system-mediated bronchoconstriction and respiratory symptoms in addition to primary irritation of the skin and respiratory tract. Workers were hospitalized after exposure to subtilisins in a detergent formulation plant where the "safe limit" was set at 0.12 $\mu g\ m^{-3}$. Other studies with monkeys and further reports on workers support the STEL for subtilisins.

A study with monkeys showed that exposure to 10 or 100 mg m^{-3} synthetic detergent dust together with 0.01–1 mg m^{-3} enzyme dust produced gross signs of respiratory distress, pulmonary histopathological effects, and pulmonary function impairment. In a study with guinea pigs, the no-observed-effect level for pulmonary sensitivity induced by exposure to subtilisins for 15 min. day^{-1} for 5 consecutive days was between 0.0083 and 0.041 mg m^{-3}. Animals exposed on the same regimen at higher levels developed enzyme asthma. Additional information on subtilisin exposure to guinea pigs can be found in Karol et al. (1985). Another study with workers reported that 3.7% of exposed workers experienced dose-related sensitization symptoms (enzyme asthma) on exposure to enzyme detergent powders. Symptoms included sweating, headaches, chest pains, influenzalike symptoms, coughing, breathlessness, and wheezing sufficient to incapacitate the worker.

At least one additional proposed occupational exposure limit exists for bacterial endotoxins: an exposure limit of 30 ng m^{-3} as an 8-hour time-weighted average was put forth by Palchak et al. (1988).

11.1.5. Other Guidance Related to Fermentation Facilities and Field Tests

The Food and Drug Administration has general facility requirements for the manufacture of genetically engineered biological products (FDA, 1991a) and finished pharmaceuticals (FDA, 1991b). Although these FDA requirements do not directly refer to the manufacture of biotechnologically derived products, they are broad enough to apply to such drugs and biologics. Further, these regulations have been interpreted for biotechnology facility and systems design in open literature articles to allow manufacturers to better understand how the FDA review of floor plans for such facilities are conducted (Hill and Beatrice, 1989a). Issues relevant to bioaerosol production are primarily focused on the design of a fermentation plant's HVAC (heating, ventilation, and air conditioning) system. Viable airborne particulates in the production and purification areas are controlled by several techniques. All work with spore–bearing microorganisms is conducted in a self-contained area that shares no air systems (and equipment and entrance) with any other manufacturing or testing area of the plant. The air supply to the production and purification areas should be filtered through high-efficiency particulate air (HEPA) filters at velocities sufficient to remove particulates in the air (airflow rates of at least 20 air changes per hour are normally acceptable). Each room used for inoculum preparation, fermentation, harvesting, or purification should normally have a separate air-handling system. The biological production process is to be conducted under strict aseptic control from propagation of the organism through filling and packaging. Production processes that are conducted in an open system (exposing the product to plant environmental air) should be performed under Class 100 air. In addition to the facility and system design guidance noted above, FDA has provided guidance on aspects of the plant operating procedure and validation that are examined for such facilities (Hill and Beatrice, 1989b). Systems are checked to see if they meet design specifications after in installation, and the operational limits of the equipment and systems are tested. Programs for maintaining the systems and equipment in good working condition, such as preventive maintenance and equipment calibration programs, may also be identified.

For small-scale field tests, limited information on treatment of aerosols is available through OECD and U.S. EPA documents. The OECD (OECD, 1991) has developed a set of good developmental principles (GDP) for small-scale field research with genetically modified microorganisms. These principles provide guidance to investigators on selecting micoorganisms and research sites and on designing appropriate experimental conditions. In general, the GDP address microorganisms applied in small-scale agricultural field tests as soil amendments, foliar sprays, seed treatments, or as inocula introduced into the vascular tissues of plants. The GDP notes that with the use of techniques that lead to generation of aerosols, such as foliar sprays and spray irrigation, greater dispersion may indicate the need for larger border areas. Alternatively, aerosol formation may be minimized by the use of drip application and drip irrigation. Aerial dispersal is influenced by several factors. These include mechanisms of entering the atmosphere, particle shape, ability to survive environmental stress, and ability to adhere to soil and other particles. Rafts of soil or dust particles are raised by wind when the ground is heated by solar radiation, or some microbes may adhere to insects or mites which can then be dispersed by wind currents. The microorganism's biological characteristics will affect its fate in the environment because these characteristics indicate the likely modes of movement/dispersal, ability of the microbe to adhere to particulates or animate or inanimate vectors, ability to infect vectors, and ability to survive transport. The OECD notes that positioning of a field research plot can be used to address and limit potential transport through the aerial route. For example, consideration can be given to situating the experimental site so that natural features of the landscape such as trees, hills, windbreaks, or fences can be used to influence wind currents.

The U.S. EPA (1992) also provided guidance on monitoring of small-scale field tests for environmental fate and impacts of microorganisms. A general scheme for designing a monitoring program is laid out in three phases. The first phase consists of defining the program's objectives based on available knowledge of the microbe to be released (the microorganisms profile), the environment (the field site profile), and the field test protocol (the experimental profile). The microorganism profile should address the traits that indicate the microorganism's ability to undergo aerial transport (temperature optimum, dormant forms, desiccation tolerance, shape, density, diameter, etc.). The environmental profile examines which vectors are present on the site, the soil characteristics, and ambient air conditions (wind speed, turbulence, wind direction, humidity, rainfall, solar intensity, presence of nutrients and pollutants, etc.). The experimental profile examines which meteorological parameters will be measured during the test, the sampling strategy, plot and border area designs, methods of aerosol application (concentration, rate, frequency, duration, discharge characteristics, etc.). The second phase in designing a monitoring program is to determine what is to be monitored and with what intensity. Monitoring of bioaerosols is considered as a potential monitoring endpoint. Monitoring intensity is determined by the degree of uncertainty and the potential severity of effects associated with the microbe. The final phase of a monitoring program is to develop the specific monitoring plan which consists of defining the physical layout of the field test, monitoring zones, and sample collection and analysis strategies. Quality control and quality assurance procedures are also discussed because they assist the researcher in balancing time constraints and procedural costs with the data quality needed to achieve the monitoring objectives. Documents similar to the OECD and EPA field test guidance documents are available through the USDA (USDA, 1992).

11.2. TSCA Experience

The microbial applications that the EPA has reviewed under TSCA have fallen into two groups: those for use in the environment and those for fermentation system use to produce a product. There are a number of techniques used in OPPT to estimate environmental and human exposure to microbes as a result of bioaerosol generation (1) in fermentation plants and (2) at small-scale field test sites.

11.2.1. Bioaerosol Releases from Fermentation Plants

The purpose of an exposure assessment is to identify the concentrations of microorganisms to which humans and the environment will be exposed. The first step for such an assessment of chemical substances consists of identifying the sources of environmental release within a facility, the media of release (air, water, land), estimating the magnitude of release, and identifying the controls/ treatment methods used to minimize release and the effectiveness of these controls/treatment methods.

Microorganism releases may be continuous and unintended, as results from an undetected leak from a pump or a seal. Alternatively, these releases could be continuous and intentional, as is the case with the limited releases of aerosolized microorganisms in fermentor off-gas. Microorganism releases may also be discrete (or fugitive), as those that result from sampling or on inoculation of the fermentor.

To determine the type and degree of aerosol releases from the fermentation process, sources of such releases need to be identified in a plant's operations equipment. The unit operations equipment used in a biotech process are classified into three categories: (1) seed stock preparation, (2) fermentation, and (3) product recovery. Product recovery equipment includes all harvesting, product concentration, and product purification equipment. Many of the equipment components of the process have been identified as potential sources for bioaerosol releases (Fig. 11.1).

For the purposes of the EPA review, two different exposures are considered which result from fermentation releases: (1) worker exposure and (2) environmental exposure. The worker exposure assessment consists of an estimation of the concentrations of microorganisms per unit time to which a fermentation worker is exposed. From Fig. 11.1, it can be seen that these exposures result from all releases of microorganisms occurring inside the plant. Primary sources of aerosolized microbes inside the plant include fugitive emissions from the product recovery operations (the solid-liquid separation process). In the product recovery process, bioaerosols form particularly under high-pressure situations where seals are involved. Such seals occur in equipment such as pumps, agitators, centrifuges, and cell disrupters. The occurrence and concentration of the release tends to be related to the type (design), age, and state of repair of the equipment, operation pressure, concentration of the microorganism in the process fluid, and worker technique (NIOSH, 1988; Martinez et al., 1988).

During the EPA review, an environmental exposure assessment is also provided. It consists of estimating potential nonoccupational exposure resulting from releases of microorganisms outside the fermentation plant. Such releases would include disposal of filter cakes and liquid waste. However, for the purposes of this chapter, aerosolized microorganisms will be the primary focus of discussion. The main source of these bioaerosols results from venting of the off–gas from

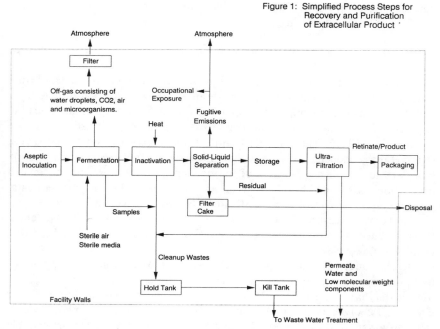

Figure 11.1. Simplified process steps for recovery and purification of extracellular product.

the fermentor to the outside atmosphere. Fugitive emissions from inside the plant could also contribute to estimated environmental exposure.

11.2.1.1. Worker Exposure Assessment

Estimating exposure to workers inside a fermentation plant is a two-step process. First, an understanding of the fermentation equipment and plant design allows for a primarily qualitative understanding of potential releases. Second, this more qualitative understanding is used to quantitatively estimate worker exposure.

An assessment of occupational exposure consists of an estimate of the number of workers exposed to the substance being reviewed, days yr^{-1} of exposure, worker activities that could lead to exposure, level of exposure and engineering controls, and personal protective equipment used to reduce exposure. The main difference between an assessment of exposure to genetically engineered microorganisms and an assessment for a toxic chemical is that the level of exposure is estimated as colony-forming units (cfu) day^{-1} rather that mg day^{-1} (unless an estimate of microbial toxins, allergens, etc. is required). This level applies to the daily amount of exposure to the microorganism that occurs through the inhalation of bioaerosols. The focus of this discussion will be on the approach OPPT uses to determine that level.

11.2.1.1.1. Worker Activities and Exposure Data

The approach involves first analyzing the fermentation process in detail from start-up to packaging and including waste treatment and disposal to determine both the fermentation equipment/design and worker activities that could lead to exposure. The potential areas for worker exposure in a fermentation plant are noted in Table 11.2. The National Institute for Occupational Safety and Health (NIOSH) has conducted a control technology assessment of enzyme fermentation processes to establish a baseline of information on the equipment and related occupational safety and health programs and practices used in enzyme fermentation processes. As part of this investigation, NIOSH collected area aerosol samples for viable process microorganisms, enzymes, and total dusts around potential omission sites. The results of their investigation are presented in (NIOSH 1988; Martinez et al., 1988).

NIOSH collected area samples of airborne microorganisms around various unit processes of three different fermentation facilities. Each of the three plants used a *Bacillus* strain (either *B. subtilis* or *B. licheniformis*) which produced either a proteolytic or a carbohydrase enzyme. An Anderson two-stage viable sampler was used that had a cutoff diameter of 8.0 μm with larger nonrespirable particles being collected on the top stage, and smaller, potentially respirable

Table 11.2. *Potential exposure points in a biotech facility.*

Major Unit Process	Potential Exposure Points
Feed stock preparation phase	Pipetting Remove cotton plug from flask Streaking of Petri plate Inserting inoculating loop into culture Decanting centrifuge culture in flask Mixing culture with pipette Opening lyophilized cultures Breakage or accidents
Fermentation phase	Manual inoculation of seed fermentor Transfer from seed fermentor to large fermentor Manual sampling during fermentation Use of rotary vacuum drum Cleaning and maintenance of equipment Agitator shaft seal
Harvest and product recovery phase	Processing releases Manual chemical testing during processing

Source: Battelle (1988).

particles collected on the bottom. Sampling was conducted over a 4-5–day period at each plant with sample time ranging from 2.5 to 20 min at a flow rate of 1 ft^3 min^{-1}. The sample times were selected based on exploratory surveys to obtain colony counts between 20 and 200. All air sampling plates were counted using standard colony counters. Colonial morphology was compared with that of the production strains of the same age and on the same medium. Randomly selected sample populations were streaked onto standard methods agar for isolation/identification. Confirmation of the production strain was made with standard Gram staining and/or sugar utilization profile analysis. Where possible, colonies resembling the production strain were included as a separate count. Results of the monitoring at the three survey sites are given in Tables 11.3–11.5.

11.2.1.1.2. Estimating Worker Exposure

The NIOSH data has been used for two purposes. First, the data were used to validate a qualitative model and to provide more quantitative estimates of

Table 11.3. *Plant 1 microorganism concentration. Production strain for this plant was* Bacillus subtilis.

Sample Location	N^a	Geometric Mean	Minimum Value	Maximum Value	Percent Production Strain
		Microorganism Concentrations (cfu m^{-3})			
Background—outside	NA^b	123	0	1,493	7
Background—cafeteria	3	270	208	313	42
Background—meeting room	5	701	602	782	51
Background—locker room	2	33	7	134	NC^c
Background—office	2	528	371	750	NC
Clean room	18	4	0	73	76
Dumpster	9	2,400	1,161	4,657	NA
Fermentor agitator shaft	20	339	84	2,298	48
Filter press—closed	17	3,906	988	23,588	NA
Filter press—closing	4	8,578	5,320	11,601	NA
Filter press—open	7	10,599	4,484	28,990	NA
Incubation room	6	332	298	382	31
Main laboratory	2	147	137	154	34
Quality control lab	8	159	91	295	35
Sample port—closed	8	196	62	490	19
Sample port—open	3	1,668	1,312	2,902	77
Scrubber	30	345	0	1,702	38
Seed agitator shaft	20	1,634	1,057	4,231	23

[a]Number of samples.

[b]Not available (counts of the production strain were not made, but the production strain was noted as the predominant strain on the plates)

[c]Not counted.

Source: NIOSH (1986a).

Table 11.4. *Plant 2 microorganism concentrations. Production strain for this plant was* Bacillus licheniformis.

Sample Location	N^a	Microorganism Concentrations (cfu m^{-3})			
		Geometric Mean	Minimum Value	Maximum Value	Percent Production Strain
Background—filter press	6	382	35	667	NC^b
Background—paint shed	14	121	4	272	NC
Background—paint shed north	6	478	217	638	NC
Between aging tanks	13	532	90	895	NC
Clean room	6	3	0	12	33
Control room	30	271	108	693	3
Fermentor agitate shaft	38	285	39	1015	NC
Outside incubation room	6	309	134	563	NC
Incubation room	8	221	32	435	NC
Rotary vacuum drum filter	45	345	61	2028	22
Sample port	6	705	336	983	17
Seed agitator shaft	24	326	104	766	NC

[a] Number of samples.

[b] Not counted.

Source: NIOSH (1986b).

Table 11.5. *Plant 3 microorganism concentration. Production strain for this plant was* Bacillum licheniformis.

Sample Location	N^a	Microorganism Concentrations (cfu m^{-3})			
		Geometric Mean	Minimum Value	Maximum Value	Percent Production Strain
Background—drop tank	21	206	21	875	1–3
Background—laboratory	7	20	5	34	NC^b
Background—outside	7	51	32	81	NC
Background—room adjacent to Incubator	NA^c	2	0	5	NC
Centrifuge	32	953	357	5855	5
Clean room	6	0	0	0	NC
Fermentor agitator shaft	28	257	85	620	1–3
Rotary vacuum belt filter	42	915	209	6572	NA
Sample port of main fermentor	8	373	151	555	1–3
Agitator shaft of seed fermentor	30	142	24	723	1–3

[a] Number of samples.

[b] Not counted.

[c] Not available (counts of the production strain were not made, but the production strain was noted as the predominant strain on the plates).

Source: NIOSH (1986c).

worker exposure. The qualitative model was developed by Battelle to produce a comparative rating system for evaluating worker exposure during the different unit processes in a biotechnology fermentation plant (Battelle, 1988). The model was intended to integrate three important factors in an exposure assessment: (1) point of release (where in the process), (2) type of release (continuous or discrete), and (3) the magnitude of release (high, medium, and low). The model was also intended to integrate the three important factors in worker exposure that would occur in a typical large-scale fermentation plant: (1) seed stock preparation, (2) fermentation (seed and large-scale), and (3) harvesting, concentration, and product purification. Application of the qualitative exposure model to the unit processes in a biotech fermentation facility is given in Table 11.6. From Table 11.6, a qualitative assessment of worker exposures can be made for different activities. Thus, the activities of a particular plant can be ranked according to whether they have a "high," "medium," or "low" potential for exposure.

Second, the NIOSH data is used to quantitatively assess worker exposure. A quantitative approach to worker inhalation exposure estimation could involve at least three different methods which are, in decreasing order of preference, (1) monitoring data for the microorganism, (2) monitoring data for analogous substance during similar processes or operations and similar worker activities, and (3) modeling approaches.

For Pre-Manufacture Notices reviewed in OPPT, the microorganism being assessed has not been manufactured commercially, and so it is unlikely that the PMN submitter will have monitoring data available. For this reason, OPPT has been relying mainly on the second approach to quantitatively estimate inhalation exposure in bioaerosols.

The data noted in Tables 11.3–11.5 are used to approximate the potential inhalation exposure for workers performing activities in various parts of a specific

Table 11.6. Qualitative exposure model.

Unit Process	Sterilization in Fermentor		Sterilization at a later stage	
	Continuous Release	Discrete Release	Continuous Release	Discrete Release
Seed stock preparation	+	+ +	+	+ +
Seed fermentation	+ + +	+ + +	+ + +	+ + +
Production fermentation	+ + +	+ + +	+ + +	+ + +
Cell concentration	+	+	+ + +	+ + +
Cell lysis	+	+	+ + +	+ + +
Debris removal	+	+	+ +	+ +
Product purification	+	+	+	+
Miscellaneous procedures	+	+	+	+

Note: + + + indicates exposure is possible and large numbers of viable microorganisms involved in the process; + + indicates exposure is possible and small numbers of viable microorganisms involved in the process; + indicates exposure is unlikely.

fermentation plant. These estimates are based on the type of plant equipment proposed and on the biological characteristics of the microorganisms proposed for production. An example which illustrates how exposure can be dependent on equipment can be seen when exposures around different separation equipment from Tables 11.3 and 11.4 are compared: Microbial counts for filter press apparatuses (Table 11.3) are higher than with rotary vaccuum filter separation equipment (Table 11.4).

As stated, the data in Tables 11.3–11.5 can be used to estimate a worker's inhalation exposure to bioaerosols for various activities during a fermentation process such as sampling. A worker's inhalation exposure during sampling can be estimated from the area monitoring data taken around the sampling port. This estimation assumes that these measurements are representative of the worker's exposure during sampling, that the biological characteristics of the microorganism to which the worker is exposed are similar to the characteristics of *Bacillus licheniformis* for which data were gathered for Table 11.4, and that the fermentation sample port design is similar to that noted in Table 11.4. From Table 11.4, the mean concentration of bacilli around the sample port of the main fermentor was determined to be 705 cfu m^{-3}. The sampling time was 6 min and the percent of total colonies that were identified as the production strain was 17%. Assuming a breathing rate of 1.25 m^3 h^{-1} (CEB, 1991), a worker's inhalation exposure to the production strain for one sampling operation is estimated to be 14 cfu (i.e, 705 cfu m^{-3} × 17% × 1.25 m^3× 6/60 h = 15 cfu).

11.2.1.1.3. Test Methods for Assessing Bioaerosol Effects on Workers

The EPA has developed protocols for exposure of mammalian species to aerosolized microbial products (microbial pesticides and bioremediation microorganisms). Two recent references (George et al., 1991; Sherwood et al., 1988) provide guidance on testing protocols for exposure to bioaerosols by aerosol inhalation, intranasal inoculation, and intratracheal inoculation. These protocols are designed to examine the potential for infectivity, pathogenicity, or toxicity by the pulmonary route using mice as human surrogates. Mice were challenged with *Pseudomonas aeruginosa, P. cepacia, Streptococcus zooepidemicus,* and influenza virus in these tests. Animals were exposed to aerosols for 10–30 min or recieved 0.05-ml suspensions injected at the bottom of the trachea or into the nostril cavity. Tests were run for approximately 14 days, and the gastrointestinal tract organs and lungs were examined for microorganisms. Clearance patterns and LD$_{50}$ values were determined when possible. Pseudomonads induced morbidity or mortality probably through expression of virulence factors or lipopolysaccharide (endotoxin) toxicity. For most situations, intratracheal and intranasal exposures can be used as replacements for aerosol inhalation methods: They are less expensive, it is easier to control experimental conditions, the dosing is more precise, and the results are easier to interpret.

A tiered testing scheme for evaluation of another endpoint, respiratory allergenicity, has been proposed by researchers not affiliated with EPA (Sarlo and Clark, 1992). This scheme consists of four levels, or tiers, of testing: (1) analysis of structure-activity information of potential allergen to determine hypersensitivity potential, (2) tests to determine ability of potential allergen to haptenate carrier molecules under in vitro conditions, (3) guinea pig injection tests to determine immunogenicity, and (4) guinea pig inhalation tests which determine relevant routes of exposure and safe chemical exposure levels. Such a tiered-testing scheme, although developed for low-molecular-weight chemicals, is relevant to the determination of microbial allergen effects. As with the mammalian protocols for infectivity/pathogenicity/toxicity, there are studies available that support the use of intratracheal techniques as substitutes for the guinea pig tier 4 inhalation tests applied to biological allergens such as enzymes (tier 4 tests) (Ritz et al., 1993).

11.2.1.2. Environmental Exposure Assessment

The second exposure assessment done as part of the OPPT review of a fermentation product is an environmental exposure assessment which estimates nonoccupational and environmental exposures resulting from plant releases, including aerosol releases. The major source for bioaerosols released to the environment is the fermentor off-gas (see Fig. 11.1). Estimation of environmental exposures can be done by analyzing the processes leading to bioaerosol formation during the fermentation process, estimating the magnitude of bioaerosol releases from the fermentation process, and estimating the environmental exposure resulting from these releases.

11.2.1.2.1. Formation of Bioaerosols in the Fermentation Process

In industrial fermentation practice, a broth containing nutrients is "seeded" with a quantity of production strain microorganisms derived from a laboratory fermentor. For aerobes, air is bubbled through the broth to permit growth while the broth is stirred with a rotating agitator to provide mixing and improved contact with the air bubbles. The growth process consists of four stages. During the lag phase, the microorganisms adjust to their new environment and begin reproduction. In the log phase, growth occurs at an exponential rate, producing metabolites and consuming nutrients. Growth then plateaus at the stationary phase as the nutrient loading is no longer sufficient to support rapid reproduction and the build-up of metabolites adversely affects growth and viability. As the influences of these two factors increases, the death phase is reached (at which the death rate is greater than the reproduction rate).

In a typical operation, several layers of foam bubbles are formed at the surface, producing a froth. These bubbles are larger than those which rose through the bulk liquid phase; they range in size from about one-half to several inches in diameter. The behavior of these bubbles tends to follow the same behavior reported in the literature on bubble drop formation. These studies show that as bubbles rise to the surface of a liquid, the liquid film forming the top layer of the bubble thins as the liquid forming the film drains down the surface of the film back toward the liquid phase. When the film thins to the point where it can no longer support the internal pressure of the gas, the bubble bursts. For bubbles smaller than about 1 mm in diameter, the depression in the liquid surface fills in so rapidly after bubble rupture that a vertical jet of liquid is formed at the center of the drop. Upon disintegration, the jet forms a primary aerosol droplet of about 100 μm and a few smaller satellite particles. For bubbles greater than about 5 mm in diameter, only droplets resulting from the fragmentation of film are produced. These droplets are less than 10 μm in size and increase in number as the particle size decreases, forming a plateau in the particle-size distribution at about 0.1–0.3 μm.

Blanchard and Syzdek (1982) have shown that droplets formed from dilute suspensions containing micron-size bacteria may be enriched by a factor of 10^1–10^3 times that of the bulk liquid. Droplet enrichment does not occur during the rise of the bubble through the bulk liquid but is due to a concentrating effect at the gas-liquid surface.

The properties of the liquid at surface of the fermentor govern the formation of aerosols in the off-gas. Pure water, the major liquid component of the broth, provides the baselines. The addition of the nutrients and other additives comprising the broth serve to increase viscosity and decrease surface tension, factors which increases foam formation. The other major impurities are the microorganisms themselves and their metabolites.

The result of these trends is that the production of aerosols in the bioreactor head space

1. is governed by the rheological properties of the fermentor fluid
2. changes with time as these properties change
3. predominantly results in the formation of an aerosol which exhibits essentially the same number size distribution in the growth (log) stage in which foaming occurs,
4. exhibits a number concentration ranging from a low of about 10^8 particles m^{-3} to about 10^{10}–10^{11} particles m^{-3} at the peak of the growth curve and then decreases
5. can be controlled to the level exhibited in the pregrowth period with antifoaming agents
6. is essentially not affected by agitator speed until a certain speed is reached, at which point aerosol production increases by a factor of 100 for broth only (no antifoaming) and by a factor of 1000 for a high cell density broth.

Control of bioaerosols is usually achieved by gas filtration systems. Sterilization of exhaust gases and air is accomplished by removing the bioaerosols from the gas stream. Filtration may be applied to primary containment to control process vent gases and in secondary containment systems to remove bioaerosols from the work space air before it is vented to the outside. Secondary containment provides for cleaning of the workplace air which was contaminated by uncontrolled process emissions resulting from leaks or poor work practices. Before the advent of highly effective filtration units, exhaust gas incineration was practiced. However, this control option has largely been abandoned in favor of filtration because of cost.

In gas filtration, particles are removed from gases by direct interception, inertial impaction, diffusion, electrical attraction, and gravitational settling; only the first three mechanisms have significance in providing microbiological containment. The most important mechanism for relatively large particles that are smaller than the filter pore size is inertial impaction. The inertia of the particle causes it to impact the internal filter surface rather than follow the gas stream lines. Thus, particles in the $0.1-1$ μm range may be collected by filters with pores in the $1-4$ μm range. Particle retention will be a function of velocity in the pores and tortuosity of the pores, as well as the relative size of particles and pores. In general, the higher the velocity, the higher the retention of particles captured by inertial impaction. Unless the filter pores are 0.2 μm or smaller, bacteria captured at higher flow rates may penetrate the filter under low or intermittent flow conditions. Therefore, validation is essential if larger pore filters are used. Furthermore, if a filter becomes wet, the inertial impaction mechanism no longer applies and particles smaller than the filter pore can move through the filter suspended in the wetting liquid. The microorganisms captured by the filter are then killed by a subsequent operation, usually by heat during steam sterilization of the system.

There are two major types of construction of filter media used in filter fabrication: depth and membrane. Both are used to remove bacteria from gases for purposes of containment. Membrane filters may be thought of as thin films containing numerous pores, which usually (but not always) possess tortuous paths. Microporous membranes used for gas filtration usually have pore size ratings of $0.1-0.45$ μm. Depth filters are usually a random array of fibers or particles with tortuous paths between them and a wider distribution of pore sizes. The limiting pore diameter of traditional depth filters such as packed fiberglass is typically $2-10$ μm (Conway, 1984). Particles tend to be captured within the depth filter as well as on the surface. Glass fiber depth filters generally have higher particle holding capacity and often have lower differential pressure drop than membrane filters, but they generally are more difficult to validate to guarantee high efficiency or sterile filtration. Therefore, the more open depth filters tend to be used as prefilter for a membrane filter.

Although depth filters can perform satisfactorily in gas sterilizing service, the current trend is toward specification of membrane filters where absolute containment of microorganisms is required. However, microfiber depth filters are still being specified for many large-scale fermentation applications. Depth filters can be designed to capture a very high fraction (over 99.99%) of bacteria-sized particles. Many HEPA filters are depth filters. HEPA filters must achieve 99.97% or better retention of 0.3–μm particles (as measured by a test such as ASTM D2986-71). Bacteria and yeast are larger particles and are more easily filtered from gas streams. The current generation of sterilizing depth filters for gases are made of glass microfibers that have been treated to be hydrophobic. These filters generally must be kept dry, but they last longer than polymeric membrane filters. When operated properly (i.e., dry), sterilizing depth filters can retain bacteria-sized particles to the limits of the detection system (Battelle, 1988).

Hydrophobic filters are preferred to hydrophilic ones for gas filtration. Hydrophilic filters can provide reliable particle removal when dry, but moisture in the gas stream can significantly alter filter properties. This may result in a higher pressure drop, reduced flow through the filter, and reduced filtration efficiency. Furthermore, the hydrophilic glass fiber filters, which were one of the earliest sterilizing filters and are still widely used today, will pass bacteria when wet (Bruno and Szabo, 1983). It does take some time (several hours to days) for bacteria to grow through such filters, but this has been identified as a source of contamination in several systems. The most common hydrophobic membrane filters are made of polypropylene, polyvinylidene fluoride (PVDF), polytetrafluoroethylene (PTFE, e.g., Teflon, Gore-tex), or treated glass microfibers. The cost of these materials increases with chemical and thermal stability, with polypropylene being the least expensive, and PTFE and treated microfibers the most expensive. Many of the commercially available filters are steam sterilizable, but the heat eventually alters the polymeric membranes and they must be replaced. Polypropylene will soften at sterilization temperatures, and particular care must be taken so as not to deform a polypropylene filter while hot. Manufacturers' recommendations should be followed (Battelle, 1988).

Off-gas filters and vent filters are two important applications in biotechnology. Fermentor off-gases especially require special consideration in filtration. The exhaust pressure of these gases from the fermentor is low (4-11 psi), and a filter in the path of these gases can cause serious pressure drops. This may upset oxygen transfer in the fermentor (Meltzer, 1987). Also, these gases are warm and moist, and condensation may form even on a hydrophobic filter, thus adding to the flow resistance. Therefore, cyclone separators, demisters (for heavy droplets), and coalescing filters (for finer aerosols) are applied before the filter. These devices do not affect the pressure drop significantly. The filter housing may also be steam-traced to sterilize the gases as they pass through the filter. Caution should be exercised in steam tracing a filter however, because there have been instances of filter failure due to heat destruction. Heat radiation from the filter housing may deform the base of the filter cartridge. Filter integrity is lost by failure of the filter seal to the filter holder and/or membrane failure.

Air filters are also used to decontaminate air venting from a tank. Air leaves or enters the tank, depending on whether liquid is let into or out of the tank. The vent filter sterilizes both the incoming and outgoing air; a vent filter, besides being sterilizable and hydrophobic, should be non-fiber-releasing to avoid contaminating the product. Membrane filters are ideal for these applications. Vent filters have to be replaced only occasionally because they are rarely subject to large loadings (Meltzer, 1987).

A typical fermentor exhaust system consists of one or more devices to remove the bulk of the particulate and water (e.g., a condenser, a cyclone, and/or a coalescing filter) followed by one or more sterilizing filters. There are many design variations in use. The system is more likely to include a condenser if the fermentation is thermophilic. The condenser, cyclone, and coalescing filter usually drain back into the fermentor. A heater often precedes the sterilizing filter in larger fermentors.

A condenser is recommended when the fermentor is operated above 45°C. The condenser serves two functions: (1) It removes excess moisture from the exhaust gas that could otherwise condense downstream and (2) it serves as additional volume in which to break foam. Condensate and broken foam drain by gravity to the fermentor.

The cyclone serves as a foam trap and is sometimes installed on fermentors used for nonrecombinant fermentations, although the use of cyclones is not typical design practice. There do not appear to be any standard design criteria for the cyclone. Data for one commercial unit showed an 89% removal of aerosolized *E. coli* (Marsh, 1984).

A sanitary demister called Turbosep (Domnick Hunter) has been developed specifically for fermentor exhaust gas. It has recently been tested for several months by pharmaceutical manufacturers in Europe. Droplet removal efficiency higher than 99.9% is claimed at flows ranging from a few standard cubic feet per minute (scfm) to 4000 scfm. This wide range of flow rates requires a range of different demister size units. The manufacturer claims that Turbosep is so efficient that the coalescing filter is not needed.

Coalescing filters will remove a large fraction of the aerosol from the fermentors. Pall and Kirnbauer (1978) claims a 99.99+% removal of droplets in the 0.1–0.5 μm range as measured by the dioctylphthalate (DOP) test (ASTM D 2986-71) with their Reverse Ultipor Coalescing filter. The DOP test involves a homogeneous smoke mixture of 0.3-μm-size DOP droplets that are passed through the filter. An industrial laboratory found a reduction in aerosolized *E. coli* from fermentation broth of 98.87-100% percent through a coalescing filter. The average reduction for five experiments with challenges of 820 to 59,000 organisms per standard cubic foot was 99.97% (Marsh, 1984).

The final containment is usually provided by a sterile filter. The NIH RAC

Guidelines for large-scale operations (BL1-LS and higher containment) call for a HEPA filter or equivalent. Although it might be argued that the combination of a cyclone and coalescing filter are equivalent to a HEPA filter, current design practice for fermentations requiring containment uses one or two sterilizing grade filters. One industrial firm using two filters in series has monitored the gas between filters and found no viable microorganisms in the exhaust gas after the first sterile filter (Battelle, 1988).

In addition to removing microorganisms and particulates, the filter must be capable of operating at the design flow rate and differential pressure. This means that it must be large enough so that possible fouling will not reduce its capacity too much during operation. The safety factor used for design will vary with the design philosophy. With modern hydrophobic filters, water condensation on the filter will not reduce the gas flux much. Occasionally, conservative designers will specify parallel filters so that the fermentation may continue if one filter plugs.

The need for a heater before the sterilizing filter appears controversial. With hydrophilic depth filters (glass fiber), it is essential to heat the gas above the dew point to maintain proper operation. With the hydrophobic depth filters, gas heating is usually recommended. Some people recommend heating with hydrophobic membrane filters, whereas others claim it is not necessary. In large systems, there is usually a heat exchanger in the gas line immediately before the sterilizing filter. In small systems, a steam-jacketed filter housing will suffice. Separate heat exchangers appear to be used for gas flows over about 350 scfm ($10 \ N \ m^3 \ h^{-1}$). Typically, the gas is heated 10–15°C to reduce the relative humidity from 100% to about 60–70%.

All tanks included in the primary containment system should be vented through a sterilizing filter. These include product holding tanks associated with such downstream operations as cell harvesting and rupture, feed tanks used to hold inoculum to be added to the fermentor, sample vessels, and hold tanks used for containment of waste materials or to pick up spills. Because of the intermittent gas flow, a filter that relies on the direct interception mechanisms should be used. A hydrophobic membrane rated at about 0.2 μm in a cartridge (or possibly a capsule for small vessels) would be suitable. Tank vent filters are sized based on the maximum expected flow rate and minimum expected pressure drop. Then a safety factor (usually $2 \times m$) is applied to protect against changes in operating conditions or premature plugging of the filter.

Filter maintenance consists of periodic testing and replacement. In flowing systems such as fermentor air supply or fermentor exhaust, an increase in pressure drop indicates partial clogging of the filter; a decrease indicates a leak. Either condition requires replacement. A more common practice is to operate a preventive maintenance program which calls for filter change on a regular schedule. The filter manufacturer's recommendation and the steam sterilization conditions will affect the useful life and replacement schedule which does not require interruption or compromise of sterile conditions during a fermentation cycle. The filters will normally be steam sterilized before replacement, so the possibility of a release of viable organisms should be negligible. Tank vent filters may not have provisions for steaming in place. If not, the maintenance procedures should be scrutinized to ensure the filters can be changed safely and without unacceptable release of microorganisms.

Incineration is a possible alternative to sterile filtration. Incineration is generally more expensive than filtration for removing microorganisms from fermentor exhaust. Although there are no commercial facilities that have substituted incineration for filtration or containment reported in the literature, there are facilities that incinerate fermentor exhaust as a method of odor abatement. For example, at least one manufacturer mixes fermentor exhaust with inlet air to the plant's boiler. Also, with particularly hazardous materials, such as work with highly infectious organisms, the containment system may consist of sterile filtration followed by gas incineration. These very hazardous organisms are unlikely to be used in commercial fermentations.

An important aspect of air filtration, especially in the biotechnology industry, is integrity testing because even the smallest penetration of filters by microorganisms may be undesirable. Integrity testing of the filtration material is performed by both the manufacturer and the user to ensure that it has no leaks. Both the manufacturer and the user commonly test the filter material as a quality control and customer acceptance measure (Robertson and Frieben, 1984; Leahy and Gabler, 1984). However, such testing does not ensure a leak-free system. It is necessary to test the complete filtration system in-place for such assurance, so that leaks caused by improper sealing of the filter media to the filter housing can be detected.

The user may test the filter upon installation and at periodic intervals. When the filter is the primary containment barrier, particularly on fermentor exhausts, a test of the filter integrity may be performed before and after each fermentation. Bubble-point (ASTM F316) or diffusion tests are commonly used to test the integrity of a filter. Both are nondestructive and can be performed in situ. Membrane filters may be tested by an air diffusion test. These tests correlate well with bacterial challenge tests (Conway, 1984; Pall and Kirnbauer, 1985). Conway (1984) describes the Forward Flow Integrity Test, a diffusion test with a high level of sensitivity. Many manufacturers use the DOP test to assign ratings to their filters and to assure quality control. Another test uses sodium chloride aerosols of various particle sizes as the test medium (Rubow, 1981). Portable aerosol testing devices are available for integrity testing (for example, Dominick Hunter's Integ 1).

11.2.1.2.2. Estimates of Release

The number of microbes present in fermentor off-gases have been estimated. These values range from 1.8×10^7 cfu m^{-3} to 7.8×10^2 to 1.9×10^5 cfu m^{-3}

s^{-1} (Reilly, 1991). Assuming 24-h day^{-1} off-gas emissions, the untreated off-gas was estimated to release 2×10^8 to 1×10^{11} cfu day^{-1}. This estimate is reasonable for fermentor head space aerosols and for worst case estimates on facilities lacking exhaust components used to reduce foam and aerosols (e.g., mechanical foam breakers and condensers); these units may reduce the microbial concentration in the unfiltered, exiting off-gas to significantly lower levels (Wyza and Leeson, 1991). In addition, some microorganisms die rapidly after release: some nonspore formers, anaerobes, those with low tolerances for dehydration and UV light, and so forth.

11.2.1.2.3. Modeling Environmental Releases

Releases of microorganisms from fermentation plants to the environment outside the plant have been estimated using a modification of the Turner point source model, developed originally to estimate chemical emissions (Turner, 1970). This model is useful for estimating fugitive emissions from a fermentation plant. The Turner model uses a sector averaging form of the Gaussian algorithm to estimate concentrations resulting from a point source release. The equation used is:

$$C = \left(\frac{2.03 \times (Q)}{(X)(\sigma_z) \times (\mu)} \right) e^{(-0.5)(H/\sigma_z)^2} \tag{11.1}$$

where C is the concentration in ambient air (organisms m^{-3}), Q is the release rate (organisms s^{-1}), H is the release height, X is the receptor distance from source (m), σ_z is the vertical dispersion coefficient (m), and μ is the mean wind speed (m s^{-1}). Using the assumptions

$H = 3$ m,
$X = 100$ m,
$\sigma_z = 5$ m (neutral atmospheric stability),
$\mu = 5.5$ m s^{-1},

Equ. (11.1) can be simplified to

$$\text{Concentration} = Q(6.165 \times 10^{-10}).$$

Releases are usually given as organisms yr^{-1}. Organisms yr^{-1} is converted into units of organisms s^{-1} by multiplying by 0.03170979.

$$\text{Concentration} = Q_{yr}(1.955 \times 10^{-11}).$$

The model assumes that the wind blows continuously in the same direction outside the plant. Because this is unlikely, a new assumption is made that the wind blows continuously in one direction 25% of the time.

$$\text{Concentration} = Q_{yr}(4.89 \times 10^{-12}).$$

To convert this to an annual exposure rate, the following equation is used:

$$\text{Exposure} = C \times IR \times D \times F,$$

where C is the concentration from the Turner model, IR is the inhalation rate, set at 1 m^3 h^{-1}, D is 24 h day^{-1}, and F is 365 days yr^{-1}. Using the above parameters, annual exposure can be calculated:

$$\text{Annual exposure (organisms yr}^{-1}) = Q_{yr}(4.28 \times 10^{-8}).$$

11.2.1.3. Regulatory Issues Under TSCA

In some cases, PMN fermentation applications have been subject to certain restrictions through the use of a TSCA Section 5(e) Consent Order. The use of such Consent Orders has been explained by Siegelbaum (1991). Section 5(e) Consent Orders are written by EPA to prevent unreasonable risks before they occur. The Consent Order established between the agency and the manufacturer submitting the PMN places controls and places limitations on activities associated with new chemical substances which will enable EPA to make a determination that those activities, as limited by the terms of the Consent Order, will not present unreasonable risks to humans and the environment. For fermentation applications, controls often include (1) procedures to ensure cell kill prior to release of cellular debris from the plant and/or (2) worker protective gear to limit inhalation of bioaerosols in the plant.

To issue a Consent Order under Section 5(e) of TSCA, the agency must be able to support the following legal findings:

1. That the information available to EPA is insufficient to permit a reasoned evaluation of the health and environmental effects of the new microbial substance AND

2. Either

   (a) that the manufacture, processing, distribution in commerce, use, or disposal of such substance, or any combination of such activities, may present an unreasonable risk of injury to health or the environment, OR

   (b) that such substance is or will be produced in substantial quantities, and such substance either enters or may reasonably be anticipated to enter the environment in substantial quantities or there is or may be significant or substantial human exposure to the substance.

Modification or revocation of the Order is based on information provided by the PMN submitter or through other sources which would change EPA's original determination. For example, Consent Orders for fermentation applications have noted that after submission of certain types of health effects tests, restrictions on cell kill procedures noted in the Order may be removed.

In addition to this discussion of Consent Orders under Section 5(e) of TSCA, exemptions for use of certain fermentation microorganisms under Section 5(h) (4) of TSCA were discussed in Sec. 11.1.3 of this chapter. These exemptions are also partially based on limiting bioaerosol emmissions from fermentation facilities.

11.2.2. Bioaerosol Releases Resulting from Intentional Use of Microorganisms in the Environment

11.2.2.1. Small-Scale Tests under TSCA

The potential for greatest exposure to bioaerosols would seem to be subsequent to a release of microorganisms for intentional use in the environment; for example, deliberate applications to an agriculultural field. However, this has not been the case with many of the applications EPA has seen so far. The reason for this is that the field application methods have been designed to minimize aerosolization of the organisms. For example, when Monsanto tested a *Pseudomonas aureofaciens* engineered with a LacZY marker system on a planting of wheat, the inoculation was accomplished by pumping the bacterial suspension through plastic tubing with a peristaltic pump. The suspension exited from a small-grain drill and dripped into the trough containing the seeds. This system produces essentially no aerosols.

Similar arrangements have been made in a number of other field tests involving recombinant rhizobia. Techniques used to monitor the field test site for aerosolized bacteria during applications have been fairly simple, consisting of placing several open Petri dishes around the test site during application of the organisms. Although this type of test has a low probability of detecting aerosolized bacteria unless a fairly large release takes place, it has been considered satisfactory: the engineered organisms being tested were low risk and the method of application minimized inadvertent releases.

In one instance, there were aerosolized rhizobia detected after a spray applications in small-scale field tests (total area less than 1 acre). For details of this release, see Sec. 11.2.2.4.

11.2.2.2. Worker Exposure Assessment

Worker exposure to bioaerosols during field application can be estimated from existing data on pesticide exposure. In one study, a worker's respiratory exposure to a herbicide applied with ground sprayers was measured to be 36 μg h^{-1} (Abbot et al., 1987). The concentration of the active ingredient in the herbicide was 0.7%. Based on this information, the total aerosol exposure is estimated to be 5 mg h^{-1}.

In the absence of data on biaerosol exposure, this data can be used as a surrogate to provide a rough estimate of potential worker exposure to microorganisms. Assuming that a worker applying a product containing 10^9 cfu ml^{-1} in a similar manner receives a similar exposure, worker exposure to microorganisms from bioaerosols is estimated to be 5 × 10^6 cfu h^{-1}. As expected, this exposure estimate is higher than that estimated for closed systems. This approach has been helpful in the TSCA PMN program for estimation of potential exposures to genetically engineered microorganisms during field application. The calculations are given below.

$$\frac{36\ \mu g\ h^{-1}}{0.7\%}\left(\frac{1\ mg}{10^3\ \mu g}\right) = 5\ mg\ h^{-1},$$

$$(10^9\ cfu\ ml^{-1})(1\ l\ kg^{-1})(1000\ ml\ l^{-1})(1\ kg\ 10^6\ mg) = 10^6\ cfu\ mg^{-1},$$

$$5\ mg\ h^{-1}(10^6\ cfu\ mg^{-1}) = 5 \times 10^6\ cfu\ h^{-1}.$$

11.2.2.3. Environmental Exposure Assessment

Atmospheric dispersion of released microorganisms can be estimated using the Microbe-Screen computer model (EPA, 1987). Microbe-Screen was derived from Tox-Screen, a chemical fate estimation program. The model was changed to account for the biological characteristics of of microorganisms by estimating growth and decay in both water and soil, and transport between air, soil, and water. Microbe-Screen can model microbial releases to four types of surface water bodies: lakes, rivers, estuaries, and oceans. In addition, the program can also model microbial movement in soil, accounting for such things as fate and transport associated with eroded soil. The air release module of Microbe-Screen makes estimates of dispersion using a modified Gaussian Plume equation for particulates. Gravitational settling and die-off rates in the atmosphere are important input parameters for this part of the model. The model does not estimate microbial growth in the atmosphere because of an assumption of insufficient nutrients. The source of an aerosol release can be turned "on" or "off" to simulate multiple release events. To date, the surface water and soil movement estimation portions of the model have been tested against actual field test data and found to correlate reasonably well. However, the air-release module has not yet been validated.

11.2.2.4. TSCA Section 5(E) Consent Orders

Aerosolized rhizobia were detected after spray applications during field tests which began in 1988. These tests were conducted under a TSCA Section 5(e) Consent Order near Sun Prairie, Wisconsin (McClung and Sayre, 1993). The general purpose of Consent Orders under TSCA was discussed in Sec. 11.2.1.3. This Consent Order specified the collection of data such as monitoring data for rhizobia in the air, soil (lateral and vertical movement in the soil), and rhizosphere. It also specified the collection of effects data such as alfalfa yields.

The *Rhizobium meliloti* being tested were genetically modified by the addition of antibiotic resistance markers and/or *nif* genes to enhance nitrogen fixation in the legume crop alfalfa. A strain comparison test was conducted in which approximately 5×10^{12} rhizobia were applied by spraying suspensions of each rhizobial strain on alfalfa seeds placed in open furrows by using a carbon dioxide propelled bicycle sprayer which was calibrated to deliver 10 ml per linear foot. Immediately following the spraying, furrows were manually covered with soil using garden rakes. In a second strain competition test, approximately 5×10^{11} rhizobia were suspended and sprayed into each of several 6-ft furrow rows using a hand-held spray bottle. The furrows were then immediately. covered with soil. For bioaerosol monitoring, selective agar plates were mounted on posts located in all four compass directions at various distances from the perimeter of the test plots on day 0, 1, 2, 3, 4, and 6 after initiation of the strain comparison trial. No colonies appeared on the vast majority of plates regardless of compass direction or distance. A total of 13 colonies appeared on one type of selective medium over a cumulative exposure of 6 h on day 0 for all compass directions and distances even though it was moderately windy on the application day. Later, samplings were for 2-h exposures only. On day 6, the number of colonies on the same selective medium had dropped from 13 at the 4-ft distance to one colony at both the 100- and 200-ft distances. Therefore, there was little aerial dispersion resulting from application of the rhizobia. Also, aerial dispersion monitoring taken during at termination when the fields were plowed resulted in no detectable dispersal of inoculant from the test site.

Disclaimer

This document has been reviewed by the Office of Pollution Prevention and Toxics, USEPA and approved for publication. Approval does not signify that the contents necessarily reflect the views and policies of the Agency nor does mention of trade names or commercial products constitute endorsement or recommendation for use.

References

Abbott, I. A., Bonball, J. L., G. Chester, T. B. Hart, G. J. Turnbull, et al. 1987. Worker exposure to a herbicide applied with ground sprayers in the United Kingdom. Amer. Ind. Hyg. Assoc. J. 48(2):167–175.

Batelle, 1988. Large-scale DNA Processing Facilities—Volume 2 of 4: Containment Technologies and Volume 4 of 4: Worker Exposure Modelling. Final Report (Sept, 1988) written under EPA Contract No. 68-03-3248. Batelle, Columbus, Ohio.

Batelle, 1991. Support document for preparation of engineering assessments for genetically-engineered microorganisms. Final Report (October 31, 1991) written under EPA Contract No. 68–10–0003. Battelle, Columbus, Ohio.

Blanchard, D. C., and L. D. Syzdek. 1982. Water-to-air transfer and enrichment of bacteria in drops from bursting bubbles. Appl. Environ. Microbiol. 43:1001–1005.

Bruno, C. F., and L. A. Szabo. 1983. Fermentation air filtration upgrading by use of membrane cartridge filters. Biotech. Bioeng. 25:1223–1227.

CDC. 1974. Classification of etiological agents on the basis of hazard, 4th edi. CDC Office of Biosafety, Atlanta, GA.

CDC & NIH. 1984. Biosafety in microbiological and biomedical laboratories, 1st ed. CDC and NIH, Atlanta, GA.

CEB (Chemical Engineering Branch). 1991. Preparation of engineering assessments - Volume 1: CEB Engineering, Chemical Engineering Branch of the Economics, Exposure, and Technology Division, Office of Pollution Prevention and Toxics, U.S. EPA, Washington, DC.

Conway, R. S. 1984. State of the art in fermentation air filtration. Biotech. Bioeng. 26:844–847.

EPA. 1986. Policy statement 51 FR23313.

EPA. 1987. Biotechnology model: Microbe-Screen. Contract # 68–02–4246), USEPA/ OTS/EED, Washington, DC.

EPA (Office of Pesticides and Toxic Substances). 1991. Summary of the biotechnology Science advisory Committee's subcommittee on the proposed TSCA Biotechnology Rule. USEPA/OPTS.

EPA. 1992. Monitoring small-scale field tests of microorganisms. Office of Prevention, Pesticides, and Toxic Substances (EPA 700 R–92–008), Washington, DC.

FDA. 1991a. In Code of Federal Regulation, Title 21, parts 600–680. U.S. Government Printing Office, Washington, DC.

FDA 1991b. In Code of Federal Regulations, Title 21, parts 200–211. U. S. Government Printing Office, Washington, DC.

George, E., M. Kohan, D. Whitehouse, J. Creason, C. Kawanishi, R. Sherwood, and L. Claxton. 1991. Distribution, clearance, and mortality of environmental pseudomonads in mice upon intranasal exposure. Appl. Environ. Microbiol. 57(8):2420–2425.

Hill, D., and M. Beatrice. 1989a. Biotechnology facility requirements, Part I: Facility and systems design. Biopharm, 20–25.

Hill, D. and M. Beatrice. 1989b Biotechnology facility requirements, Part II: Operating procedures and validation. Biopharm 28–32.

Karol, M., J. Stadler, and C. Magreni. 1985. Immunotoxicologic evaluation of the respiratory system: animal models for immediate- and delayed-onset pulmonary hypersensitivity. Funda. Appl. Toxicol. 5:459–472.

Leahy, T. J., and F. R. Gabler. 1984. Sterile filtration of gases by membrane filters. Biotech. Bioeng. 26:836–843.

Maigetter, R., F. Bailey, and B. Miller. 1990. Safe handling of microganisms in small- and large-scale BL-3 fermentation facilities. Biopharm 22–29.

Marsh, M. M. 1984. Attachment II, Recombinant DNA Advisory Committee Large-Scale Working Group, National Institutes of Health, Minutes of Feb. 7, 1984 meeting.

Martinez, K., J. Sheehy, J. Jones, and L. Cusick. 1988. Microbial containment in conventional fermentation processes. J. Appl. Ind. Hyg. 3(6):177–181.

McClung, G., and P. Sayre. 1993. Ecological risk assessment case study: Risk assessment for the release of recombinant rhizobia at small-scale agricultural field sites. Prepared for the Risk Assessment Forum, Office of Research and Development, U.S. EPA. (EPA 600/in press).

Meltzer, T. H. 1987. Filtration in the pharmaceutical industry. Marcel Dekker, New York.

Milewski, E. 1990. EPA Regulations covering release of genetically engineered microorganisms. Pp. 319–340. In J. Nakas and C. Hagedorn (eds.), Biotechnology of plant-microbe interactions, McGraw-Hill Publishing Co., New York.

National Cancer Institute. 1974. National Cancer Institute safety standards for research involving oncogenic viruses. U.S. Government Printing Office, Washington, DC. [DHEW Publication No. (NIH) 75–790]

NIH. 1986. Guidelines for the use of recombinant DNA molecules, Federal Register 51(88):16958–16985.

NIH. 1991. Revision of the NIH guidelines subcommittee meeting Minutes from 15 Octoberr 1990 and 7 December 1990. Recombinant DNA Tech. Bull. 14(4):411–437.

NIOSH, 1986a. In-depth survey report: Control technology assessment of enzyme fermentation processes. Assessment of enzyme fermentation processes at Gist-Brocades., USA, Inc. (Kingstree, S. Carolina). U.S. Department of Health and Human Services, National Institute for Occupational Safety and Health. U.S. Government Printing Office, Washington, DC. [DHHA (NIOSH) Publication No. ECTB 116–1b]

NIOSH, 1986b. In-depth survey report: Control technology assessment of enzyme fermentation processes. Assessment of enzyme fermentation processes at Novo Biochemical Industries (Franklin, N. Carolina). U.S. Department of Health and Human Services, National Institute for Occupational Safety and Health. U.S. Government Printing Office, Washington, DC. [DHHA (NIOSH) Publication No. ECTB 116–15b.]

NIOSH, 1986c. In-depth survey report: Control technology assessment of enzyme fermentation processes at Miles Laboratories (Elkhart, Indiana). U.S. Department of Health and Human Services, National Institute for Occupational Safety and Health. U.S.

Government Printing Office, Washington, DC. [DHHA (NIOSH) Publication No. ECTB 116–16b.]

NIOSH, 1988. Control technology assessment of enzyme fermentation processes. U.S. Department of Health and Human Services, National Institute for Occupational Safety and Health. U.S. Government Printing Office, Washingtin, DC. [DHHA (NIOSH) Publication No. 88-114].

OECD. 1986. Recombinant DNA safety considerations. OECD, Paris.

OECD. 1991. Safety Considerations for the use of genetically modified organisms. OECD, Paris. [DSTI/STP(91)14]

OSHA. 1992. CFR Part 1910.1000. Table Z–1-A. Limits for Air Contaminants, p. 27.

Office of Science and Technology Policy. 1986. Coordinated framework for regulation of biotechnology; Announcement of policy and notice for public comment. Federal Register 51:23302–23393.

OECD. 1991. Safety considerations for the use of genetically modified organisms, Part 1: Elaboration of criteria and principles for the good industrial large-scale practice. OECD, Paris.

Plachak, R., R. Cohen, M. Ainslie, and C. Hoerner. 1988. Airborne endotoxin associated with industrial-scale production of protein products in gram-negative bacteria. Am. Indust. Hyg. Assoc. J. 49: 420–421.

Pall, D. B., and E. A. Kirnbauer. 1978. Bacterial removal prediction in membrane filters. 52nd Colloid and Surface Symposium held at the University of Tennessee, Knoxville, TN, June 1978.

Reilly, B. 1991. Analysis of environmental releases and occupational exposure in support of proposed TSCA 5(h) (4) exemption. Draft EPA Report.

Ritz, H., B. Evans, R. Bruce, E. Fletcher, G. Fisher, and K. Sarlo. (in press 1993) Fundamentals & Applied Toxicology. Respiratory and immunological responses of guinea pigs to enzyme-containing detergents: a comparison of intratracheal and inhalation modes of exposure. Fundamentals and App. Toxicol.

Robertson, J. H., and W. R. Frieben. 1984. Microbial validation of vent filters. Biotech. Bioeng. 26:825–835.

Rubow, K. L. 1981. Submicron aerosol filtration characteristics of membrane filters. Ph. D. Thesis, University of Minnesota, Minneapolis.

Sarlo, K., and E. Clark. 1992. A tier approach for evaluating the respiratory allergenicity of low molecular weight chemicals. Fund. Appl. Toxicol. 18:107–114.

Sayre, P. 1990. Assessment of genetically engineered microorganisms under the Toxic Substances Control Act: Considerations prior to small-scale release. Pp. 405–414. In P. M. Gresshoff, L. E. Roth, G. Stacy, and W. E. Newton (eds.), Nitrogen fixation: Achievements and objectives. Chapman and Hall, New York.

Sherwood, R., P. Thomas, C. Kawanishi, and J. Fenters. 1988. Comparison of *Streptococcus zooepidemicus* and influenza virus pathogenicity in mice by three pulmonary exposure routes. Appl. Environ. Microbiol. 54(7):1744–1751.

Siegelbaum, H. 1991. Regulation through Consent Orders. Speech given at the EPA Conference "TSCA Section 5 and 8 Conference". Washington, DC, 4–6 June.

Turner, B. 1969. Workbook of atmospheric dispersion estimates. U.S. EPA/Research Triangle Park, NC. Public Health Service, Research Triangle Park, North Carolina.

USDA. 1992. Guidelines recommended to USDA by the Agricultural Biotechnology Research Advisory Committee December 3–4, 1991. Document No. 91–04 available through the USDA Office of Agricultural Biotechnology.

Wyza, R. B., and A. Leeson, 1991. Generation of bioaerosols in fermentation equipment and their control. Final report written under EPA Contract No. 68–10–0003, WA 0–10. Battelle, Columbus Division, Columbus, OH.

12

Safety and Containment of Microbial Bioaerosols

Barbara Johnson and I. Gary Resnick

Introduction

Guidelines for the safe use of pathogens in various laboratory and industrial settings have evolved to provide an acceptable level of protection for the laboratory worker, the public, and the environment. Considerable data exist to support the efficacy of these guidelines. However, there is little established regulatory criteria on which to base a biosafety program. In addition, the guidelines are primarily concerned with clinical laboratory settings as is the majority of the historical safety data. Therefore, these guidelines must be interpreted by a biosafety professional to ensure that the inherent level of risk, less mitigative measures, is understood by management. Management, in consultation with workers and government regulators, in turn decides whether the risk is acceptable. This is often an iterative process for novel laboratory operations such as aerosol studies with pathogens or toxins.

The information that follows is a compilation of established biosafety procedures used throughout the biological containment community. In addition, the authors share their experiences in developing and implementing a successful biosafety program for aerosol studies with pathogens and toxins.

Management

A successful biosafety program requires an informed and committed management body. It is imperative that management possess a basic understanding of microbiology to appreciate potential modes of disease transmission, host susceptibility, and environmental fate of pathogenic agents. The commitment of management to the safety program must be commensurate with safety program resource requirements. Facility design, construction and maintenance, training, safety equipment, and operational controls can be very expensive for work witn eiioiogi-cal agents. Without the commitment of management the biosafety program will not achieve an acceptable level of performance.

Role of Management

A well-established and delineated chain of management is required to address safety issues at each level of the organization. The laboratory director or senior laboratory manager/scientist should bear overall responsibility for the safety of the laboratory. It is the duty of the laboratory director to establish and staff a safety program to address all aspects of safety, with particular emphasis given to biosafety. To assist management in this effort a full time safety officer with significant training in microbiology and experience working in containment is extremely desirable. The safety officer should report directly to and be rated by the laboratory director. A less desirable alternative is to have a senior scientist perform this function as a collateral duty.

Senior management provides on a generic or laboratory-wide basis the compo-nents of a biosafety program. However, the responsibility for adherence to the program must rest for each project with the project scientist. Ultimately, the success of the program rests on the dedication and preparation of each individual working in the laboratory.

The diverse support requirements required for a biosafety program sustain the need for an institution level biosafety committee (IBC). An IBC with appropriate organizational representation provides a forum for coordination of required sup-port activities such as immunization and medical surveillance programs, physical security and risk assessment. In addition, an IBC provides a convenient forum for public review and input to programs.

Implementation and Documentation

Development of an implementation plan and documentation of plan execution are vital aspects of biosafety program development. Implementation requires a well-managed and coordinated approach with trained and motivated workers, both in the laboratory and out. Documentation of implementation milestones (e.g., training, resolution of deficiencies) is of great importance to protect the rights of employees and to preclude unfounded liability claims against the organi-zation. Plan documentation should include an installation level biosafety program document, project specific safety reviews, and detailed protocol level procedures.

Oversite

Unbiased oversite can provide an added measure of assurance that a biosafety program is being executed as recorded, and appropriate documentation is being maintained. Oversite by government agencies can foster greater acceptance by local communities. However, considerable resources can be expended toward this goal, with little improvement in safety, if trained oversite personnel are not available. The authors have found it desirable to utilize the IBC for credible program oversight (i.e., invite representatives from government agencies to participate in the IBC).

Biosafety Levels

Risk assessment is the basis for determining what safety precautions should be taken in a given situation. Estimating risk has both objective and subjective components, with the ultimate assessment reflecting the level of risk which is perceived to be acceptable. While these assessments can be made methodically, there will be unique aspects in each situation which potentially alter the recommendations for mitigation of the risk.

The appropriate biosafety level (BSL) for performing work with microorganisms is determined by considering both the characteristics of the organism being used and the procedures involved in the investigation. Some factors to consider include: pathogenicity, virulence, infectious dose, route of entry, persistence in the environment, potential or intentional creation of aerosols, and possibility of auto-inoculation. In general these factors are assessed in a two-step process involving an estimation of the potential for exposure of target organisms (i.e., laboratory workers, support employees, or the public) and determination of the impact of that exposure. The risk assessment process requires the input of professionals trained in diverse disciplines. A knowledge of engineering principles is needed to estimate the potential for organisms escaping physical containment systems. The probability of a viable pathogen reaching a potential host, once it escapes containment, is best assessed by a microbiologist. Specialized medical training is needed to evaluate the impact of viable pathogens that have entered a host with consideration given to the host's immune status. A critical risk assessment is required to guide establishment of appropriate mitigative procedures.

The CDC and NIH have jointly published a universally accepted set of guidelines (Richardson and Barkley, 1988) which describes BSL-1 through BSL-4, and recommends safety practices, equipment and facility designs which promote safe working conditions in various biosafety levels. The CDC/NIH Guidelines are highly recommended as a source of information and as a starting point for composing biological safety programs and manuals tailored for specific institutions. Each subsequent BSL incorporates the practices of the prior level and adds more stringent practices. However, application of the CDC/NIH guidelines does not eliminate the need for an in depth risk assessment of all proposed work with pathogens. This is particularly true when performing novel operations such as aerosol studies.

Biosafety Level 1

BSL-1 is suitable for work done with defined, characterized strains of microorganisms of no known or minimal potential hazard to healthy adults. Work is done using standard microbiological practices, often on an open bench top. While no general personal protective equipment (PPE) is required, situations may arise when specific types of PPE are desired (i.e., an animal handler may desire donning a dust mask and gloves prior to handling animals and cages). All biological materials must be decontaminated prior to disposal. If the site of decontamination is separate from the laboratory, materials must be transported in leak proof bags to the site.

Biosafety Level 2

BSL-2 involves work with indigenous organisms of moderate risk to man, which are usually associated with diseases of varying severity. By using good microbiological techniques much work can be done on an open bench. BSL-2 principals differ from those of BSL-1 in several aspects: (1) the laboratory director may limit access to the lab when work is being done, (2) laboratory personnel are trained in handling pathogenic agents, and are under the direction of competent scientists, and (3) procedures which potentially create infectious aerosols are performed in a high efficiency particulate air (HEPA) filtered biological safety cabinet (BSC) and/or with the proper PPE. The universal biohazard warning sign and pertinent information should be posted in labs where work with infectious agents is ongoing. A medical surveillance program, including serum banking and immunization, spill and accident plan, and a biosafety manual should be prepared and instated.

Biosafety Level 3

BSL-3 containment work may involve indigenous or exotic agents which can cause serious or lethal disease upon infection via the respiratory tract. All procedures involving infectious organisms are performed in BSC by individuals wearing the proper PPE, or combinations of both. No work is done on the open bench. Special building engineering aspects are discussed later in the chapter. The laboratory director establishes entry and exit requirement policies in these labs. Work surfaces are decontaminated daily or immediately following a spill. PPE includes laboratory over-clothing, gloves, dust masks or respirators (especially for work with infected animals) and eye protection. Vacuum lines should be protected by $0.22\mu m$ or HEPA filters, and liquid decontaminant traps. Needle and syringe use should be limited, recapping, destroying and nonessential handling should be minimized.

Biosafety Level 4

BSL-4 or "maximum" containment is required for agents in which the risk of lethality is high. Many of these agents are exotic, with no known, or limited prophylaxis and cures. Vaccines should be provided to personnel when available. In addition to previously discussed precautions, all work will be performed in a Class III BSC, or in a Class II BSC by personnel equipped with a one-piece positive pressure suit with supplied breathing air. All materials leaving the laboratory are sterilized by autoclave, and liquid effluent is sterilized in a holding tank by chemical or heat methods. Both supply and exhaust air are HEPA filtered. Ventilation, plumbing, and vacuum systems for BSL-4 laboratories are separate from the central building systems. Maximum containment areas are constructed such that all penetration, joints and seams are thoroughly sealed. For more specific details pertaining to practices, PPE, and facility design appropriate at the different biological safety levels CDC/NIH Guidelines and 32 CFR Parts 626 and 627 (US Code, Title 32) are recommended.

Aerosolization Studies

Studies involving the intentional aerosolization of pathogens or toxins are not specifically covered in most biosafety reference material. Work involving aerosolization of pathogens poses a greater risk than conventional procedures. Aerosolized materials assume a range of sizes, including that of droplet nuclei. Droplet nuclei are comprised of single bacterium surrounded by moisture or organic material. Generally these droplets assume a size less than 2 μm in diameter. Particles in the 1–5 μm diameter range effectively penetrate into lung tissue (Hatch, 1961). When not adequately contained, microorganisms and toxins carried as droplet nuclei can access the respiratory system in numbers sufficient to result in infection, and topical or systemic damage. Often, the natural route of infection of many organisms includes, or is limited to the respiratory tract.

For these reasons, the intentional aerosolization of biological materials necessitates an increase in the normally assigned BSL for work with a particular organism. For example, work involving nonproduction quantities of *Bacillus anthracis* are normally performed at BSL-2. When large quantities of organisms are produced, or risk of aerosol production is high or intentional, work should be conducted in accordance with BSL-3 precautions. Studies involving the infection or exposure of animals to infectious materials will also generally be conducted at a BSL higher than regularly assigned. There is no single rule that applies to all situations or all organisms when assigning a BSL to projects involving aerosol production. The most logical and conservative approach is to identify the regularly assigned BSL, perform a comprehensive risk analysis which takes aerosol exposure or release into account, and determine what additional safety precautions can be taken to mitigate the risk.

When aerosol production is intentional, facility and primary barrier designs can and should be augmented. Chambers patterned after Class III BSC are completely sealed and highly adaptable for aerosol containment. An airtight aerosol fixture, placed inside the Class III BSC, contains the aerosol challenge. Downstream from the aerosol test fixture the aerosol is exhausted through two "in-line" HEPA filters prior to exhaust into the building ventilation system. There should be at least one HEPA filter in the building exhaust, as work involving intentional aerosolization of microorganisms or toxins will most likely be conducted under BSL-3 or BSL-4 conditions. This type of design provides additional protection and simplified decontamination. Technically, only the area from the aerosol generator through the interior of the secondary chamber and into the HEPA filters will become contaminated. The interior of the Class III BSC should not be exposed to infectious aerosol provided there are no leaks in the aerosol delivery system or secondary chamber.

Before conducting aerosol studies all BSC, secondary containment chambers and delivery equipment should be tested for leaks by pressurizing the system with detectible gases (e.g., a freon substitute or nitrogen) and monitoring for the escape of gas. Following positive pressurization with 2 in. water gauge pressure a soap solution should be applied to all gloves, seams, joints and areas where containment breaches or material weaknesses may occur. Personnel should observe for leaks, make repairs, and retest the system until it is completely airtight. No testing should commence until the system can be certified as airtight. Testing should be repeated at least annually and when ever magnahelic pressure varies significantly, gloves are replaced, modifications to the structure of the cabinet are made or when the cabinet is moved.

Additional PPE is highly recommended for studies involving pathogen and toxin aerosolization. Personnel should wear respiratory protective equipment such as positive air pressure respirators, goggles to protect the ocular mucosa and wrap around surgical style gowns over laboratory clothing. Individuals performing aerosol studies should be vaccinated against each agent if possible, and serum samples should be periodically collected and maintained as part of an immunological surveillance program.

Training

Providing technical and safety training to personnel is central to an effective biological safety program. It is the responsibility of the employer to provide training which will promote safety and safe working conditions in the workplace (Richardson and Barkley 1988). The training can be performed and documented by an appointed safety officer, principal investigator or experienced technician. Suggested topics for training are listed below.

Hygiene

Universal hygiene policies prohibiting eating, drinking, application of cosmetics or lip balm, and smoking in laboratory areas should be adhered to always. Hands should be washed frequently, throughout the day, following manipulations of potentially infectious materials, handling animals, and prior to exiting the lab. Fingers, pens, and other objects should not be put in the mouth, eyes, or facial area. Mouth pipetting of chemicals, samples and biological materials should be strictly prohibited. Food should not be stored in the lab or in refrigerators used for storing laboratory materials. A variety of personal protective equipment can be used in conjunction with good hygienic practices to limit the risk of disease transmission by the oral and ocular route.

Practices and Techniques

Standard microbiological practices and techniques should be adhered to at all times, and supplemented when they are deemed insufficient to control the hazard associated with the agent or procedure involved. Supplements to standard practices should be made following a risk assessment of the situation. Examples of modifying a technique or practice based on risk (of infection) include, but are not limited to the following policies (a) restricting the recapping or clipping of needles used for work performed with human blood (US Code Title 29, Part 1910.1030), (b) posting a biohazard sign on laboratory doors when infectious materials or animals are present, (c) restricting traffic flow through the work area when infectious materials are present, (d) reducing aerosol generation by not "blowing out" the last droplets from pipets or using pipets designed to minimize the creation of unintentional aerosols (Hazel and Halbet, 1986), and (e) transporting samples in and out of the lab in secondary unbreakable sealed containers.

Specialized Training

For persons working with infectious materials or in areas where such work is done, specialized training describing signs and symptoms of the diseases caused by the materials present should be provided. In addition to signs and symptoms of disease, route of transmission, procedures for decontaminating personnel and work areas and accident reporting should be discussed. Optimally, the occupational health providers could provide this training to employees or to individuals responsible for training employees. First aid and self aid training should be provided with emphasis on emergency care and decontamination of potentially exposed co-workers in need of assistance.

Storage of Pathogens

The storage of stock quantities of pathogenic or infectious materials should be limited to a secure area central to the laboratories where manipulations will be performed. Centralization minimizes the distance the material must be transported, and reduces the potential for contaminating larger non-laboratory areas, such as hallways. Access to pathogens storage areas should be restricted, and the removal or deposit of materials should be logged and recorded in a central note book. Similarly, when not in use, pathogens in the laboratory should be maintained in a contained, labelled area such as a freezer or refrigerator. Preferably laboratory or freezer doors can be locked when pathogenic materials are present.

Safety Plans

Formal safety plans and procedures should be made for special circumstances and emergencies relevant to each laboratory or facility. These plans should be included as part of a biological safety manual which is reviewed and understood by all employees on an annual basis.

The biological safety manual should describe operations and procedures for all work done in BSL-2 or higher levels. The manual should outline in detail standard operating procedures for all protocols involving pathogen, chemical and radiological work. Examples of topics include: culture and harvest of organisms, infection and handling of animals, PPE requirements, decontamination and sterilization procedures, spill clean-up, chemical hygiene, handling and disposal of hazardous materials, radiological safety, transportation of infectious materials in and out of the building, and emergency evacuation plans.

Evacuation in the case of a fire, biological release, or other serious mishap should be planned and rehearsed. Building maps highlighting current location and the two nearest exists should be posted in all rooms and intersecting hallways. Evacuation drills should be held on an "as needed" basis.

The formation and training of an emergency response team (ERT) for spill containment and decontamination is prudent as few HAZMAT and rescue services are prepared to render assistance in the event of a biological emergency. The laboratory director should participate with the ERT in the assessment of the scope of the accident and the remediation plans. Optimally, the leader of the ERT would have training in biological and chemical safety. Members of the ERT will require training in topics including but not limited to selection and use of proper PPE, risks associated with decontaminants, mitigation of hazards, containment and decontamination of spills, large scale sterilization procedures (formaldehyde fumigation, sodium hypochlorite washdown), entry and exit from contaminated areas, and signs and symptoms of intoxication or infection. ERT members should meet quarterly to participate in exersizes, discuss strategies for hypothetical accidents, practice donning and removing protective equipment and overgarments (tyvex suits) and discuss potential improvements to current practices.

Primary Barriers (Safety Equipment)

Primary barriers are the second system of protection available to the worker, the first being training. Primary barriers include various types of safety equipment whose function is to contain biological materials or isolate the worker (and other workers in the laboratory) from the agent. Primary barriers differ from secondary barriers in that the latter are mainly concerned with architectural and engineering features. The following are examples of primary barriers: biological safety cabinets (Class I, II and III), sealed centrifuge rotors. sealable containers (preferably leakproof and non-breakable), animal cages with filter membrane covers and bonneted HEPA filtered animal cage rack. The following types of personal protective equipment which meet or exceed OSHA standards (US Code, Title 29, Part 1910) also act as primary barriers: gloves, gowns, lab coats, scrubs, coveralls, aprons, shoe covers, hair bonnets, respirators, one piece suits with or without positive pressure or breathable air (ANSI 288, 1989), NIOSH approved disposable or cartridge respirators, goggles and face shields.

The selection of primary barriers is determined following an assessment of the risk involved and the assignment of an appropriate biosafety level for the project. Barriers are not an alternative to good microbiological technique, but a means to mitigate specific risks and provide increased safety in the work environment.

Biosafety Cabinets

BSC are universally used for providing primary containment during work with infectious materials. They provide effective containment in operations assessed as BSL-2 and BSL-3, when used and maintained properly. BSC can be constructed around centrifuges to further mitigate risks associated with unintentional aerosolization. The National Sanitation Foundation has set accepted industry performance standards to evaluate proper function of BSC (NSF, 1976). BSC should be tested and certified upon installation, following modification or moving and at least on an annual basis. There are three classes of BSC (Class I, II, III), and several subclasses which vary in airflow velocities, characteristics and design (NRC, 1990). Cabinets can be ducted directly (hard plumbed) or by "thimble connections" into the building ventilation system for complete exhaust to the outside environment, or ducted to allow recirculation of HEPA filtered air back into the laboratory. HEPA (High Efficiency Particulate Air) filters are designed to capture 99.97% of 0.3um particles upon challenge (Chatigny, 1986). This size particle is more difficult to capture than particles of greater or lesser size. HEPA filters trap large particles (bacteria) by interception, and small particles (viruses) by impaction and diffusion.

There are many differences between types of cabinets, one of the most striking involving airflow characteristics. While all classes of BSC HEPA filter the exhaust air, Class I BSC are not designed to HEPA filter room air prior to contact with the interior work surface. Workers are protected from contaminated material, but specimens are not protected from "contaminated" room air. This may pose problems for cell culture or when isolation of organisms is desired. Room air entering Class II and III BSC is drawn through HEPA filters prior to contact with the work area. These systems protect both the work and the worker. Class III BSC or gloveboxes provide the greatest level of protection to the worker, as they are completely enclosed. Work is conducted through glove ports. These cabinets are often custom designed and can be equipped with dunk tanks or pass through autoclaves for the decontamination of materials or wastes prior to removal and entry into the laboratory (Kuchne, 1973).

Enclosed Containers

The use of sealed rotors and centrifuge cups provide protection against accidental aerosolization, and may be more economically feasible than constructing a BSC around the centrifuge. Other methods for decreasing contamination resulting from the production of aerosols involves the use of sealed containers such as screw top centrifuge tubes and screw top tubes equipped with washers. Sealable nonbreakable containers also limit the potential for contamination due to spills and splashes. When transporting potentially infectious materials within or between laboratories, a non-breakable leakproof secondary container should be utilized.

Personal Protective Equipment

PPE should be chosen based on the operations performed and risks present. For example, gloves should be selected based upon the organism or biological material, as well as the diluent in which the organism is suspended and the procedures involved with work. While latex gloves may be suitable for working with biological samples in culture media, nitrile is more suitable for work with some solvents, and butyl rubber for work with caustics, acids and bases (Lynch, 1982; Thomas, 1970). Light weight wire mesh or leather gloves, are advocated to protect personnel working with non-anesthetized animals.

Respiratory protective equipment should also be chosen with care. Included under the title of respiratory protective equipment are surgical masks, dust/mist respirators, dust/mist/fume respirators, high efficiency respirators, half and full face cartridge respirators, HEPA filtered positive air pressure respirators, self contained breathing apparatus and one-piece suits with canisters or supplied air. NIOSH has made recommendations for choosing respirators based on the hazard analysis of the work (NIOSH, 1992). A study conducted using a biological aerosol evaluated the protective efficacy of several respiratory protective devices and found 33%, 22%, 8% and 0.05% leakage of bacteria in surgical masks, dust/mist respirators, high efficiency respirators, and full face bonneted positive air pressure HEPA filtered respirators. Leakage around the face seal area was shown to be responsible for 28%, 19% and 4% of the bacterial penetration in the first three types of devices tested (Johnson et al., 1992). This study emphasized the importance of attaining and testing for a tight face seal, and demonstrates the effectiveness of several commonly used devices.

One-piece suits with supplied breathable air or with HEPA filter canisters can be used to form a barrier between the worker and the organism being studied. This type of suit can be especially useful for ERT personnel responding to a containment breach when work involves aerosolization of toxins or pathogens. The suits are chemically decontaminated prior to exiting the laboratory. Suits should be checked carefully for holes, rips or weak spots prior to entering the laboratory.

Secondary Barriers

Laboratory architectural and engineering design form the basis for secondary protective barriers. While primary barriers protected personnel in the laboratory and immediate area, secondary barriers protect personnel in other areas of the building, and protect the environment from exposure to biological materials. Types of secondary containment present in a facility depends on the facility design. Facility design should be based on the anticipated scope of work to be conducted. The following are examples of secondary barriers, and will be discussed in relation to biological safety levels later: directional airflow, HEPA filtered exhaust air, separate ventilation systems, limited access zones, airlocks/airbreaks, change rooms, vector and pest-proof construction, sealed enclosures and penetrations, and solid and liquid waste sterilization systems.

General

At all biosafety levels it is desirable to have surface materials which are impervious to damage by acids, bases, alkalis and other chemicals. The surfaces should be seamless and easy to clean. Sinks for handwashing and emergency eyewash stations should be available within the laboratory. BSL-2 facilities should be pest-proof with well screened windows which access the building exterior. Optimally, windows would not open to the outside as this could interfere with, if not completely abrogate, the air balance within the laboratory. Air passage from open windows, doors and even air ducts can adversely effect the efficiency of a BSC and cause contamination of work area. Additionally, an autoclave should be readily available for sterilizing waste prior to its removal from the laboratory.

Penetrations made for conduit, gas lines, water pipes, light switches, fixtures, wall mounted autoclaves, etc. . . must be sealed. The degree of rigor by which sealing is performed will depend on the intended use of the laboratory. At BSL-3, it is appropriate to "seal by sight," so that no penetrations are visible to the eye, or seal to a lesser degree of stringency than at BSL-4 during a smoke containment test. At BSL-4, penetrations must be sealed to provide a room which is airtight, and capable of passing pressure drop or smoke containment tests.

BSL-3 laboratories should have self-closing locking doors to ensure that access is limited to authorized individuals. Change rooms should be supplied with lavatories, laboratory clothing, lockers and showers which enables workers to change into work clothing prior to entering the laboratory and shower out prior to leaving the laboratory.

Ventilation

Facilities which have administrative areas, non-restricted work zones (BSL-1) and zones of increasing restriction (BSL-2,3 or 4) require directional airflow be present in the laboratory design. Ventilation systems should be designed so that air flows from non-contaminated areas to potentially contaminated areas. A gradient should exist so that the air pressure in areas of greater contamination should be negative to those of non-contaminated areas to prevent the potential release of microorganisms into the building. For example, at BSL-3, it is desirable to have a 1.0–2.0 in. water gauge pressure differential across the entry door.

Entry into a BSL-3 area is normally through an airlock or airbreak. This design provides an additional level of protection to the building environment by channeling building air. Airlocks are designed so only one set of doors can be opened at a time, and are often equipped with an interlock system. Doors are designed to form an airtight seal. Airlock doors for BSL-4 conditions are often inflatable with the air in the ante room exhausted through a HEPA filter. There should be a -1.0 in. water gauge pressure in the airlock enclosure, as compared to the entry corridor. The containment laboratory on the other side of the airlock is progressively more negative in pressure than the airlock area. This system is particularly useful if operations are to be performed in one piece suits, which are cumbersome and required chemical decontamination after use. Airbreaks have a similar design, but are not required to be airtight. Air from hallways can be pulled in under the doors into the working laboratory. Because laboratories are negative to the ante room and outer hallways, inward flow of air is ensured.

Ventilation systems in BSL-3 laboratories are often designed to provide HEPA filtration of exhaust air. This is required for BSL-4, as is HEPA filtration of supply air. HEPA filtered building exhausted air provides another level of protection against accidental building contamination. Room air from containment laboratories is never recirculated back into other parts of the building. Air from BSL-4 laboratories is contained in ducts and pulled by fans through HEPA filters prior to release into the environment. It is preferable that the ducts, fans and other constituents for ventilation of containment laboratories be in an independent system from the system which serves the rest of the building.

Supply and exhaust fans in containment laboratories should be interlocked.

In the event of system or equipment malfunction this will prevent the supply fan from supplying air to the room if the exhaust fan is disabled. Essentially, an interlock prevents the room from becoming positive to the corridor. An emergency back-up generator for lights, fan motors, breathable air supply system and other key pieces of equipment should be included in the design.

Disinfectant Barriers

When work is conducted at BSL-3 and BSL-4, the laboratory should be equipped with double door autoclaves with an interlock mechanism. This type of autoclave is mounted in the wall common to an active containment room and a non-containment area. Only one door can be opened at a time. This facilitates the transfer of sterilized materials from containment areas into general laboratory areas. Sterilized materials can then be disposed of or routinely cleaned for reuse. Other methods of removing materials from containment areas include passage through a "dunk tank" filled with an appropriate decontaminating agent. It is advisable to install a FAX and computer in BSL-3 and BSL-4 laboratories, as data transport is otherwise cumbersome.

As alluded to, all materials exiting a containment laboratory must be sterilized or carefully packaged prior to leaving the laboratory or entering the environment. Exhaust air is sterilized by HEPA filtration. At BSL-3 when aerosol studies are in progress, and at BSL-4 it may be desirable to incinerate exhaust air after HEPA filtration. Materials are autoclaved or chemically decontaminated, and waste water must be treated. BSL-3 effluent water can be chemically or heated treated prior to release into an approved sanitary water system. This is accomplished by the inclusion of holding tanks for effluent. The tanks can accommodate handling of waste and decontaminating chemicals or they can be equipped with coils to heat the contents to high temperatures capable of sterilizing waste prior to release.

Decontamination and Sterilization

Proper techniques for decontamination and sterilization of materials contaminated with microorganisms or toxins is a necessary component in the prevention of laboratory acquired infection and contamination of the environment. Decontamination is defined as processes which will make a contaminated material safe to handle by inactivating or removing microorganisms. It does not imply or ensure total destruction of the biological burden. Sterilization is a more stringent form of decontamination. Sterilization is microbiologically defined as a process which consistently produces negative growth following treatment of biological indicators (biological indicators routinely contain 10^6 organisms). When working with moderate or high risk organisms, sterilization is strongly advised.

Some methods for decontamination include: chemical treatment (e.g., with ethylene oxide, vaporized hydrogen peroxide and formaldehyde gas), autoclaving, and UV irradiation. There are positive and negative aspects concerning each method. Decontaminating procedures are chosen based on the efficiency of the procedure for the biological material and the surroundings to be decontaminated, ability to safely monitor or mitigate personnel exposure levels, and expense of the system.

Chemical

There are a wide variety of chemical compounds available for decontamination. Not all chemicals work on all biological materials. Following are some examples of decontaminants and their properties. Phenolic compounds are highly effective against mycobacterium, vegetative bacteria and enveloped viruses, but relatively ineffective against bacterial spores. Care should be used in handling phenolics as exposure can cause skin depigmentation, necrosis, and systemic poisoning (Brown et al. 1985; Conning and Hayes, 1970). Quartenary ammonium compounds are cationic, and effective against rickettsia, lipophilic viruses and vegetative bacteria. Exposure can result in minor contact dermatitis (Shumunes and Levy, 1972). Sodium hypochlorite (chlorine, bleach) is active against almost all biological microorganisms (at a concentration of 2.5% free chlorine) including bacterial spores, vegetative cells, viruses, fungi, and most toxins. While dilute solutions of chlorine may be a pulmonary or ocular irritant, concentrated solutions (or gas) is an asphyxiant in poorly ventilated areas. Of the chemicals discussed, each can be used for surface decontamination, while only sodium hypochlorite can also be used for decontaminating liquids. Each requires at least 10–30 min contact time to work. Specific details on contact times and microorganism susceptibility have been described (Vesley and Lauer, 1986).

Heat

Steam autoclaving has long been considered the preferred method for sterilizing biologically contaminated materials. Gravity displacement autoclaves operate at 121° C at 15 lbs/in^2 (at sea level). Sterilization time is usually 30–60 min (Lauer et al. 1982), and can be verified using biological indicators (*B. stearothermophilus*). Autoclaving inactivates microorganisms as well as most proteinaceous toxins. The limitations of autoclaving include chamber size and insulated loads. Materials which are "insulated" include animals carcasses, large volumes of laundry and large volumes of fluids. Large volumes of fluids can be handled more effectively by chemical decontamination followed by longer term heating in a sewage treatment tank. Carcasses should be incinerated. Risks to personnel operating autoclaves include burns due to steam, super-heated liquids, and hot surfaces.

Ethylene Oxide

Ethylene oxide (EtO) sterilization is another sterilization process which requires containment in a chamber. EtO is effective against all bacteria (including spores), viruses, fungi and molds. Currently, EtO is the prefered method for use in the sterilization of contaminated electronic equipment. Its efficacy against toxins is not known. In addition to the specialized chamber requirements for EtO use, a rigorous personnel monitoring program must be instituted. EtO is a desiccant, mutagen, suspect carcinogen, and causes ocular, dermal and respiratory irritation (Glaser, nd).

Paraformaldehyde

Large scale decontamination can be achieved using formaldehyde gas in a sealed room. Gas is generated by heat sublimation of powdered or flake paraformaldehyde. To decrease the risk of explosion, paraformaldehyde can be dissolved in H_2O at a concentration of 0.3 grams/ft^3 area. The room temperature and relative humidity must both be approximately 70% prior to commencement. An exposure time of 4 hr to overnight is required before treatment with gaseous ammonium bicarbonate for neutralization. Following 1 or more hours aeration, the level of residual formaldehyde can be tested and biological indicators collected and analyzed for growth prior to the resumption of work. Formaldehyde is a suspect carcinogen, highly toxic and very irritating (NIOSH 1976). However, it is currently the most effective proven method for large scale decontamination.

Ultraviolet Irradiation

Ultraviolet irradiation at a wavelength of 255 nm has been used for inactivating viruses, mycobacteria, and bacteria on bench top surfaces and in airborne states. Its efficacy for inactivating toxins has not been demonstrated. There is controversy and conflicting data concerning the decontaminating ability of UV radiation. The organism, time of exposure, distance from the UV source and its susceptibility to damage by UV radiation all influence the success of decontamination. Personnel exposure to UV radiation can result in damage to the skin and cornea (Duke-Elder and MacFaul, 1972). If UV radiation is to be used, a calibrated photoelectric UV intensity meter should be utilized to measure UV emissions upon installation of bulbs and at least every 6 months thereafter. Bulbs should be cleaned weekly as a layer of dust will interfere with emission (at least 40 microwatts/cm^3 is required at the surface for decontamination.

Vaporized Hydrogen Peroxide

Vaporized hydrogen peroxide (VHP) is a relatively new method of decontamination currently under evaluation. VHP is generated by a programmable unit which can be used to decontaminate BSC, rooms and other equipment. The benefits of this technology include ease of operation, rapid decontamination time, harmless breakdown products (H_2O and O_2), minimal chance for personnel exposure to hazardous chemicals and potential use with large sized equipment. Possible problems include personnel exposure to VHP leading to ocular, mucosal and lung irritation or damage, and initial expense for equipment purchase. This technology has been shown to successfully decontaminate a variety of bacteria (spores and vegetative), fungi, viruses and toxins (Klapes and Vesley, 1990; Johnson et al., 1993; Rickloff, 1990). It appears that physiologic levels of salts in diluent solutions reduce the decontaminating efficiency of VHP. Additional data is needed to determine the limits of this technology with regard to various work practices before an accurate assessment can be made.

Animal Use and Holding

The housing and handling of animals in studies involving infectious agents and aerosolization pose unique safety risks. If animal use is to be included in studies, the quality of animal care before, during and after the study must meet standards provided by the institution conducting the work, and other guidelines (NIH, 1985; US Code, Reg. 9). Recommendations for "animal" biosafety levels have been published (Richardson and Barkley, 1988).

Housing

A point to note is the importance of maintaining separate rooms, or facilities for animals involved in testing and animals maintained for breeding or those in quarantine. It is also beneficial to have a "clean"/"dirty" corridor design so used or contaminated materials exit the animal area without passing areas which house breeder colonies, quarantine, or animals not on studies. The rooms should be designed for easy cleaning, include a pest control program, have floor drains either sealed, filled with water or an appropriate disinfectant, and should be able to contain animals which have escaped from their cages or handlers. While escape is rare when proper handling and care is observed, it is a possibility which should not be completely dismissed.

Handling

Animal facility workers should be forewarned of the potential for developing allergies and contact dermatitis after prolonged work with animals and animal waste. Workers should have access to gloves and dust/mist respirators, and be instructed to always wash their hands thoroughly after working with animals or dirty materials. Optimally exhaust air from animal facilities should not be recirculated throughout the building, at the very least recirculation of air could disseminate allergens to sensitized workers. Discussed below are recommendations for work with animals in the four animal biosafety levels (Richardson and Barkley, 1988).

Precautions

At animal BSL-1, in addition to standard and special BSL-1 practices it is also advisable that doors to the facility open inward and are self closing. Cage materials should be handled in such a way as to minimize the creation of dust and aerosol. Normally at BSL-1 organisms used are not harmful or are of little threat to healthy adult humans, but they may be of threat to animals of the same or different species (i.e., infectious canine hepatitis or parvo viruses, murine hepatitis virus, sendai virus, and some vaccine strains of viruses). For this reason, and others, it is important to separate different species and infected animals from non-infected animals. No special containment is required for animal BSL-1.

At animal BSL-2 all of the above mentioned precautions are followed as are laboratory BSL-2 practices. Due to the potential to create aerosols when changing bedding materials, it is advisable to autoclave the bedding while still in the cage, and perform normal cleaning following sterilization. Work performed with infected animals (inoculation, lavage, necropsy, examination), tissues, eggs and insects should be conducted in a BSC. The facility design and use of primary containment equipment during work with infected animals is important but must be augmented due to hazards intrinsic to work with animals. The risk of aerosol transmission of pathogens increases as organisms can be shed in nasal or oral secretions (sneezing and coughing), in the urine, feces and other body fluids. Inoculation of pathogens by animal bites should never be underestimated. Exposure to organisms may be via contact with residue from the animals coat. To mitigate risk in these scenarios the proper PPE should be chosen, and procedures employed to minimize each form of exposure. Examples of microorganisms used in animal studies which require animal BSL-2 precautions (work excluding use of non-human primates) include but are not limited to: *Corynebacterium diphtheria, Mycobacterium leprae, Mycrosporum sp.,* and *Cryptococcus neoformans).*

In many situations, the BSL at which work is conducted increases during studies involving the use of animals due to the increased risk of pathogen aerosolization and injection (via bite). Work with organisms normally used in accordance with BSL-2 recommendations are often worked with at BSL-3 when animal or vector use is involved (i.e., lymphocytic choriomeningitis virus, *Coxiella burnetii, Neisseria meningitidis, Mycobacterium tuberculosis* and *Clostridium botulinum).*

Animal BSL-3 practices include those described for animal BSL-2, laboratory BSL-3 and housing animals in partial of full containment systems. This can be accomplished by placing regular cages inside of HEPA filtered bonnets or bonneted racks, in ventilated (laminar flow) cabinets, or by using cages of solid construction with filter media acting as a barrier between room air and air inside the cage. Entry into areas where infected animals are housed is through two sets of doors (airlock or airbreak), and there should be a pass-through autoclave in the animal facility for sterilization. The other side of the autoclave opens into a "clean" or non-contaminated area for unloading autoclaved materials. The autoclave should have interlocking doors to prevent inadvertent opening of both doors at the same time. The ventilation system should be providing directional airflow into the animal room, and not recirculate the air within the room unless it is HEPA filtered. In addition to previously described PPE, more efficient respirators such as positive air pressure respirators or high efficiency respirators may be desired. Over-clothing should be of the wrap around variety such as a surgical gown, and not the lab coat which buttons down the front.

Work conducted at animal BSL-4 must incorporate all the safety practices of animal BSL-3, and laboratory BSL-4. This includes conducting all work not performed in a Class III BSC in a one piece positive pressure ventilated suit. Supply air is single HEPA filtered, and exhaust air is double HEPA filtered. Organisms requiring animal BSL-4 maximum containment include: marburg, machupo, lassa and ebola viruses, and herpes virus simiae.

Conclusion

Safety equipment, operational procedures and facility features can be employed to provide an acceptable level of protection for laboratory workers, the public and the environment. Risk assessments must be performed for all proposed work, with pathogens or toxins, to guide selection of mitigative equipment and procedures. Aerosol studies with pathogens and toxins present unique safety consideration. The success of a biosafety program is a measure of the commitment of management and workers to the program.

References

ANSI Z88, 1989, subcommittee on respiratory protection against infectious aerosols.

Brown, V. K. H., V. L. Box, and J. J. Simpson. 1975. Decontamination procedures for skin exposed to phenolic substances. *Arch. Environ. Health,* 30:1-6.

Chatigny, M. A. 1986. pp. 144–172, Primary barriers, in B. M. Miller, (ed.), *Laboratory Safety: Principles and Practices,* American Society for Microbiology, Washington D.C.

Conning, D. M. and M. J. Hayes. 1970. The dermal toxicity of phenol: an investigation of the most effective first-aid measures. *Br. J. Ind. Med.,* 27:155–159.

Duke-Elder, S. and P. A. MacFaul. 1972. Injuries. Non-mechanical injuries, pp. 912–933. In S. Duke-Elder (ed.), *System of Opthamology,* vol 14, part 2., C. V. Mosby.

Glaser, Z. R. Special occupational hazard review with control recommendation for the use of ethylene oxide as a sterilant in medical facilities. Department of Health and Human Services publication no. 77-200. U.S. Government Printing Office, Washington, D.C.

Hanel, E. Jr. and M. M. Halbert. 1986. Pipeting, pp 204–214. In B. M. Miller (ed.), *Laboratory Safety: Principals and Practices,* American Society for Microbiology, Washington, D.C.

Hatch, T. F. 1961. Distribution and deposition of inhaled particles in respiratory tract. *Bacteriol. Rev.,* 25:237–240.

Johnson, B., B. G. Harper, A. J. Mohr, D. R. Winters, and I. G. Resnick. 1993. Potential use of vaporized hydrogen peroxide in the inactivation of toxins., Abstr. Annual American Society for Microbiology, 93 General Meeting, Atlanta, Ga., Q263, 1993.

Johnson, B., D. D. Martin, W. S. Huff, and I. G. Resnick. 1992. Efficacy of selected respiratory protective devices (RPD) challenged with bacteria, Abstr. 35th Annual American Biological Safety Association Conference, 1992.

Klapes, N. A. and D. Vesley. 1990. Vapor-phase hydrogen peroxide as a surface decontaminant and sterilant. *Appl. Environ. Microbiol.* 56:503–506.

Kuehne, R. W. 1973. Biological containment facility for studying infectious disease. *Appl. Microbiol.,* 26:239–243.

Lauer, J. L., D. R. Battles, and D. Vesley. 1982. Decontaminating infectious laboratory waste by autoclaving. *Appl. Environ. Microbiol.* 44:690–694.

Lynch, P. 1982. Matching protective clothing to job hazards. *Occupat. Health Safety,* Jan:30–34.

National Institutes of Health, Guide for the Care and Use of Laboratory Animals, DHHS publication no. (NIH) 86-23 (revised 1985), U.S. Government Printing Office, Washington, D.C.

National Institute for Occupational Safety and Health. 1976. Criteria for a recommended standard-occupational exposure to formaldehyde., Department of Health and Human Services publication no. 76–142., U.S. Government Printing Office, Washington D.C.

National Institute of Occupational Safety and Health. 1992. Recommended guidelines for personal respiratory protection of workers in health care facilities potentially exposed to tuberculosis. U.S. Department of Health and Human Services, Centers for Disease Control, NIOSH, DHHS (NIOSH), Atlanta, Ga.

National Research Council, 1990, *Biosafety in the Laboratory,* National Academy Press, Washington, D.C., pp. 25–28.

National Sanitation Foundation, NSF Standard No. 49 for biohazard cabinetry, 1976 (revised June 1987), NSF, Ann Arbor, Mich.

Richardson, J. H. and W. E. Barkley, W. E. (ed.). 1988. *Biosafety in Microbiological and Biomedical Laboratories.* U.S. Department of Health and Human Services, Centers for Disease Control and National Institutes of Health, HHS Publication no. (CDC)88-8395, 2nd Ed.

Rickloff, J. A. 1990. Use of vaporized hydrogen peroxide for the biodecontamination of enclosed areas., Abstr. Interphex USA Conference, New York, N.Y., 1990.

Shumunes, E. and E. J. Levy. 1972. Quartenary ammonium compound contact dermatitis from a deodorant. *Arch. Dermatol.* 105:91–95.

Thomas, S. M. G. 1970. The use of protective gloves. *Occupat. Health,* 22:281–284.

U.S. Code of Federal Regulation 9, Subchapter A, Parts 1-3, Laboratory Animal Welfare Regulations, U.S. Government Printing Office, Washington, D.C.

U.S. Code of Federal Regulations, Title 29, OSHA, Part 1910, Occupational Health and Safety Administration Safety and Health Standards.

U.S. Code of Federal Regulations, Title 29, OSHA, Part 1910.1030, Occupational Exposure to Bloodborne Pathogens.

U.S. Code of Federal Regulations, Title 32, Department of the Army, DOD, Parts 626-627, Biological Defense Safety Program.

Vesley, D. and J. Lauer. 1986. Decontamination, sterilization, disinfection and antisepsis in the microbiology laboratory. p. 182–198. In: B. M. Miller (ed.), *Laboratory Safety: Principles and Practices,* American Society for Microbiology, Washington, D.C.

Index*

* Boldface page numbers refer to illustrations or tables.